ROADSIDE GEOLOGY of OHIO

Mark J. Camp

2006
Mountain Press Publishing Company
Missoula, Montana

ROADSIDE GEOLOGY IS A REGISTERED TRADEMARK OF
MOUNTAIN PRESS PUBLISHING COMPANY

Library of Congress Cataloging-in-Publication Data
Camp, Mark J., 1947-
  Roadside geology of Ohio / Mark J. Camp.—1st ed.
    p. cm.
  Includes index.
  ISBN 0-87842-524-1 (pbk. : alk. paper)
  1. Geology—Ohio—Guidebooks. I. Title.
QE151.C36 2006
557.71—dc22

                                    2006020781

Printed in the United States of America
BY FC PRINTING, SALT LAKE CITY, UTAH

Mountain Press Publishing Company
P.O. Box 2399 • Missoula, Montana 59806
(406) 728-1900

This volume is respectfully dedicated to the geologists, historians, and naturalists who collected the data and made this work possible, and to the citizens of Ohio.

MICHIGAN
LAKE ERIE
PENNSYLVANIA
INDIANA
WEST VIRGINIA
KENTUCKY
O H I O V A L L E Y
TILL PLAINS
ALLEGHENY PLATEAU
Ashtabula
Painesville
Jefferson
Chardon
Cleveland
Lorain
Sandusky
Port Clinton
Toledo
Wauseon
Bryan
Napoleon
Bowling Green
Defiance
Fremont
Norwalk
Elyria
Medina
Ravenna
Warren
Akron
Youngstown
Fostoria
Tiffin
Findlay
Paulding
Van Wert
Upper Sandusky
Ashland
Wooster
Canton
Lisbon
East Liverpool
Bucyrus
Mansfield
Lima
Kenton
Wapakoneta
Celina
Marion
Mt. Gilead
Millersburg
Carrollton
New Philadelphia
Steubenville
Bellefontaine
Mt. Vernon
Sidney
Delaware
Coshocton
Cadiz
Marysville
Greenville
Piqua
Urbana
Troy
Columbus
Newark
Cambridge
St. Clairsville
Dayton
Springfield
London
Zanesville
Eaton
Xenia
Lancaster
New Lexington
Woodsfield
Caldwell
Middletown
Circleville
McConnelsville
Hamilton
Lebanon
Washington Court House
Logan
Wilmington
Marietta
Chillicothe
Athens
Hillsboro
McArthur
Cincinnati
Batavia
Waverly
Pomeroy
Jackson
Georgetown
West Union
Portsmouth
Gallipolis
Ironton
20
80/90
24
75
23
2
90
11
422
71
80
76
250
224
77
30
127
68
40
70
50
33
32
52
7
N
0 10 20 30 40 miles
0 10 20 30 40 50 kilometers

CONTENTS

PREFACE

In the early 1700s Ohio country lay destined for greatness. Thousands of people venturing westward followed its trails, rivers, and the shoreline of Lake Erie searching for a better future west of the Alleghenies. A few carved their homesteads out of the wilderness that was the Ohio Valley; many settled near the mouths of the Great Miami, Muskingum, and Scioto Rivers. Although they may not have realized it, they were interacting with geology. Those that cleared the forest and planted crops could thank the glaciers for having left a mantle of enriched soils. Ancient bedrock slabs and fieldstone became chimneys, hearths, and mill dams. Water was harnessed to power grist- and sawmills, structures that were often built of local stone. Early settlements like Defiance, Fort Recovery, and Marietta formed around military installations sited at vantage points along rivers. Chillicothe, Cincinnati, Cleveland, Columbus, Dayton, Steubenville, Youngstown, and Zanesville are among other pre-1800 communities that owe their location to geology.

Early travelers little realized that vast mineral resources lay hidden below the surface. However, seams of coal on the banks of the Ohio River and rusty, heavy cobbles and salt springs in the southeastern hills caught the eye of those schooled in the natural sciences. In 1803 Ohio became the seventeenth state and the first to form out of the Northwest Territory. The importance of geology was evident even in these early years. For example, the federal government held in reserve three Ohio salt springs so that they would be available to the entire nation.

The digging of canals opened up trade routes as Ohio was settled in the 1800s, allowing tons of building stone to move across the state. The coming of railroads accelerated the pace of development and particularly stimulated the coal and building stone industries. As forests were cleared, coal became the favored fuel for locomotives and stone was in demand for bridge and culvert construction. Iron smelters appeared in eastern Ohio. The Cleveland-Youngstown-Pittsburgh corridor became the world's greatest steel belt and Cleveland became a major iron port. The discovery of oil also made it a world-class refinery center. Ashtabula, Conneaut, Fairport Harbor, Huron, Lorain, Sandusky, and Toledo also became major ports as industrial growth focused on the North Coast. Railroads carried coal north from southeastern Ohio and iron ore south from Lake Erie ports to the steel belt.

The ceramic industry flourished in several areas of the state where clay or shale was common, but nowhere did it flourish more than in Ohio River towns between East Liverpool and Steubenville and the Crooksville-Roseville-Zanesville area of central Ohio. The discovery of natural gas and oil along a flexure of rock between Lima and Toledo around 1900 was an exciting, albeit wasteful, time in Ohio history. The resulting availability of cheap energy brought the glass industry to northwestern Ohio. As you can see, most of Ohio's economic success and industrial growth was tied closely with geology. This part of Ohio history is a common theme in this book.

Truly, geology is a subject that is all-encompassing. It is behind much of what we do. Just as it underlies the development of Ohio's industrial and geographic growth, it also defines its scenery. The flat plains of northwestern Ohio owe their makeup to the ice sheets that covered the region and to waves and currents that washed the shores of Lake Erie's predecessors. The Lake Erie coast attracts people to its pleasant views and numerous recreational opportunities. Its shape and composition are dependent on wave and current action, but the lake is not very old geologically speaking. The Ohio River defines Ohio's southern boundary; surprisingly it's a relatively recent addition to Ohio, too. The hills of eastern and southeastern Ohio—some of the oldest landscape in the state—represent areas that glaciers and glacial lakes left relatively untouched. Streams produce their rugged beauty. Anyone who has been to Hocking Hills State Park or other scenic areas within the uplands has witnessed its natural beauty composed of rock shelters, high cliffs, and waterfalls. Ohio has many great park systems ranging from its only national park, Cuyahoga Valley National Park, to a wonderful series of state, county, township, and municipal parks. Most owe their redeeming features to geology. This book looks at many of them.

Geology, to scientists, includes trying to understand what various natural materials mean and how they came to be. To many, *geology* means collecting minerals, rocks, and fossils. Ohio is known for some spectacular mineral occurrences, especially in the northwestern counties. Important fossils have brought fame to such locations as Cincinnati, Rocky River, and Sylvania. Indeed, the hills of southwestern Ohio yield abundant, well-preserved fossils wherever its gray bedrock is exposed. Representative and spectacular specimens are in many museum and university collections around the world. If you're interested in finding specimens, this book will guide you in your hunt.

Geology is also about people—the geologists, naturalists, and all who have picked up a rock or fossil and pondered its history. Interwoven amongst the description of Ohio's geology are the stories of pioneers who helped solve its mysteries. Some of the names you'll recognize; others will be new to you. These names, however, only represent a select few who have contributed to the story. This book is based on the research and studies of many others, and I could not have written it without them.

For the purposes of this book, I divided Ohio into four areas—the western Ohio till plains, the Allegheny Plateau, the Lake Erie shoreline, and the Ohio River valley—based on bedrock geology, glaciation, and geographic location. These regional chapters follow an introductory chapter, which provides the basic background for understanding the geologic makeup of Ohio. Following an introductory section, each regional chapter includes road guides of selected highways. The road guides run north to south and east to west but are written so they serve readers traveling in any direction. Just read them in reverse order if you are traveling in the opposite direction. The road guides also include selected topics that focus on sites of particular geologic significance (black subheads). Don't forget to read these as you pass through the towns and regions they describe. There are a number of side trips as well.

The oldest bedrock in the state covers the southwestern corner of Ohio, stretching as far north as Dayton. The Cincinnati Arch defines the edges of this bedrock in this corner of the state. The strips of bedrock adjacent and parallel to the arch become increasingly younger—Silurian to Mississippian age—stretching from Cincinnati to the Michigan border and to the east. With the exception of stream cuts, quarries, excavations, and road and railroad cuts, the soil hides our view of the bedrock. The western Ohio till plains region was also glaciated by pre-Illinoian, Illinoian, and partially by Wisconsinan ice. It includes various till plains, the Huron-Erie lake plain, and a small portion of the Allegheny Plateau.

Ohio's eastern half is underlain by younger bedrock of Mississippian through Permian age, and only its northern and northwestern parts were glaciated. Bedrock appears on hillsides, in stream cuts, and wherever the soil is thin. Eastern Ohio is a hilly upland, the Allegheny Plateau, and it is a strikingly different landscape than relatively flat western Ohio. The plateau serves as a gradual transition to the Appalachian Mountains far to the east.

Ohio's northern and southern borders are defined by major geologic features. The north borders Lake Erie, so the focus of its geology is the constantly changing shoreline, plus glimpses of a history that far pre-dates the lake. Bedrock is visible, especially in the stretch from Sandusky to the Cleveland area. The Ohio River marks the southern border. The river exposes bedrock in the bordering bluffs and provides a medium for geologists to discuss recent and current geologic activity.

Though you may encounter bedrock of many ages along roads in all four chapters, I discuss Ohio's most common formations in detail in the chapter where you will encounter them the most. I discuss Ordovician, Silurian, and Devonian formations in the Till Plains chapter and Devonian, Mississippian, Pennsylvanian, and Permian formations in the Allegheny Plateau chapter.

Ohio is rich in roads. Obviously I could only include a select few. I included most interstate and U.S. highways, but only a few state routes have separate road guides. The geology of routes near the interstates and U.S. highways is usually similar and can be easily deduced from the road guides and geologic maps. I also include an Additional Reading section and a glossary. If you want more information or are unfamiliar with a term, be sure to consult these sections.

I tried to describe sites and features that are easily seen from the road, but some places are off the road and require permission to visit. Others exist only in history and require your imagination. Much of Ohio's roadside is private property. Please be considerate and do not trespass; seek permission. Please obey all traffic rules and be cognizant of traffic. Be considerate of others who share the road. Never enter quarries, mines, and caves without permission and an experienced guide. One careless or thoughtless act by an individual can ruin a site for all responsible visitors. These places can also be dangerous.

The teachings, guidance, and encouragement of Mary Bowermaster, Jane Forsyth, Cal Gettings, Craig Hatfield, William Kneller, Aurele LaRocque, Doran Snyder, and Richard Whisler set the foundation for me to compile *Roadside Geology of Ohio*. I thank all my colleagues in the geology, natural sciences, and Ohio history fields; many of them I have never known. Without them this book would have been impossible. I thank the Ohio Department of Natural Resources, Division of Geological Survey, for its pioneering work in Ohio geology, continued advancement in its study, review of manuscript portions, numerous discussions, and use of historical photos and maps from its extensive library and database. I also appreciate the support the Ohio Academy of Science has offered geologists and other scientists throughout its history. I deeply appreciate the travel support provided by the Department of Earth, Ecological, and Environmental Sciences at the University of Toledo. I would like to compliment the staff of Mountain Press for their suggestions and editing and seeing this project to fruition.

I hope you find *Roadside Geology of Ohio* a useful travel companion as you drive the roads of the Buckeye State and discover its geologic splendor.

TIME	BEGAN MILLIONS OF YEARS AGO	TYPICAL GEOLOGICAL EVENTS AND MATERIALS IN OHIO
CENOZOIC ERA		Ice cover and forests. Mastodons, giant beavers, and other Ice Age animals present. Glacial tills, outwash, lake and river sediments, and soils deposited. Glacial sediments not found in southeastern Ohio.
Quaternary Period	2	
Tertiary Period	65	No sediments or fossils from this time found in Ohio.
MESOZOIC ERA *Cretaceous Period* *Jurassic Period* *Triassic Period*	145 200 251	No sediments or fossils from this time found in Ohio.
PALEOZOIC ERA *Permian Period*	299	Great swamp forests; sail-back reptiles common. Coal, shale, sandstone, limestone, and clay deposited in repeated cycles. Deposits restricted to Athens, Belmont, Jefferson, Meigs, Monroe, Morgan, Noble, and Washington Counties of southeastern Ohio.
Carboniferous Period — *Pennsylvanian Subperiod*	318	Great swamp forests. Fossil plants common; fossil amphibians and reptiles occasionally found. Coal, shale, sandstone, conglomerate, limestone, and clay deposited in repeated cycles. Major coal and clay mines have been developed in deposits of this time. Eastern and southeastern counties.
Carboniferous Period — *Mississippian Subperiod*	359	Shallow seas. River deltas spread sand into Ohio from eastern and northern uplands. Sandstone and shale. Northeastern, central, and extreme northwestern Ohio counties. Mississippian rocks well exposed in Hocking Hills State Park in south-central Ohio.
Devonian Period	416	Shallow seas. Fish, trilobite, brachiopod, and coral fossils common. River deltas spread sand and mud into Ohio from eastern uplands. Sandstone, shale, and limestone deposited. Berea sandstone, an important building stone, formed during this period. Central Ohio counties from Erie to Adams County and Defiance, Fulton, Henry, Lucas, Paulding, Putnam, and Wood Counties.
Silurian Period	443	Shallow seas. Barrier and patch reefs common. Seas withdraw from Ohio late in period, forming evaporites. Dolomite, limestone, shale, gypsum, and rock salt. Bedrock under much of western Ohio, including Bass Islands, from this time.
Ordovician Period	488	Shallow seas. Bryozoans and brachiopods common. Shale and limestone. Ohio's most fossiliferous rocks and oldest exposed bedrock are from this time. Restricted to southwestern counties.
Cambrian Period	542	Shallow seas. Sandstone and dolomite known from drilling.
PRECAMBRIAN TIME	4,600	Igneous and metamorphic rocks. Known only from deep drilling.

Geologic timescale. Bedrock from the periods in the brown font appear at the surface in Ohio.

INTRODUCTION TO OHIO'S GEOLOGY

A Geologic Calendar

We are about to embark on an excursion deep into time, long before human-kind, to when foot-long trilobites and squidlike animals with long conical shells wallowed in the bottom mud of the shallow, warm sea covering ancestral Ohio; to when giant cockroaches, salamander-like amphibians, and sail-back reptiles roamed coastal swamps in eastern Ohio; to when rivers flowed north into a val-ley that Lake Erie fills today; and to when great sheets of ice scraped and gouged the early Ohio landscape. This is the geologic story of Ohio, of how the soils, rocks, and landscapes of the Buckeye State came to be.

Weaving a story about the geology of Ohio demands relying on an al-most unbelievable amount of time—what some people call *deep time.* Deep time spans back through 4.5 billion years, to the formation of the earth. Why do geologists deduce that the earth is so old? By observing and interpreting the geologic processes occurring today, which vary from steady and slow to intermittent and rapid, geologists know that the earth must be very old. Take for example the Grand Canyon in northern Arizona; at places the Colorado River has carved a mile into the earth's surface. Can you imagine how long the river took to carve this spectacular natural landmark? Have you noticed any great change in the depth of a river valley near your home during your lifetime? How about in your parents' lifetime, or your grandparents'? Probably not. Indeed, floods may cause sudden changes, but overall they have little effect on the depth of a river valley. It takes a lot of time for river valleys to deepen.

Landscape development generally is an extremely slow process; negligible in terms of human lifetimes. The Colorado River carved into thousands of feet of rock, which itself had formed over a long interval of time. Modern-day depositional rates of pebbles, sand, and clay and precipitation rates of calcite and other minerals, the particles that make up the sedimentary rocks of the Grand Canyon's walls, are quite variable and depend on many things. Among the higher rates at which sediment is accumulating on the earth today is 2 to 3 centimeters per year, which is how quickly the Mississippi River delta is forming. Far from the continents in the deep oceans, a centimeter of mud may take 10,000 years to accumulate. So you might begin to see how long it took the rocks in the Grand Canyon to form before they were cut by the Colorado River.

Sedimentation alone is still not the whole picture since the sediments had to be compacted and cemented together to form the canyon's colorful rock layers. This took time as well. Besides that, the ancient seas covering this area and depositing much of the region's sediments withdrew at times, exposing the former sea bottom to the forces of erosion. Ancient rivers and streams and wind swept over the surface, carrying away tremendous volumes of sediment. Unconformities, gaps in the geologic record where rock layers are missing, mark these times of erosion. Sometime between 10 to 5 million years ago, the Colorado River arrived and began contributing to the erosion and deposition of the area, including the long process of canyon cutting. Ohio's sedimentary rocks formed as a result of the same cycles of deposition and erosion.

Although the observation of rock layers (including the fossils they contain)—how they are related to other rock layers above, below, or adjacent to them—set the basis for determining geologic time, it was the discovery of radioactivity in 1896 that led to sophisticated methods of dating. These methods have allowed geologists to assign actual ages to rocks and events. When certain elements, like uranium and thorium, form and become part of the makeup of a rock, they immediately begin emitting tiny particles of matter—radiation—and eventually turn into stable elements like lead. These emissions do not occur haphazardly but proceed at regular rates. The emission process is known as *radioactive decay*. By knowing a given decay rate and comparing the amounts of original radioactive material with the new stable elements, geologists can calculate the age of a rock. Geologists have used this technique to generate the numerical values on the geologic timescale, which they use to tell and better understand the earth's story.

The enormity of geologic time is difficult to comprehend. The geologic timescale serves as a calendar for geologic events; our yearly calendar appears minuscule in comparison. Geologists divide geologic time into three basic units—eras, periods, and epochs—much like our calendar consists of years, months, and days. One major difference is that geologic time deals in millions to billions of years rather than 365-day intervals.

Plate Tectonics and Ohio's Origins

Earth's rocks reveal a complicated early history. Much of the evidence is hidden below the surface in deeply seated rocks that have been subjected to numerous episodes of melting, baking, folding, uplift, and erosion. Much of the evidence of the planet's early years has long since been erased by its continuing evolution. Geologists, like detectives, continue to unravel the mysteries of early Earth by piecing together clues in the rocks and structures found around the globe. Ohio began developing at a point on the earth's surface some 4.5 billion years ago.

As the planet cooled and differentiated into its internal zones—the inner core, outer core, mantle, asthenosphere, and lithosphere—the roots of Ohio were set. The lithosphere began forming as basaltic lava oozed from a series of interconnected fractures, called *ridges*, across the earth's surface. At some point early in the history of the earth, an atmosphere formed and chemical reactions between the hot rocks, colliding meteorites, and atmospheric gases led to precipitation and ocean formation. Continents, exceptionally thick masses of lithosphere (60 to 150 miles), formed amongst the oceans. Ridges, associated with lava extrusion, are found mainly in ocean basins where the lithosphere averages about 40 miles thick. Intersecting the ridges are deep oceanic troughs called *trenches* and other fractures not associated with lava extrusion. The combination of the ridges, trenches, and fractures divides the earth's surface, or lithosphere, into segments called *plates*.

Present-day Earth has six major plates and several smaller ones. The movement of these plates is referred to as *plate tectonics*. The extrusion of lava, or new lithosphere, at the ridges pushes older lithosphere away, producing what geologists call *seafloor spreading*. It's not a readily visible movement, resulting in only inches of movement per year. The oldest lithosphere eventually reaches the trenches and descends, or is subducted, back into the earth, eventually melting and becoming new magma.

Continents are generally composed of deep-seated granitic rocks with a veneer of sedimentary rocks, all of low density, unlike the seafloor, which is composed of denser basaltic rocks. When plates collide, continents, unlike the seafloor, resist being dragged into the depths of the earth. The rocks that compose the continents tend to pile up at the earth's surface, crinkling, cracking, and sliding against each other. This collision results in mountains and often causes an adjustment of plate motions—a slowing of spreading rates or a seaward shift of a trench. Granitic magmas generated at this time push upward into the sedimentary layers as intrusions.

The ridges and subduction zones are centers of volcanic and earthquake activity. Through geologic time, however, some ridges and trenches ceased to function and were replaced by new ones elsewhere. Such is the case below Ohio. A once-active ridge lies buried in the Precambrian basement far below the

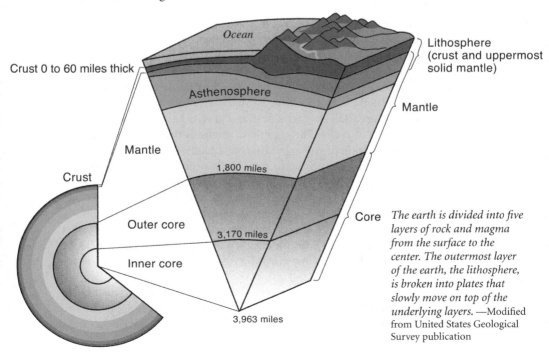

The earth is divided into five layers of rock and magma from the surface to the center. The outermost layer of the earth, the lithosphere, is broken into plates that slowly move on top of the underlying layers. —Modified from United States Geological Survey publication

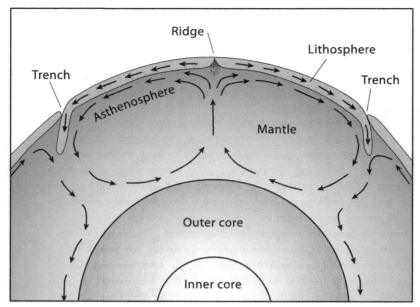

Lithospheric plates move, or spread, from oceanic ridges and rifts to subduction zones, or trenches, at rates of ¾ to 4 inches per year.
—Artwork by Mountain Press

Buckeye State. Today the closest active ridge to Ohio is in the middle of the Atlantic Ocean.

Plate tectonics has been responsible for the geography of the earth through time. Ever since humans have been around there has been little change in the positions of continents or oceans on the global surface; for example, North America has been separated from Europe by the vast Atlantic Ocean. This has not always been the case.

Ohio's Basement

No native Precambrian rocks are present at the surface anywhere in Ohio, though others have been carried to this state by glaciers. They do, however, underlie the entire state a few thousand feet below the surface in western Ohio and over 13,000 feet down in southeastern Ohio. To indirectly see these deeply buried rocks, geologists resort to geophysical techniques. For example, geologists have transmitted electronically generated waves into the subsurface and noted changes in the earth's magnetic field and gravity below Ohio that reflect changes in the types of rocks. Through these methods they have been able to interpret the composition of rocks, fractures and faults, and folds, all of which are basically undetectable from the surface. Wells drilled through glacial material and the underlying Paleozoic sedimentary rocks have penetrated these ancient rocks as well, and have allowed geologists to directly observe samples brought to the surface. These ancient rocks, from 1.5 billion to 800 million years old, are a mixture of igneous, metamorphic, and sedimentary rocks, what geologists call the *basement*. Even older igneous and metamorphic rocks lie below the basement, but until we can gather data on them, information remains speculative.

The first well to tap Precambrian rock dates to 1912. It was in the oil field surrounding Findlay. At 2,770 feet down a reddish granite gneiss appeared and continued to the bottom of the hole, another 210 feet. By 1932 three other wells in northwestern Ohio provided proof of similar Precambrian rock. This area of basement rock was named the granite-rhyolite province. By the late 1960s ninety-four wells hinted at the complexity of Ohio's basement. Wells in eastern Ohio tapped various schists and gneisses, metamorphic rocks that form under intense baking and pressure; wells in western Ohio showed unmetamorphosed volcanic rocks and several types of granite, igneous rock types. The area of basement rock under eastern Ohio was named the Grenville province. New geophysical data and continued drilling brought many more discoveries from 1985 to present, including the identification of many previously unknown faults.

One interpretation of Ohio's basement is that the granite of western Ohio is part of a larger chunk that pushed upward from deep below the surface some 1.5 billion years ago. This granite began to pull apart about 1.3 billion years ago, forming a deep rift, or downward-slipping block of crustal rock, in which volcanoes formed. Eventually, this surface rift filled with sediment and basalt lava flows, what is now deep below southwest Ohio. About 1 billion years ago the rifting ended. The reason is still unclear, but plate motion led to a cessation of the upward warping and tensional cracking of this part of ancient North America, which is also referred to as Laurentia.

During this time period the margin of ancient North America may have corresponded with eastern Ohio. An unnamed continent collided with this coastline between 1.2 and 1 billion years ago, pushing up subsurface rocks to form the ancient Grenville Mountains. The Grenville Mountains extended from present-day eastern Ohio northeast to Labrador and southwest to Texas. They were composed of highly deformed schists and gneisses, types of metamorphic rocks, and granites and granodiorites, types of intrusive igneous rocks. These rocks, the stubs of what were tall mountains, are now deep below the eastern half of Ohio. Originally, the metamorphic rocks were deposited as coastal sands and calcareous muds along small continents and volcanic islands that were folded, faulted, and metamorphosed during a series of small collisions that preceded the collision with the larger, unnamed continent. All the collisions combined formed a supercontinent called Rodinia. The granites under eastern Ohio are the solidified magmas associated with the igneous activity of this event. The boundary between these rocks and the granite-rhyolite basement of western Ohio, which remained unaffected by this mountain building, lies roughly parallel to I-75. The Grenville Mountains were worn down to gentle hills over 300 million years of exposure to streams, rivers, and other erosional agents.

Geologists are only beginning to unravel this hidden history and how it relates to important mineral occurrences and earthquake activity in Ohio. This book, however, is about geology along the road and thus focuses on younger rocks seen at the surface. In Ohio, these are the rocks of the Paleozoic Era.

Paleozoic Time
North American Mountain-Building Events
Though relatively flat, much of Ohio's geology has been dictated by mountains, the eroded roots of which now exist to the east. Three separate mountain-building events, related to the collision of continents, built three different ranges in Paleozoic time. These mountains shed massive amounts of sediment

through tens of millions of years, creating most of Ohio's rocks that are visible at the surface today.

A chain of volcanic islands located along a trench east of Laurentia, and perhaps some unknown landmasses, slowly slid underneath eastern North America some 475 million years ago in the vast sea of Ordovician time. Laurentia was mostly underwater at this time, and its eastern edge was around 440 miles east of present-day Ohio. As the plates collided, a belt of volcanoes on the edge of the Baltica (European) plate ran into the eastern border of Laurentia, causing a belt of folded and contorted rocks to rise out of the sea. These mountains were the Taconic Highlands, and they dominated the geology of eastern North America throughout late Ordovician and early Silurian time. The Taconic Highlands extended about 125 miles on either side of the present Atlantic Coast and stretched from Georgia to Newfoundland. A vast volume of debris was eroded from these mountains to form the surrounding sedimentary rocks from present-day Delaware to Wisconsin. Some geologists estimate that the mountains may have reached heights of 13,000 feet. As time passed, streams flowing down their slopes poked fingers of sediment in all directions, including toward Ohio.

About 375 million years ago in Devonian time, after the Taconic Highlands had been eroded away, the cycle repeated itself when another section of Baltica and another chain of volcanic islands collided with Laurentia far to the northeast of Ohio, forming the Acadian Mountains, an even higher and larger range. This range paralleled the present-day Atlantic Coast from South Carolina north. The eastern margin of what was to become North America was as much as 400 miles farther east than it is today. Streams flowing from the Acadian Mountains carried sediments westward once again, building a great sequence of overlapping deltas. Devonian- and Mississippian-age shales and sandstones mark this influx of erosional debris. Deep below the surficial rocks of eastern Ohio there are thin sheets of ash that also date to this time—the result of volcanic eruptions to the east caused by this mountain-building event.

The culmination of mountain building in eastern North America began about 318 million years ago during Pennsylvanian time. A number of volcanic

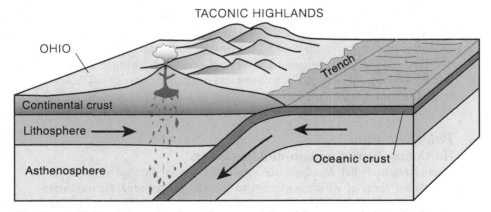

The Taconic Highlands formed as the seafloor was subducted down a trench that existed about where the Atlantic Coast is today. Volcanic islands at this trench were melded to Laurentia, which became North America. —Artwork by Mountain Press

island chains, microcontinents, Africa, and South America collided with the supercontinent that included present-day North America and Europe. The process lasted about 65 million years. The closest this collision came to Ohio was in the southern Appalachian region, where northwestern Africa slid into North America. Once again the resulting mountains—the Appalachians—shed debris in Ohio's direction. Deltas of major rivers formed across eastern Ohio, continually changing shape with fluctuating sediment supply and uplift from the ongoing mountain building. Through time, the depositional environments in Ohio alternated between river channels, floodplains, coastal swamps, uplands, brackish estuaries, and shallow seas. These different environments are reflected in the Pennsylvanian- and Permian-age rocks of southeastern Ohio. These rocks compose the hilly part of Ohio known as the Allegheny Plateau, and this region marks the transition from the flat till plains of western Ohio to the deformed rocks of the Appalachian Mountains in Pennsylvania and West Virginia.

The Ordovician World

Sedimentary rocks, sandstones, shales, and limestones to name a few, underlie Ohio's surface. The deposition of these rocks began as a shallow sea covered ancient North America in Cambrian time, about 542 million years ago, long before the Taconic Highlands developed. The oldest sedimentary rocks of Paleozoic age in Ohio were derived from the initial flooding of the eroded, barren landscape made up of Precambrian rocks. The highest part of this ancient landscape covered much of present-day eastern Canada; even still, it remained a lowland and was a source of sand, silt, and mud during early Paleozoic time. Initially, the Cambrian sea spread this sediment as a layer of sand and pebbles, which became sandstone. After all of the coarse eroded debris settled out of the sea, the chemical precipitation of calcite and, locally, dolomite began and continued into late Ordovician time. All of this early rock is buried far below Ohio's surface and is only seen in deep oil and gas wells.

The oldest rocks at the surface in Ohio come from late Ordovician time, about 445 million years ago. A shallow, warm sea still stretched across what would become Ohio and most of central and western North America. The Taconic Highlands stretched from what is now Georgia into Canada. Rushing streams carried pebbles, sand, and mud from these uplands into present-day Pennsylvania and West Virginia, forming fingerlike deltas where they joined the sea. Ohio was too far from the mountains to receive coarse sediments, but mud occasionally clouded the Ordovician sea. Ohio's closeness to the Ordovician equator, hence milder weather, favored the deposition of limestone on the seafloor, which forms from seashells and chemically precipitated calcite (calcium carbonate is precipitated more favorably in warm water; it dissolves more readily in cooler water). But shale, formed from clay sediments of the eroded highlands, was also deposited in great quantities, often settling in low spots of the ocean floor.

The Silurian World

As the Taconic Highlands in eastern North America gradually wore away, less clay reached western Ohio. As a result, limestone and dolomite were the dominant rock types that formed in the shallow sea that still covered Ohio. This change marked the beginning of Silurian time. On the southern border of the subsiding Michigan Basin in northwestern Ohio and along the Cincinnati Arch running from Cincinnati to Findlay, massive reefs flourished in warm Silurian seas from around 435 to 420 million years ago since Ohio lay much closer to the equator. Colonial corals, large brachiopods called *pentamerids*, and clams are characteristic fossils of this age. Today, Silurian bedrock is mainly dolomite, a rock that formed when groundwater rich in magnesium percolated through limestone. Fossils from this time are commonly molds and casts and often lack detail because mineral crystals, especially dolomite, began growing again after the fossils were created. Today, openings in the original Silurian rocks hold crystals of celestite, calcite, fluorite, and other minerals.

By 425 million years ago the Silurian sea began to withdraw from Ohio, eventually exposing the sea bottom everywhere except for Ohio's northern counties. A large evaporite basin formed across present-day Michigan, the site of Lake Erie, and northeast Ohio. The sea continued to exist in this area as it withdrew elsewhere, but it was abnormally shallow. This shallowness and a drier climate led to extreme evaporation. The remaining water became saltier and saltier. Instead of the normal precipitation of calcite in the water, dolomite, gypsum, and halite crystallized and accumulated on the sea bottom to form some 2,250 feet of strata.

The Devonian World

About 416 million years ago, while Ohio lay exposed as land, streams carved into the landscape, removing tremendous volumes of rock and sediments that had been deposited. This erosion created the major unconformity, or gap in the sediment record, that marks the transition into Devonian time in Ohio. Salt and gypsum continued to crystallize in the Devonian remnant of the late Silurian sea. Northeast of ancestral Ohio, the Baltica plate containing the European continent was slowly colliding with the North American plate, setting the stage for the uplift of the Acadian Mountains. The shifting of the plates brought the ocean back to Ohio about 400 million years ago; slowly, low-lying areas became fingers of the sea. Sea level rose. By 375 million years ago the state was once again submerged.

Ohio was even closer to the equator than it had been in earlier Paleozoic time, and warm tropical conditions led to a proliferation of sea creatures in the shallows and the deposition of fossiliferous limestones. River deltas spread westward toward Ohio beginning around 375 million years ago, depositing lobes of sediment eroded from the uplifting mountains to the east. Ohio was far enough west of the mountains in Devonian time that only mud clouded the sea, forming shales, until delta growth brought sand to northeastern and central Ohio.

Plants carpeted the land west of the Acadian Mountains to a shoreline in present-day southeastern Pennsylvania, way east of Ohio. This was the first significant land vegetation that developed late enough in time to have left behind plant fossils, but only the larger plants have been found as fossils in Devonian shales of Ohio. Fossils include 100-foot-high trees, members of the progymnosperms (primitive conifers) called *Archaeopteris*, which grew near the mountains. Scattered fossil logs of these trees, referred to as *Callixylon*, washed into the muddy sea covering Ohio. Suggestive of a long period of transport, some of the logs found in Adams and Delaware Counties are encrusted with crinoid fossils (sea-dwelling echinoderms); geologists speculate that the crinoids might have lived on the underside of the logs as they floated in the sea. By about 365 million years ago sandstone was forming in several delta lobes that extended from the present Lake Erie coast toward the Ohio Valley. This became the Berea sandstone, which was an economically important stone.

The Mississippian, Pennsylvanian, and Permian Worlds

By around 360 million years ago at the start of the Mississippian Subperiod, eroded sediments from the Acadian Mountains were still spreading across northern and eastern Ohio. The Cincinnati Arch, including land from Conneaut to Toledo and Toledo to Cincinnati, was exposed; sand was accumulating in a shallow sea that covered the southeastern part of the state. Throughout Mississippian time the shorelines fluctuated as one delta expanded and another shrank; deposits shifted from deltaic to nearshore marine deposition as rivers, and then ocean waves, assumed control. The sands were moved about by waves and currents, forming cross-beds (intersecting layers) and ripple marks; mud settled in low spots. Only the most robust shells and skeletons retained their structures in these deposits to be preserved as fossils. The Cuyahoga and Logan formations of this time contain an assemblage of fossil clams, snails, and brachiopods mainly preserved as casts and molds. The Mississippian sandstones are relatively resistant to erosion and form many cliffs and knobs throughout northeastern, central, and south-central Ohio.

Beginning around 318 million years ago the Appalachian Mountains started to rise. Again, Ohio remained far removed from the mountains, but erosional sediments continued to pile up across Ohio; all of eastern North America from the present-day Atlantic Coast to Kansas was land. However, sea level rose and fell many times from late Mississippian through Permian time.

During times of higher sea level, slim fingers of seawater slipped into Ohio, depositing shale and limestone layers containing fossils of spiny brachiopods, snails, clams, and corals. At other times the primary deposition was sand in stream channels, silt and mud on the surrounding floodplains, organic-rich mud and coal in swampy areas, and nonmarine limestone in lakes and ponds. As is the case in modern deltas, the environments of deposition were constantly changing; what was a stream channel at one point in time became a floodplain at another, and later still, a shallow sea. The sedimentary layers of this type of environment follow certain patterns of order, which is often a repeating sequence. Some geologists refer to these repeating layers as *cyclothems*. In passing from one cyclothem to another, often the site becomes one of erosion, not deposition. These breaks in deposition are unconformities that mark the top and bottom of a typical cyclothem. Other geologists mark the boundaries of what they call *sequences* by paleosols (ancient buried soil horizons), not unconformities.

The Pennsylvanian landscape across Ohio was dominated by lush vegetation and many swamps. Essentially, Ohio was at the equator at this point in time and had a warm, humid climate. The terrestrial strata commonly yield

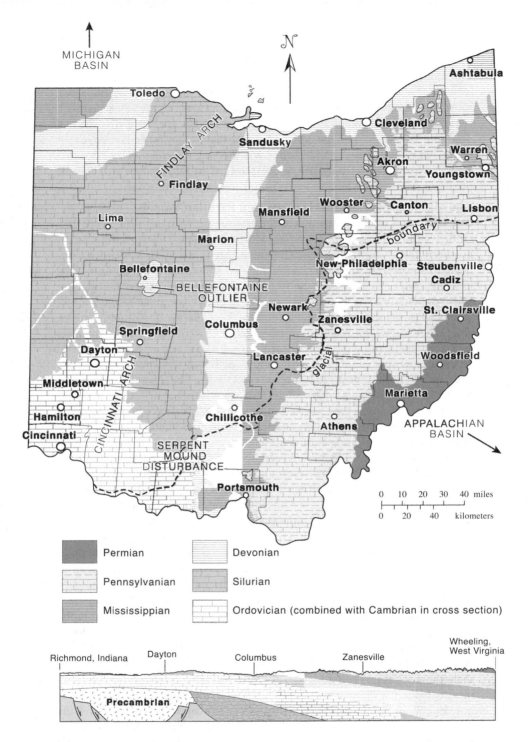

Bedrock geology of Ohio. A geologic map shows the distribution of rock formations at the surface of the earth. It tells what rocks are underfoot and what to expect elsewhere. Because much of the bedrock is covered by sediment, geologists must carefully observe rocks exposed at the surface, or outcrops, and use their knowledge of these rocks and the fossils they contain, and geologic structure, to project what lies below covered areas. Their projections are often confirmed by records from area wells and borings. —Modified from Ohio Department of Natural Resources, Division of Geological Survey publication

abundant fossils of pteridophytes (true ferns), pteridospermophytes (seed ferns), lycopods (scale trees), sphenophytes (horsetails), and the *Cordaites*, an extinct group of conifer-like trees. The youngest rocks, not sediment, in Ohio date to Late Pennsylvanian–Early Permian time around 300 to 295 million years ago. Plant fossils are also common in these rocks. These strata formed under strictly terrestrial conditions and thus lack marine fossils but contain some interesting fossil amphibian and reptile remains.

The Missing Record

The rise of the Appalachian Mountains and formation of the supercontinent Pangea in late Paleozoic time changed eastern North America forever. Pangea began to break apart 190 million years ago along the present-day Atlantic Coast, and the Atlantic Ocean began to form. Back in Ohio, and across North America, sediments accumulated here and there in stream channels and lakes, but erosion was the primary geologic process of this time. The lack of a widespread sheet of sediments meant that not much of a sedimentary record was preserved. Thus bedrock representing latest Paleozoic, Mesozoic, and most of Cenozoic time, between 290 million and 300,000 years ago, is not present in Ohio, representing a roughly 290-million-year gap in knowledge. Besides local smatterings of Permian sediments that weren't eroded away, the next sediments that left a lasting record in Ohio came from great continental ice sheets during the last 2 million years.

Arches and Sedimentary Basins

Sediments in the ancient seas that covered Ohio during early Paleozoic time accumulated as flat layers, but few layers remained that way. Some areas of the earth's crust remained stable or rose while others sank, producing what geologists call *arches* and *basins*. Beginning in late Ordovician time, about 450 million years ago, rock layers in southeastern Ohio began to slowly sink. These rocks became the western edge of the Appalachian Basin, in which sediments eroded from rising mountains to the east accumulated. The Paleozoic strata on the western edge of the basin that didn't sink are known as the Cincinnati Arch. This arch runs from northern Kentucky to Clark County, Ohio. The basin sank sporadically during Paleozoic time, tilting rock layers in central and eastern Ohio to the southeast. Pennsylvanian rocks outline the edge of the Appalachian Basin in southeastern Ohio. This basin extends from about Steubenville south along the Ohio Valley to Pomeroy and continues into West Virginia and southwestern Pennsylvania. Because this region continued to sink into latest Paleozoic time, it is the only place in eastern North America where strata of Permian age accumulated in great enough thicknesses to escape later erosion.

On the west side of the Cincinnati Arch, in Indiana, Paleozoic strata tilt westward into the Illinois Basin, another area where sinking of the earth's crust allowed the accumulation of great thicknesses of sediment. In northwestern Ohio, strata tilt similarly into the Michigan Basin to the northwest from a higher area called the Findlay Arch, which extends from Toledo south to Springfield. Erosion of the crests of the arches when they were periodically above sea level stripped away one layer of rock after another, leaving the oldest rock in a strip down the center of each arch and parallel bands of increasingly younger rocks on either side of them. The erosion of these arches is the reason the state geologic map has such distinctive north-south bands of bedrock.

Both the Cincinnati and Findlay Arches are broad features as well as long, extending across most of Ohio, eastern Indiana, and northern Kentucky; thus the tilt, or inclination, of the rock strata is not readily noticeable in a localized area. The tilt only becomes significant when geologists trace certain rock layers and find that they occur deeper and deeper below the surface the farther east and west they are of the arches. On the center of the Cincinnati Arch in downtown Cincinnati or the Findlay Arch in Toledo, the incline is less than 1 degree, so essentially the rock layers are flat.

Sedimentary arches and basins are extremely important to the modern world because many contain oil and gas and sometimes coal. Oil and gas exist across the state and are associated with the Cincinnati and Findlay Arches and Appalachian and Michigan Basins. Coal lies within the Appalachian Basin, greatly benefiting Ohio's economy. Arches and basins are also important because they influence what kind of landscape develops at the surface. In southeastern Ohio the arches and basins are the reason that more erosion-resistant late Paleozoic sandstones occur at the surface and the Allegheny Plateau, an upland, exists; in western Ohio, the softer early to mid-Paleozoic limestones, which also underlie the strata of southeastern Ohio, are at the surface and have been eroded flat.

The Pleistocene Ice Age

Sometime around 2.5 to 2 million years ago, a cooling of the world's climate led to continental glaciation in the Northern Hemisphere. An ice mass centered on the present-day Hudson Bay area gradually extended icy fingers across the northern section of the continent, at times covering all but southeast Ohio, which is perched on the Allegheny Plateau.

Many questions remain unanswered: How many ice sheets came and went? When and why did they start and stop? What was the weather like? The modern landscape contains clear evidence of the last two ice sheets and a separating interglacial time, but little is clear about earlier advances and retreats because much of that evidence was reworked by succeeding glaciation or has been eroded. Scientists reason that the weather was both cooler and wetter than what we know today. It seems that low summer temperatures not low winter temperatures were instrumental in keeping ice cover in glacial times. At least the last glacial episode ended with an abrupt change in climate and a great melting of ice that sent tremendous torrents of water down Ohio's drainage system, which the glaciations drastically altered.

Before 240,000 years ago, what's called the pre-Illinoian ice sheet covered western Ohio, reaching as far south as northern Kentucky. During the interglacial time that followed its retreat, pre-Illinoian till and other glacial sediments gave rise to soils as woodlands returned. Unfortunately, geologists know very little about this time in Ohio's history. The only pre-Illinoian deposits that remain in Ohio are in the Cincinnati area.

About 230,000 years ago another glacial stage began with the arrival of Illinoian ice, so named because its deposits were first recognized in that state. The

Illinoian glaciers reworked (eroded and redeposited) most of the pre-Illinoian sediment, transforming it into various younger tills. Illinoian ice eventually covered all but the southeastern counties of Ohio, lapping onto the edges of the Allegheny Plateau. Illinoian ice, its deposits, and its meltwater modified numerous Ohio streams and rivers. Outwash, laid down by meltwater streams, formed high terraces that flank major southern valleys, including those of the Great Miami, Little Miami, and Scioto Rivers. Streams that flowed northward were reversed. Geologists believe that Illinoian ice disappeared by 125,000 years ago, but the ages and stratigraphic relationships of some deposits are still in question. After the ice retreated the climate moderated, and a long interglacial period began when large volumes of Illinoian sediments were eroded away.

About 117,000 years ago ice again entered Ohio. The younger glacial episode is called the Wisconsinan, and it lasted until about 10,000 years ago. Since the ice that covered Ohio came from the present-day Great Lakes region, it is often referred to as the Huron-Erie ice, or Huron-Erie lobe. The history of earlier ice ages that the earth experienced suggests that ice ages tend to last much longer than a couple million years, leading many scientists to deduce that the earth may only be in a warmer interval of the Quaternary Period. Northern North America may see a return of the ice sheets in the geologic future.

Geologists use several methods to date glacial materials and events. Since ice does not deposit material in well-defined layers and fossils are rare, geologists resort to other techniques to unravel Pleistocene history. Radiometric dating—measuring the radioactive decay of radioactive elements—is still among the most important ways of dating Quaternary sediments. Other techniques used in Ohio include the analysis of paleomagnetism, recognized by reversals in the magnetic polarity of glacial sediments, and changes in the amino acid structure of conifer wood.

Another obstacle geologists face is distinguishing and correlating tills; for example, how do you tell an Illinoian till from a Wisconsinan till? Geologists can decipher this mystery with several methods. They may look at the orientation of elongate pebbles and small cobbles embedded within the till—the till fabric. Tills of different age may have come from ice traveling in different directions, and the long axes of the pebbles in the till fabric indicate this. The composition of the larger particles in a till, perhaps eroded from an area with distinctive bedrock somewhere north of Ohio, may also indicate to geologists from which direction an ice sheet came. Geologists can examine the scratches or striations of underlying bedrock in order to determine the direction of ice advance, which might point to a particular glacial lobe. Other techniques involve the particle shape of a till and the distribution of particles of different sizes.

Ohio's preglacial drainage system trended in a northerly direction. Many of Ohio's current streams and rivers, which run in a southerly direction, developed as a result of the ice sheets, which dammed many rivers and streams and forced them to cut new paths and/or flow in other directions. Draining glacial lakes also cut many new channels that persist throughout Ohio today. I discuss these topics in more detail in the following chapters.

Lake Erie developed along a large bedrock valley system—connected with the Gulf of St. Lawrence—that existed before the ice sheets came. Lake Erie's first glacial predecessor developed about 16,000 years ago and was followed by many more. As climate varied, so did the water level of the different glacial lakes; some were much higher than present-day Lake Erie, and some were much lower. I discuss Lake Erie's glacial history in the Allegheny Plateau and Lakeshore Ohio chapters.

Glacial Erosion and Deposition

Geologists estimate that ice thicknesses ranged from several hundred to several thousand feet across Ohio during the maximum extent of the different glacial stages. The thickness and mass of the ice allowed it to push, scrape, and pluck loose material from the ground over which it passed. Most of this loose material remained near the ground as the ice moved, but internal ice movements pushed some of the debris higher in the ice sheet, allowing the ice to carry materials great distances. Just as a sheet of sandpaper wears away wood, the ice sheets abraded the underlying land surface. The ice planed off bedrock at the surface in the direction of its movement. Harder particles, frozen in the sole of the ice sheet, left scratches, or striations, on the bedrock surface, again in the direction of ice movement. Occasionally, gouges or grooves formed as a result of gushing streams of meltwater that flowed under the ice.

Sediment deposited directly from glacial ice consists of debris of all sizes, shapes, and types mixed together. It is called *till* and typically is gray or tan. Till is the parent material of the soil over much of glaciated Ohio. One could make a rather impressive rock collection from the various rocks contained within Ohio's tills. Indeed, most of the rocks are unlike any that compose the bedrock of Ohio; they are foreign rocks—granite, gabbro, quartzite, schist, and gneiss. Many come from the Precambrian bedrock of Canada. Ohio farmers are all too familiar with the larger versions of these glacially transported rocks called *glacial erratics.*

Glaciers plastered most of their till on the surface to make a flat and rather featureless deposit called *ground moraine,* or *till plain.* Where the rate of ice

Wisconsinan-age till forms the upper slopes of a quarry near Sylvania. Boulders and cobbles are glacial erratics.

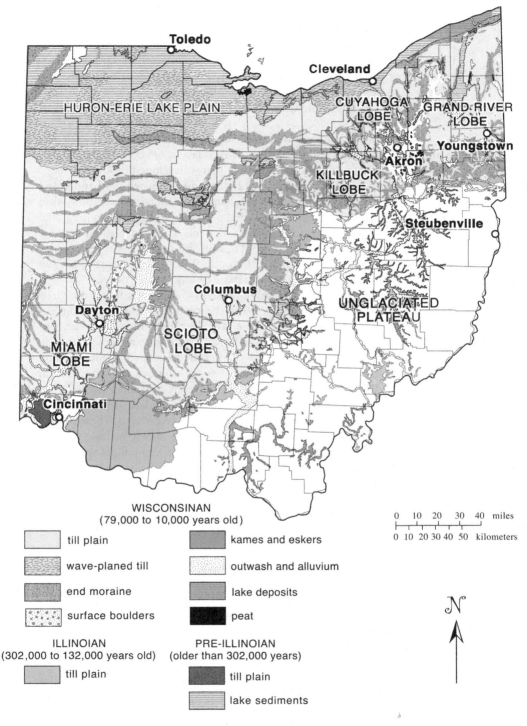

Generalized glacial geology of Ohio. —Modified from Ohio Department of Natural Resources, Division of Geological Survey publication

advance matched the rate of melting at the edge of an ice sheet, till accumulated in a hilly ridge, much like a pile of gravel forms at the end of a conveyor belt. Between Lake Erie and the Ohio River, a series of end moraines mark various stoppages of the Wisconsinan ice sheet; they are separated by till plains. The Wisconsinan ice removed evidence of earlier Illinoian moraines except in areas of southern and central Ohio where Illinoian ice had spread farther south than the Wisconsinan ice.

Glacial meltwater deposits a different-looking material called *outwash*. It is mainly sand and gravel and forms well-defined layers, or strata, because it settles out of water rather than ice. Outwash plains developed beyond the edge of melting ice sheets as numerous twisting streams deposited sand and gravel in broad sheets. Between lobes of ice, meltwater streams often combined and poured down valleys, alternately widening, deepening, and filling in preglacial valleys. Long fingers of outwash, called *valley trains*, were deposited in these valleys and extend into southern Ohio. The outwash in Ohio, now mainly preserved in terraces, is not as deep to the south; it is deeper in northern Ohio where it meshes with the till plains near its northerly starting point.

Meltwater streams also deposited outwash in low areas on top of the ice or as fan-shaped piles along its margins. When the ice finally melted, this sediment was lowered onto the ground to form sandy hills called *kames*. Outwash also filled the valleys of meltwater streams on the ice surface and was deposited along tunnels carved through the ice. It survived after the ice melted as long winding ridges called *eskers*. A string of gravel pits on today's landscape often marks an esker's course. Many Ohio eskers have been completely quarried away.

Huge blocks of ice broken from the edge of receding ice sheets were stranded and surrounded by outwash, sometimes even covered. These chunks of ice persisted after the main ice sheet disappeared, slowly melting away to become low spots in the outwash plain called *kettles*. Many filled with water to become kettle lakes. Temporary shallow lakes also formed in ground moraine as the ice sheets blocked the drainage of some rivers and streams. Most of the lakes formed in tributary valleys of major north-south rivers when outwash in the main valleys piled up and dammed the tributaries. The evidence for these lakes is Ohio's numerous flat plains that are underlain by silts and clays.

Windblown Sediments

After each ice sheet retreated, strong winds blew across the adjacent ice-free land, depositing silt winnowed from tills and outwash. This material, called *loess*, became soil and reached thicknesses of 3 to 10 feet in southwestern Ohio and along the edges of the glaciated Allegheny Plateau. Geologists recognize at least three distinct windblown loess horizons. Each loess marks an interglacial time. Except under the microscope, loess is virtually indistinguishable from other soil and is not a prominent feature in Ohio.

In some areas, wind blowing across sandy surfaces created ripples and sometimes built sand dunes. Sand dunes also formed near sandy beaches and offshore bars that were exposed by lowering water levels of the various glacial lakes, particularly in the Lake Erie Basin. Many of these dunes are still visible, for example, at Oak Openings Preserve Metropark near Whitehouse in northwestern Ohio.

Sedimentary rock strata along Ohio 125 west of Georgetown. The thicker beds that jut from the cliff face are limestone; the inset beds are shale. These rocks formed in a sea that covered Ohio some 450 million years ago.

Record in the Rocks

Most sedimentary rocks begin as deposits of ordinary sediments such as mud, sand, and gravel, which are common today, and calcareous mud, which is now rare. Given geologic time and the weight of overlying material, mud, sand, and gravel solidify into shale, sandstone, and conglomerate, respectively. The sediments accumulate in nearshore sea areas, stream or river channels, deltas, or deeper oceanic basins. As sediments continue to build up, older sediments are covered and compressed. Groundwater moving through the buried sediments precipitates mineral cements, like calcite or quartz, which bind the layers of particles. Along seashores where smaller amounts of land-eroded sediment accumulate and water temperature is warm, calcite may precipitate from the water to form limestone. Because animals and plants also frequent this environment, fossils are often found in limestone. Dolomite, gypsum, and rock salt form where extreme evaporation, warm temperatures, reduced precipitation, and/or restricted sea circulation cause seawater to become supersaturated with minerals—dolomite, gypsum, or halite—that settle out of the water.

Geologists divide sedimentary rocks into formations, which are simply distinctive rock units that they recognize from one place to another and can plot on maps. Members are distinctive and recognizable parts of formations, but they are usually too thin to plot individually on a map. As the name suggests, groups are two or more closely related formations that geologists consider a single unit for convenience of mapping. Many sedimentary layers that occur elsewhere in North America are missing from the geologic record of Ohio.

These gaps, or unconformities, represent undetectable amounts to hundreds of millions of years of geologic history. They suggest times when sediments didn't accumulate or were being eroded away.

The rocks of Ohio are like the pages of a novel; each layer tells us something about Ohio's past, and when geologists can interpret them in the appropriate order, a story unfolds. Where unconformities occur, the story is more diffi-cult to follow. Rocks contain all kinds of clues about their history. Sedimentary rocks commonly have horizontal or slightly tilted layers, called *strata*. These layers indicate to geologists that sediments settled out of water and accumu-lated, grain by grain, on top of one another on an ocean floor, river bottom, or some other area of sediment settling. This process is called *deposition*. Tilted rock layers result when areas of an ocean floor settle at different rates; currents scour out some areas while sediments slump into depressions; or up-and-down and sideways tectonic forces bend or crack overlying sedimentary layers.

Some Ohio sandstones consist of many intersecting layers that are called *cross-beds*. In a cross section of a sedimentary rock, cross-beds are inclined, or they intersect one another, in relation to the parallel layering of other sedi-ments. They represent deposition that occurred in an area dominated by water or wind currents. Careful analysis of cross-beds allows geologists to determine

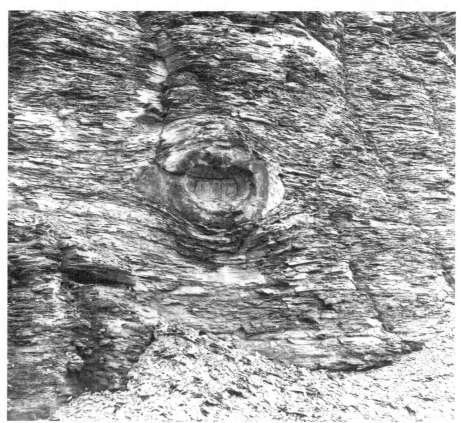

A typical siderite concretion in the Devonian-age Ohio shale of Copperas Mountain west of Chillicothe. This specimen broke in half, exposing a core that was filled with mud.

ancient current or wind patterns. Rocks with cross-beds usually have ripple marks. Both water currents and wave motion in nearshore areas or streams or wind blowing across sandy terrain form ripple marks. Some are useful in determining the direction of the currents. Shales and limestones sometimes have mud cracks, which form when muddy sediments are exposed above water and bake in the sun on tidal flats and floodplains. Concretions and nodules are common in Ohio's Paleozoic rocks. They are balls or saucers of calcite (calcium carbonate), pyrite (iron sulfide), quartz (silicon dioxide), or siderite (iron carbonate) that differ in composition from the surrounding sedimentary rock.

Concretions and nodules form from the successive buildup of concentric shells of minerals deposited by groundwater around a nucleus—a sand grain, crystal, or fossil—embedded within shale, or sometimes other sedimentary rocks. They develop as the sediment hardens into rock, often causing a local bending of the strata above and below the concretion. As a concretion dries out cracks can develop, only later to refill with mineral crystals and form a septarian concretion, which has unique crisscrossing ridges. Concretions can range from 1 inch to 8 feet in diameter and possess smooth surfaces. Nodules are generally smaller than concretions and typically have a rough outer surface. Concretions or nodules that have a hollow interior lined with crystals are called *geodes*.

Ohio's rocks are a treasure trove of fossils that provide geologists with glimpses of the creatures and plants that lived in the vast shallow seas, swamps, lakes, rivers, and forests that once covered the state. Some fossils form when sediments cover a shell, bone, tooth, or piece of wood. Groundwater fills openings in these hard remains, precipitating certain minerals that turn the organic matter into inorganic minerals. Plant fossils often form when leaves and stems are compacted between sedimentary layers, leaving only a thin film of black carbon outlining their shapes. Other fossils are molds of the original organisms, such as the imprint a shell might make when pressed into mud or sand. Other molds form shapes not readily recognizable to most people because they show the inside of an organism. For example, paleontologists have determined that some curlicue sedimentary objects are the internal molds of snail shells that formed when the shell filled with sediments. Other fossils may only be a trail or track left by an animal as it crossed a muddy area. Fossils range in size from microscopic pollen grains to the skeletons of Ice Age elephants. They help geologists date rock formations because in a vertical succession of rock layers, fossils change in type and abundance from one layer to another. Knowing the proper order of strata and comparing them to areas with similar strata, geologists can use these changes in fossils to chart changes in depositional conditions and animal and plant life over geologic time.

Common Ohio Fossils

The most likely candidates for fossilization were plants and animals with some kind of hard part—a shell, stony skeleton, tooth, bone, or wood. The organism also needed to live in an area where upon its death it was quickly buried in sediment. Quick burial slowed decay and limited the amount of destruction caused by scavengers. Ohio is rich with fossils because most of its rocks formed in and around shallow seas where animals were buried quickly.

Among the common fossilized Paleozoic sea animals are fusulinids, brachiopods, bryozoans, corals, snails, clams, cephalopods, ostracods, trilobites, and crinoids. Fusulinids, now extinct, were one-celled animals that had calcareous shells; modern protozoa are related to them. Generally, they were the size and shape of grass seeds. They are found in Pennsylvanian-age marine limestones of eastern Ohio. Ostracods are also small and go unnoticed by many fossil collectors. They are members of the arthropod phylum and have elongate calcareous shells composed of two halves. Because they tolerated stagnant and extremely salty water, their fossils are sometimes the only ones that exist in certain shales and dolomites. Ostracods are still common, but only freshwater species now make Ohio home.

Paleozoic brachiopods superficially resembled clams, but unlike clams they had shell halves that were different shapes. They lived on or burrowed in the sea bottom. Today, brachiopods have decreased in diversity and abundance but still frequent the world's oceans. Paleozoic mollusks include snails and clams, which require little introduction, and nearly extinct cephalopods called *nautiloids* and *ammonoids*. Nautiloids looked much like modern squids, but unlike squids they had calcareous shells. Some nautiloids had cone-shaped shells, and a few species reached lengths of several feet, qualifying them as among Ohio's largest marine fossils. Ammonoids were not as large. They had coiled shells, much like some modern snails. Snails and clams are, of course, still common, but only one species of nautiloid still exists and there are no ammonoids.

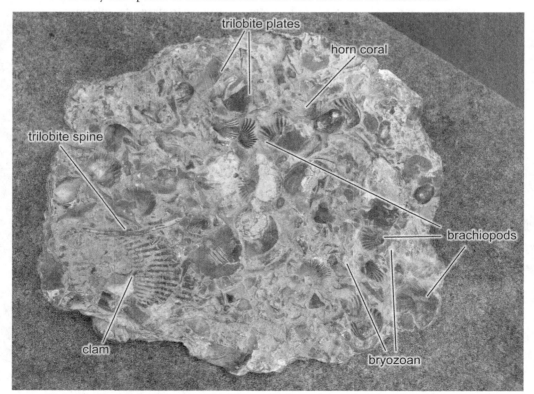

Typical fossils of Ordovician time. This Waynesville formation specimen was found near Oxford.

Bryozoans are tiny colonial animals that form calcareous mounds, many branching plantlike growths, and netlike mats on the seafloor. The Paleozoic seafloor was covered with many branching and mound-shaped colonies. Bryozoans built sturdy skeletons of calcium carbonate and are among the commonest of fossils. You need a magnifier of some sort in order to see the details of these creatures in a fossil because the animals extended from pinprick-sized openings in their stony skeletons. Corals had similar growth patterns, but they are much larger animals; individual coral animals, or polyps, ranged from ⅛ inch to 4 inches in diameter. Solitary horn corals, now extinct, looked like small cornucopias attached to the sea bottom. Paleozoic colonial corals left fossils that look like honeycombs. Corals began to dominate reefs by Silurian time. Huge barrier reefs extended for hundreds of miles across what is now the Great Lakes region, ancient predecessors of the Great Barrier Reef of modern day.

Trilobites, extinct three-lobed arthropods, frequented early to mid-Paleozoic sea bottoms—in areas similar to those that modern crabs and shrimp occupy. Since they grew by shedding old skins, one individual may have left behind several fossils. Paleozoic crinoids were stemmed echinoderms, an animal group that also contains the more familiar starfish and sea urchin. Paleozoic crinoids, or sea lilies, had flexible stems that consisted of little disks of calcium carbonate strung together by an internal fleshy stalk. The crinoid rooted its stem in sediments or cemented it to another hard object. At the top of the stem there was a rounded calcareous crown with tentacles that extended outward. Because of their delicate nature, crinoid fossils are rarely found intact; usually only segments of their stems and crowns remain as fossils.

Carbonized imprints of fern fronds, bark, stems, and fragments of Pennsylvanian- and Permian-age plants are the most common late Paleozoic nonmarine fossils. Occasionally, petrified logs and stumps of Devonian and Pennsylvanian trees appear in formations. Bones, teeth, and tracks of Paleozoic fish, amphibians, and reptiles are among the rarest and most valued finds. In later chapters I will discuss some important finds of fossil fish, including ostracoderms (extinct jawless armored fish), placoderms (extinct armored fish), acanthodians (extinct minnow-sized fish with many fins), and sharks in the Cleveland area. The Devonian black shales of this region have yielded many clues about the early evolution of fish groups.

Several sites in Pennsylvanian and Permian strata of the Ohio Valley have disclosed even rarer fossils—those of amphibians living in swamps and marshes and early land-dwelling reptiles. Fossil preservation in land areas was much less common than in the seas because the remains of dead organisms were more likely to be destroyed by scavengers, chemical decomposition, and erosion before burial and fossilization could take place. Ohio, however, had some unique and interesting denizens of the coal swamps, like the 6- to 8-foot-long sail-back reptiles, or pelycosaurs, that were fossilized. Carnivorous pelycosaurs (*Dimetrodon*) were at the top of the late Paleozoic food chain, walked on all fours, and had distinctive spiny fins down their backs. *Edaphosaurus*, another pelycosaur, had spikes protruding from its fin and fed on plants.

During Pleistocene time, Ohio was periodically covered with ice, but for a good portion of the last 2 million years the state was covered with forests, grasslands, and wetlands. Records from cores taken in bogs, marshes, and small lakes and buried logs recovered from till and outwash across the state give geologists a picture of what the landscape was like. Around 23,000 years ago when the Wisconsinan ice sheet was at its maximum southern position, southern Ohio was much like modern-day northern Canada, both in climate and vegetation. Although a tundra-type environment might be expected, fossil evidence indicates that there was a dense spruce, or boreal, forest. By 10,000 years ago the spruce gave way to pine and some hardwoods; and by 9,000 years ago a mixed-hardwood forest, dominated by oak, became the prevalent vegetation. More recently, openings in the hardwood forest gave rise to prairie plants in parts of western Ohio.

Pleistocene-age fossils lie hidden in glacial deposits, only to be exposed now and then in stream or ditch banks or during excavation and construction. Mastodon and mammoth teeth, tusks, and bones are by far the most common Ice Age fossils that have been reported, most likely because they are large and people recognize that they are something out of the ordinary. Other mammals that once frequented Ohio's countryside during this time include giant beavers, ground sloths, stag moose, short-faced bears, peccaries, musk oxen, bison, and caribou.

Rocks as Resources

Ohio's early settlers and the Native Americans before them took advantage of the state's rocks. The quarrying of Paleozoic limestones and dolomites has always been one of Ohio's top mineral industries. Ohio ranked fourth out of thirty-four producing states in lime production and seventh out of forty-nine in crushed stone (including sandstone) in 2004. Early on, people used limestones and dolomites as building stone and to make lime, but by the mid-1900s people mainly used them to produce crushed stone and lime. Quarriers pried building stone mainly from late Ordovician-age limestone; Silurian-age Brassfield and Dayton formations and Springfield and Greenfield dolomites; Devonian-age Columbus, Dundee, and Delaware limestones and Berea sandstone; Mississippian-age Cuyahoga and Logan formations; and Pennsylvanian-age sandstone formations.

Settlers built their cabins out of logs but used locally gathered rocks for chimneys and hearths. As communities grew, so did the use of stone. Just about any local stone was suitable for foundations. The tendency of limestones and dolomites to split in thin to thick layers made them suitable for the walls of public buildings like schools, churches, and courthouses. Ohioans made lime by burning limestone and used the resulting calcium oxide in mortar, plaster, and whitewash. It became a mainstay of the chemical, farm, and glass industries. Although most limestones found local use, major lime-producing formations include the Silurian-age Brassfield formation, several younger Silurian dolomites, and the Devonian-age Columbus limestone. Most formations were, or still are, crushed into aggregate and used for lime, cement, railroad ballast, road construction, and building construction. Larger blocks are used as riprap to prevent the erosion of embankments and shorelines.

Although limestone building stone production is no longer important in Ohio, the state ranks twelfth in the production of all dimension stone of the

A typical northwestern Ohio stone quarry in Silurian-age dolomite in the early 1900s. This operation crushed the limestone and burned it in a kiln to make lime.

thirty-four producing states and third in sandstone building stone as of 2004. Although all of Ohio's sandstones were used locally, the main contributors to this recent ranking are the Devonian-age Berea, Mississippian-age Buena Vista, and Pennsylvanian-age Massillon sandstones. Sandstones and conglomerates that weren't usable for construction were, and continue to be, crushed for use as aggregate, construction, foundry, and glass sand.

In 2004 Ohio ranked fifth in clay and shale production out of forty-one producing states. Major units that companies mine include late Ordovician-age shales, Devonian-age Chagrin and Bedford shales, Mississippian-age Logan formation shales, Pennsylvanian-age underclays and shales, and Pleistocene-age tills. Ohio is famous for its pottery and other ceramic products that potters made mainly from the Pennsylvanian-age rocks of eastern Ohio. Just about any town of significance boasted a pottery or two, a brick plant, or tile factory in the late 1800s and early 1900s. The ceramic and cement industries still use clay and shale but most goes to line landfills.

Coal mining in the Appalachian Basin remains an important part of the economy of Ohio. Ohio ranked thirteenth out of twenty-six producing states in 2004. Ninety-four mines reported production that year. Ohio mines are mainly in Belmont, Harrison, Vinton, Jefferson, and Athens Counties in the Pittsburgh, Middle Kittanning, Meigs Creek, Lower Freeport, and Clarion seams.

A problem with Ohio's coals as an energy source is that they contain high quantities of sulfur. The sulfur comes from three sources: pyrite, gypsum, and plant matter. Ohio coals average about 3.5 percent sulfur (2.2 percent pyritic sulfur, 0.01 percent sulfate sulfur, and 1.3 percent organic sulfur), which places them in the high-sulfur category as defined by the U.S. Environmental Protection Agency. According to the Clean Air Act of 1970 and later amendments, atmospheric emissions from coal burning must not exceed 1.2 pounds of sulfur dioxide per million Btu burned. Ohio's coals average 5.22 pounds. Physical cleaning or "washing" of the coal can remove up to 50 percent of the pyritic sulfur but has no effect on the organic sulfur content. Researchers continue looking for new, more-effective methods of cleaning Ohio's coals.

Ohio rock salt comes from mines at Cleveland and Painesville; the brine plants that distill the salt are in Licking, Summit, and Wayne Counties. Ohio was fourth out of fifteen producing states in 2004. Its salt is mainly used for ice control. Although flooding terminated gypsum mining at a quarry in Gypsum in Ottawa County in 2004, this area was the center of Ohio's gypsum production, with several underground mines and surface quarries, starting in the late 1800s.

Over 75 percent of Ohio's counties have yielded oil or gas since 1860 from over 250,000 wells. Pumping continues in about 20 percent of the wells. Oil and gas comes from some thirty formations at depths ranging between 50 to 8,500 feet. Production in the Ordovician-age Trenton limestone of northwestern Ohio dominated the industry's early years, but production of the Silurian Brassfield formation (also called the Clinton sands) of central Ohio replaced it in the early 1900s. Ohio's total production through 2004 was over 1.1 billion barrels of oil and 8 billion cubic feet of gas.

Sand and Gravel

Glacial outwash deposits are one source of sand and gravel, Ohio's second most important—behind crushed limestone and dolomite—nonenergy mineral resource. Other sources of sand and gravel are Lake Erie and the Ohio River. Sand and gravel extraction is the largest mining activity in most of the industrialized world. Builders rely on it for construction aggregate and road material. Any sand and gravel deposit close to a major market is valuable. Unfortunately, formerly good sources in Ohio may be close to not being available anymore either because mineral rights are not available or because active quarry sites are nearly depleted of usable material. Loss to urbanization is an especially difficult problem because sand and gravel must be close to a market, which may then put a gravel pit out of business as urban sprawl makes the land more valuable for other uses.

Ohio's Common Minerals

All rocks are made up of minerals, which are chemical elements and compounds. Minerals can harden, or crystallize, from magma, hot fluids, or groundwater; they can form around volcanic vents; they can develop as a result of metamorphism; and they can be derived from the weathering of rocks. Over forty kinds of minerals are native to the Buckeye State. Only gypsum, halite, and flint have been quarried and found useful. Most minerals in Ohio are scattered

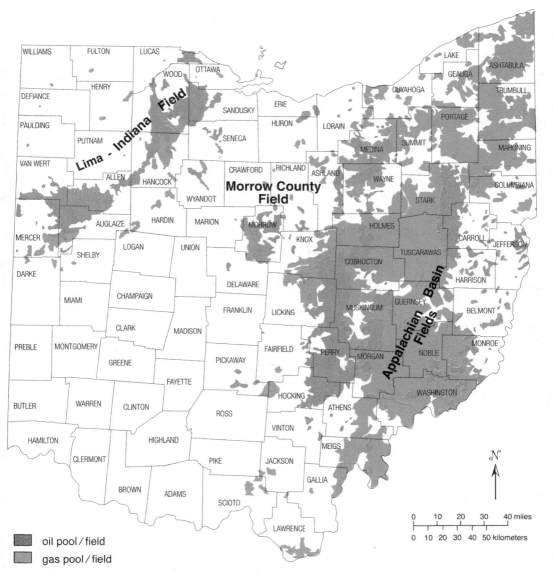

Ohio oil and gas occurrences as of 1996. —Modified from Ohio Department of Natural Resources, Division of Geological Survey publication

throughout various rock layers and are not concentrated enough to make them valuable enough to mine. It's much more economical to crush them with the enclosing limestone and dolomite to make aggregate than it is to separate them. However, many of the minerals are colorful and occur in attractive crystal clumps, making them sought after by collectors.

Without a doubt the most prolific occurrence of collectible minerals is in the Silurian-age dolomites of western Ohio. Barite, calcite, celestite, dolomite, flint, fluorite, galena, gypsum, hematite, pyrite, and sphalerite are the most

COMMON MINERALS OF OHIO

Mineral Name
(Chemical Composition)   **Occurrence**

Anhydrite
(calcium sulfate)
White to gray beds or nodules in association with dolomite, gypsum, and halite in Silurian rocks of northern Ohio.

Barite
(barium sulfate)
Clear, white, yellow to dark brown vein fillings in Ohio shale concretions and crystals associated with calcite and celestite of Silurian dolomites of western Ohio.

Calcite
(calcium carbonate)
Clear, white to dark brown crystals and crystalline masses associated with celestite, fluorite, and other common minerals and fossil replacements throughout western Ohio dolomites and limestones. Also found as vein fillings in concretions of central and eastern Ohio.

Celestite
(strontium sulfate)
Clear, white to bluish crystals filling cavities and cracks in western Ohio dolomites. Commonly associated with calcite, fluorite, and other minerals.

Chalcopyrite
(copper iron sulfide)
Brassy yellow metallic crystals filling veins in some concretions.

Dolomite (calcium magnesium carbonate)
White, tan, pink, and brown crystals in western Ohio Silurian strata and Ohio shale concretions.

Flint
(silicon dioxide)
Yellow, brown, black, and red masses associated with veins of quartz crystals. Found throughout the state but best known from central Ohio.

Fluorite
(calcium fluoride)
Yellow, brown, and purple crystals associated with calcite, fluorite, and other minerals of western Ohio dolomites.

Galena
(lead sulfide)
Lead gray, cubic, heavy crystals associated with other minerals of western Ohio dolomites. Weathered surfaces are dull gray but fresh surfaces have a silvery metallic sheen.

Gypsum (hydrous calcium sulfate)
Clear, white, and gray crystalline masses associated with younger Silurian dolomites of northern Ohio. Clear variety is called *selenite*.

Halite
(sodium chloride)
Clear to white crystalline masses associated with younger Silurian strata of northern Ohio.

Hematite
(iron oxide)
Reddish brown to gray heavy masses forming the cores of concretions and nodules in some Silurian and Pennsylvanian strata.

Marcasite
(iron sulfide)
A gray powder to crumbly crystals; a disintegration product of pyrite.

Melanterite
(hydrous iron sulfate)
Powdery white growths associated with the disintegration of pyrite. Common in Ohio shale and coal seams of central and eastern Ohio.

Pyrite
(iron sulfide)
Brassy yellow metallic cubic crystals found in Silurian and Devonian dolomites and limestones. Forms nodules in shales of Devonian to Pennsylvanian age in central and eastern Ohio. Also replaces fossils. Many people call it "fool's gold."

Quartz
(silicon dioxide)
Clear to white hexagonal crystals filling veins and geodes in Silurian, Devonian, and Pennsylvanian strata. Quartz is probably the most common of all minerals and is the main constituent of sand and silt.

Sphalerite
(zinc sulfide)
Small brownish to reddish crystals in Silurian dolomites and Ohio shale concretions.

Strontianite
(strontium carbonate)
White crystalline masses in western Ohio dolomites.

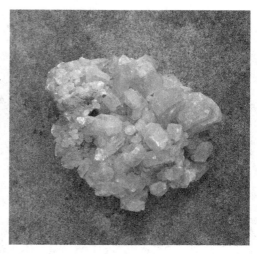

Celestite is one of northwestern Ohio's most common minerals. These crystals formed in a cavity in Silurian-age dolomite quarried at Lime City.

common; in lesser abundance are anhydrite, chalcopyrite, halite, melanterite, and strontianite. Of course, lots of other minerals, including diamond and gold, have been brought to Ohio by the ice sheets of Pleistocene time and many more lie undiscovered in basement rocks.

Other Notable Geologic Features
Groundwater and Karst

Water that soaks into the ground and does not immediately return to the surface becomes part of the groundwater, a vital natural resource. It may surprise you, but the amount of groundwater in Ohio far exceeds the amount of water in Lake Erie and Ohio's rivers and streams. The deposits of outwash on the floors of many of Ohio's stream valleys contain a wealth of good quality subsurface water in the open spaces between sand grains and pebbles. About 45 percent of Ohioans get their water from groundwater sources.

Groundwater seeps deeper than surface soil and loose sediment. It percolates down through cracks and pores in underlying sedimentary rock. Water moves through the subsurface more slowly than it does in surface streams, and typically it doesn't flow in channels. The groundwater system is more like a damp sponge that only yields water when squeezed. Groundwater is forced through the pores of geologic material by the pressure of overlying and adjacent groundwater. It moves from areas of high pressure into regions of lower pressure, sometimes even moving upslope. Groundwater moves most easily if every pore in the area is filled with water and there is no air space. Groundwater, which continually replenishes itself, occurs at different depths depending where you are in Ohio, so well drillers drill wells into the bedrock accordingly. The depth where groundwater is encountered below the surface is called the *water table*. Have you ever seen a deep hole that slowly filled to a certain depth with water even though it hadn't rained? You were essentially looking at the water table.

Where the water table is close to the surface, wetlands often develop. Where the water table intersects the side of a hill, a spring may form. Springs

Dry Cave near Bainbridge is carved in Silurian-age dolomite.

are proof that not all groundwater is pleasant tasting. Water that flows through rock may pick up dissolved minerals, particularly iron and calcium sulfates, which impart a foul taste and accompanying stink. People of the nineteenth and early twentieth centuries, however, prized such water for the treatment of whatever ailed them. Health resort towns like Green Springs, Magnetic Springs, Mineral Springs, and Yellow Springs grew up around such mineral springs.

A geologic material that readily yields groundwater is called an *aquifer*. Conglomerates, sandstones, limestones, dolomites, gravels, and sands are generally the aquifers in Ohio. Some rocks and sediments, like shale and clay, are just too fine-grained and tightly compacted to allow water to readily move through them. Aquifers sandwiched between these kinds of materials exhibit pressurized flow called *artesian flow* and often have water levels above the local water table; a few flow onto the surface as artesian springs.

Limestones and dolomites may be dissolved by groundwater, causing caves and related features to form. Ohio has a lot of caves, especially in the belt of Devonian limestones stretching from Lake Erie into southern Ohio. Groundwater commonly contains carbonic acid, a combination of water and carbon dioxide, which readily breaks down calcium carbonate, or limestone. Caves evolve from tiny vertical cracks that groundwater erodes to a complex of horizontal and vertical tubes and passageways; sometimes they consist of a number of levels at different depths. The deepest part of some caves may be at or below the water table, which explains why some commercial caves advertise underground rivers. The beautiful natural decoration of cave walls, ceilings, and floors forms when dripping groundwater precipitates calcium carbonate (calcite), the reverse of the chemical process that originally carved the cave.

The landscape above caves is often rolling due to the dissolution of the bedrock surface. Funnel-shaped holes called *sinkholes* develop over solution-widened cracks or joints. Sometimes sinkholes lead to caves below. Land dominated by sinkholes is called *karst*. Most of the drainage in karst is through caves. Valleys may still exist at the surface, but surface streams only flow short distances before emptying into sinkholes.

Rivers and Streams

The streams and rivers that stretch across Ohio's landscape are relatively young features since most of them couldn't have appeared in their present state until after the last glaciers left some 15,000 years ago. In that short time they have done a lot. Major valleys were at least partly inherited from preglacial streams; however, many others formed when postglacial streams and rivers were forced to flow parallel to the end moraines until the water found a low spot and breached them. Natural rivers tend to follow curving paths called *meanders* and are constantly changing their courses.

The strongest currents in rivers and streams stray to the outside bank when a channel curves, eating into the bank and further accentuating the curvature. The bank often forms a steep bluff because of continual undercutting and the collapse of shoreline material. On the opposite bank, where water is sluggish and drops the sediments it's carrying, sandbars and gravel bars build out into the channel. They are called *point bars*. Eventually, a narrow, long loop forms in the stream. The stream may cut through the neck of this loop, cutting off

Meandering Cranberry Creek at Glandorf. Sediments accumulate on the left bank just beyond the dolomite bridge abutment. The opposite bank is a cutbank that developed because of stronger currents.

the meander and forming crescent-shaped ponds, or oxbow lakes. These lakes occur near many rivers in Ohio. Oxbow lakes that drain become meander scars. Through a combination of meander evolution and periodic flooding, a river valley becomes broader. The area just above normal water level forms the floodplain. Some stream and river valleys have steplike margins. These are terraces and represent earlier floodplains that formed before the stream had cut down to its present level. Most terraces in Ohio are related to changes in flow conditions caused by alternate advances and retreats of the Pleistocene ice sheets. Although the flatness of floodplains make them ideal for town sites and farming, the very nature of their formation makes them susceptible to complete submergence during floods.

Flooding Rivers and Flash Floods

Ohio's varying topography, geology, and numerous rivers and streams cause regular flooding. A rise in water level to a point where it spreads onto the floodplain is a common event along all Ohio streams and rivers. Since waterways were the initial pathways into the wilds that would become Ohio, explorers and settlers were well aware of flooding. But back in those days, floods were not as dangerous as they are now. Rivers, like the Ohio River, were much shallower because there weren't dams, and floods actually made traveling easier. Also, rivers were not confined and could spread widely across floodplains, resulting in lower flood levels at any one point, and wetlands, forests, and prairies blanketed the watershed, reducing the amount of runoff in channels and lowering flood levels. Finally, drain tile wasn't in place and therefore did not provide a direct path to streams and rivers. In less than 150 years humans have drastically altered Ohio's drainage patterns.

Without a doubt the 1913 Flood is the most noted in Ohio history. Actually, it was many floods that took place in most of the drainage basins of Ohio, Indiana, and Illinois to New York. Akron, Cincinnati, Cleveland, Columbus, Dayton, Hamilton, Marietta, Piqua, Sidney, Steubenville, Tiffin, Youngstown, Zanesville, and many smaller towns experienced similar death and destruction. Many of the major communities that were affected are located on floodplains. Dayton received the most press, but it's difficult to determine who was hit the hardest. Communication was greatly disrupted, and initially officials estimated that there had been thousands of casualties. Luckily this estimate had been greatly exaggerated.

Heavy cold rain preceded the flood, falling for two days on a landscape already saturated with winter snowmelt. Streams rose to flood stage by March 24, 1913. On March 25, floodwater topped or broke through levees at many communities. Waves rushed down city streets sweeping away anything in their path. Water ripped frame buildings from foundations and floated them away. Weakened buildings crashed suddenly into swirling waters. Fires broke out, burning structures to the water line. People and animals crowded on rooftops and hills that like islands poked above the floodwaters. Many drowned or died of exposure. Rivers lost their banks as muddy torrents rushed for miles across floodplains.

The Great Miami River, which normally flowed through Dayton at less than 250 cubic feet per second, increased steadily to 250,000 cubic feet per second. The story was the same in the Tuscarawas Valley. Mineral City and

March 25, 1913, floodwaters in downtown Carey.

Sandyville became islands as rising water surrounded them. Only one bridge survived major damage in Tuscarawas County. Three days after the flood had started, water levels began to lower and the worst was over. Four hundred sixty-seven Ohioans lost their lives in the flood, and damage amounted to $143 million ($386 million in 2004 dollars) statewide. The tremendous devastation and loss of life heightened public awareness of flooding and led to a number of flood control plans with the hope that a disaster like this would never happen again. The Miami Conservancy and Muskingum Watershed Conservancy Districts are examples of such plans; each planned a series of dams and holding reservoirs within their respective watersheds. This story has been repeated over and over since 1913. Flooding of Ohio's rivers is a natural process that will continue no matter how many levees and dams we build.

Another type of flooding is characteristic of the hills of eastern and southeastern Ohio—flash flooding. These sudden floods descend on unsuspecting communities downstream after cloudbursts in the uplands overtax the holding capacity of the Allegheny Plateau's small streams. Sometimes people have little time to escape to higher ground.

Slides and Slumps

Anywhere the land surface slopes, the force of gravity becomes a player in geologic processes. Sometimes the movements are rapid, and sometimes they are imperceptible. Four types of downslope movement, or landslide, occur in Ohio: creep, slump, earthflow, and rock fall.

Creep is the slow movement of unconsolidated material downslope. Often it is controlled by daily temperature changes in the upper soil. Although it is a year-round process, freezing and thawing usually accelerates the movement.

When ice forms in the space between particles of sediment, the expansion of the ice pushes particles farther apart and decreases the amount of friction that occurs between particles. This allows the sediment to hold more water after the ice thaws. Larger masses of ice form during the next drop to freezing temperatures and particles are pushed once again. This gradually causes the sediment to move downslope. Creep occurs across the state. You can see evidence of it in tilted fence and utility poles and inclined tombstones. Slump is the slippage of intact blocks of geologic material along a curved surface. The upper part of the block moves downward and the lower part moves outward. Often more than one slump block forms, one behind the other, making a series of tilted steps along a slope. Slumps usually develop over several months, even years. Earthflows are common where water soaks into unconsolidated material, making it more susceptible to slow downslope movement. Wet but not fluid rock and sediment slowly flows down hillsides forming distinctive lobes that are slightly higher than surrounding material. Rock falls are dangerous, sudden drops of cliff rock to the ground. Rock falls occur because of instability in cliff faces caused by undercutting of the slope and the development of vertical fractures.

Landslide-prone areas include the coast of Lake Erie between Huron and Conneaut, the Cuyahoga Valley, the lower Scioto Valley, the Allegheny Plateau, and the Cincinnati area and adjacent Ohio Valley. Pleistocene lake clays and silts and Wisconsinan tills characterize the northeast coast of Ohio. They are

The Devonian-age Ohio shale is notoriously weak and fails to maintain a stable slope, constantly moving downslope in response to gravity. A bent tree illustrates its constant battle to maintain upward growth as the soil creeps into the Vermilion River valley at Birmingham.

very susceptible to undercutting by waves. Slumps and earthflows are common along the coast, particularly from Cleveland east where the shoreline bluffs are the highest. Bluff recession is occurring rapidly in Ashtabula and Lake Counties, a combination of wave erosion and slump.

Between Cleveland and Akron the Cuyahoga River cuts through numerous Wisconsinan lake silts and clays that were deposited in tributary valleys. These sediments are notoriously weak and easily fail when saturated with water, resulting in slumps. They were a continual problem for builders as they built I-80 across the valley.

The Devonian-age Bedford shale is the culprit in landslides along the Scioto River from Circleville to Portsmouth. It often slumps into the valley, carrying overlying rock and sediments onto the valley bottoms. Pleistocene lake deposits, much like those along the Cuyahoga River, experience similar failures.

The steep hills of southeastern Ohio, underlain by a mixture of Pennsylvanian- and Permian-age shales, sandstones, clays, and coals, are prone to landslides. All that's needed is a little excessive moisture or ill-advised human activity, such as clear-cutting forested slopes, filling in depressions for building construction, blasting, and oversteepening slopes. Nearly 85 percent of the slumps and earthflows in southeastern Ohio occur in the red shales of Pennsylvanian age. These beds rapidly lose strength when they become wet and flow as thick mud. A 1972 slump event in Athens forced the city to remove eight apartment buildings built on fill over this same shale. Another in the Minford clay near Jasper in 1972 disrupted the construction of Ohio 124. This fine-grained silt and mud accumulated in lakes that formed when Wisconsinan ice blocked the flow of northward-flowing rivers and streams in southern Ohio.

Thick blocks of Pennsylvanian sandstones occasionally break loose from high, steep cliffs, tumbling downslope. Rock falls are always potentially hazardous, as was exhibited in 1986 near Hanging Rock along U.S. 52. Several tons of sandstone broke loose from a 150-foot-high cliff, falling on the roadway below and unfortunately crushing an automobile. It was Ohio's first landslide fatality. Construction crews have since blasted the rocky cliff back from the highway at several dangerous points. Another significant landslide occurred in 1986 along I-70 near New Concord. A slump in fill material in a nearby ravine took out the westbound lanes of the interstate in a matter of a couple of days after heavy rains hit the area. Similar slides caused the rebuilding of Ohio 7 near Steubenville and U.S. 52 across the river from Ashland, Kentucky. In 1998 another landslide occurred along U.S. 50 just east of Athens.

Surprisingly, it's not the hills of southeastern Ohio that experience the majority of landslide problems, but instead the terrain surrounding Cincinnati. Slumps and earthflows are common in the Ordovician-age Kope formation, a mixture of limestone and shale layers, and also in glacial lake sediments. The landslide problem is directly tied to the population density of the region. Most of the failures result from human activities. An example is the Mt. Adams slide of 1973, which was associated with the construction of I-471. Tens of millions of dollars were required to rectify the situation. Downslope movement may be a more mundane part of geology, but it certainly causes more than its share of problems.

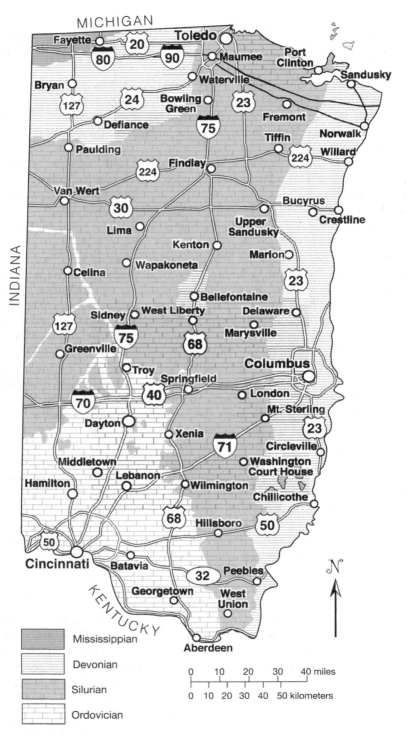

MICHIGAN

Fayette○   ⑳   ⑨⓪   Toledo○
⑧⓪   Maumee○
  Port
  Clinton
Bryan○   Waterville○   Sandusky○
⑫⑦   ㉔   Bowling○
  Green   ㉓
Defiance○   ⑦⑤   Fremont
Paulding○   Tiffin○   Norwalk
  ㉒④   Willard○
Van Wert○   ㉒④   Findlay○
  Bucyrus○
㉚   Crestline○
Lima○   Upper
  Sandusky
Celina○   Kenton○
  Wapakoneta   Marion○
⑫⑦   Bellefontaine○   ㉓
Sidney○ West Liberty○   Delaware○
⑦⑤   ⑥⑧   Marysville○
Greenville○   Columbus○
Troy○   Springfield○   London○
⑦⓪   ④⓪   Mt. Sterling
Dayton○   Xenia○   ㉓
  ⑦①   Circleville○
Middletown○   Washington○
  Lebanon○   Court House
Hamilton○   Wilmington○   Chillicothe○
  ⑥⑧   Hillsboro   ㊿
⑤⓪   Cincinnati○ Batavia○   ㉜   Peebles
  Georgetown○   West
  Union

KENTUCKY

N

Mississippian

Devonian

Silurian

Ordovician

Aberdeen○

0   10   20   30   40 miles
0   10 20 30 40   50 kilometers

INDIANA

Highways and bedrock of the western till plains. —Modified from Ohio Department of Natural Resources, Division of Geological Survey publication

WESTERN OHIO—THE TILL PLAINS

Some of the flattest land in the Midwest stretches south from Lake Erie and the Michigan border. A rolling preglacial landscape, carved in Paleozoic-age bedrock, lies buried beneath layers of Ice Age sediments. The flattest terrain was scoured by glacial ice and meltwater and marks the extent of early glacial lakes that eventually assumed the confines of present-day Lake Erie—lakes like Glacial Lake Maumee and others that formed along the southern edge of the ice of Wisconsinan time as it receded from Ohio about 16,000 to 14,000 years ago. From Allen and Van Wert Counties south to the Dayton area, the flat plains yield to a more rolling landscape with broad east-west-trending uplands that mark end moraines and separate flatter bands of ground moraine, or till plain. South of Dayton to the Ohio Valley, hills appear where ice was thinner; here bedrock pokes through the glacial blanket. The oldest rocks exposed in Ohio surround Dayton and Cincinnati; those under Toledo are slightly younger, but they are still considerably older than the strata that underlie eastern Ohio.

Ordovician Formations

Between 475 and 443 million years ago, a warm sea covered much of what was to become Ohio. Limestones were deposited in great quantities in this southern tropical-latitude environment, along with shales derived from clay sediments eroded from mountains to the east. Also, thin seams of light-colored, fine-grained sediment occur at scattered intervals, marking the fall of ash from volcanic eruptions in the rising Taconic Highlands to the east. Geologists find the ash beds very useful; they are precise time markers within the monotonous, repetitive limestone and shale beds since the ash fell across the region shortly after the eruptions. Two ash beds in the subsurface middle Ordovician strata of Ohio mark what may have been a couple of the most violent eruptions known, thousands of times greater than the 1980 Mount St. Helens eruption. Storm waves occasionally scoured the ocean bottom, leading to layers of broken shell debris geologists call *fossil hash*. Combined, these deposited materials resulted in the typical gray, fossiliferous Ordovician-age limestone and shale layers seen today along streams, and in roadcuts and railroad cuts in southwestern Ohio.

The Ordovician strata around Cincinnati were among the first rocks studied in western Ohio as early on Cincinnati emerged as an important river port and center of learning in the region west of the Allegheny Mountains. The strata, known as the Cincinnatian series, were deposited in the latest part of the

A seascape depicting the typical animals of late Ordovician time now found as fossils in the rocks of southwestern Ohio.

Ordovician period, from about 454 to 443 million years ago. The Cincinnati-area strata are the North American standard for Late Ordovician-age rocks; geologists around the world use these rocks—comparing fossils and rock types to establish whether or not similar rocks were deposited in Ordovician time. They are by far the richest fossil-bearing units in Ohio, and fossils from this area are in museums around the world. Collecting is easy; just find an exposure, obtain permission, and begin.

Isotelus, a large trilobite found in Cincinnatian rocks, is the state fossil. Small sand dollar–like animals called *edrioasteroids*—distant relatives of sea stars—are sometimes found attached to brachiopod shells in these rocks. One species, *Isorophus cincinnatiensis*, is the official fossil of Cincinnati. Among the most abundant animals were bryozoans, tiny bottom-dwelling colonial animals that secreted calcareous skeletons in the shapes of fans, branching twigs, flat mats, and hemispherical mounds in the Ordovician seas. Another common creature was the brachiopod, an animal with two calcareous valves, or shells, much like clams. Other fossils include shells of clams and snails; cone-shaped shells of nautiloid cephalopods; horn corals; the smaller but much more common trilobite; *Flexicalymene* and various types of echinoderms, including crinoids and some of the earliest starfish. Occasional small spores suggest that land plants had appeared in the uplands to the east.

The abundance and diversity of fossils, differences in bedding, and differences in the amount of limestone and shale led geologists to subdivide the Cincinnatian rocks into smaller units called *groups, formations,* and *members.* As geologists studied Ordovician rocks around the world more thoroughly, they devised other stratigraphic schemes based on variations in rock type, combinations of fossil and rock characteristics, and separation of layers by unconformities into units called *sequences*. The net result is that there is a plethora of names for Ohio's Ordovician rock units, which can be quite confusing to the uninitiated. It's not over yet. The Ordovician time period has had a history of controversy since it was established in 1879 to resolve a boundary problem between the underlying Cambrian-age strata and overlying Silurian-age strata that geologists named in the 1830s. Many geologists didn't immediately accept this new delineation. In fact, the Ohio Geological Survey did not formally use "Ordovician" until 1909. Older works on Ohio geology refer to these rocks as "Lower Silurian." Even geologists have difficulty distinguishing the formations of Ordovician time; they are all gray, fossiliferous, and composed of various combinations of limestone and shale.

Point Pleasant Formation

The oldest rocks of Ordovician time lie deep below Ohio and can only be studied from cores of rock pulled from deep drill holes. The oldest visible bedrock in Ohio is from middle Ordovician time, around 470 to 460 million years ago. It comprises about 80 feet of thick limestone beds interbedded with shale—that is, the beds alternate with one another. As a whole, the formation has about a fifty-fifty mixture of limestone and shale. Since it was first described near Point Pleasant in Clermont County, it is named the Point Pleasant formation. This formation was also known as the River Quarry beds because all of the early quarries in this formation were located at the level of the Ohio River.

Kope Formation

Some 240 feet of shale with thin beds of limestone helps form the base of the Cincinnatian rocks around Cincinnati; this unit appears above the Point Pleasant formation and below the Fairview formation. Many early studies of these rocks, later named the Kope formation, took place in quarries and stream cuts in the Cincinnati area, including Eden Park just east of I-71, where the

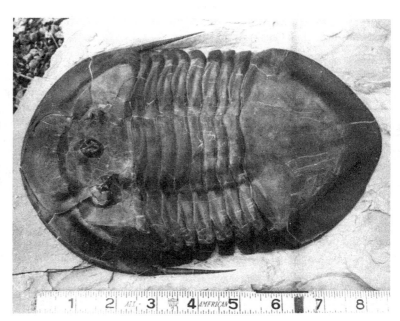

Ohio's state fossil, Isotelus maximus, an Ordovician-age trilobite that reached lengths of more than 18 inches. This specimen came from the Arnheim formation of Highland County. —Ohio Department of Natural Resources, Division of Geological Survey photo

Cincinnati Art Museum is located. Twin Lakes, just inside the park's entrance, originally was the site of a quarry. Because of the prevalence of softer shale in this formation, much of it has eroded to gentle slopes. It is prone to landslides, and as a result many older exposures are now covered in rocky erosional debris, or colluvium. Kope strata show only small amounts of fossil hash but a lot of fine-grained sediments, suggesting a deepening of the ancient sea and a shift to a more offshore environment that was below the scouring effect of even storm waves.

Fairview Formation

On the hillslopes around Cincinnati there is roughly 100 feet of shale and limestone that was deposited on top of the Kope formation. The widespread quarrying of these rocks for building stone in the late 1800s led to their early name, the Hill Quarry Beds; later they were named the Fairview formation after a cliff in Fairview Park, a former quarrying site east of I-75 off Scenic Drive. The Fairview rocks suggest a slight shallowing of the sea and a deep subtidal environment; certainly depths were no greater than 50 feet. The Fairview strata alternate between thick layers of fossil hash and fine mudstone and have a diverse assemblage of fossils, mainly brachiopods and bryozoans. The sea bottom during this time was only occasionally disturbed by strong storm waves, forming fossil hash; other times fine-grained sediments were deposited.

Grant Lake Formation

About 85 to 90 feet of shale and limestone lie above the Fairview formation around Cincinnati. The fossils in the Grant Lake formation are commonly brachiopods with thick shells, like *Hebertella* and *Platystrophia* species, and massive bryozoans; many of the fossils are broken and/or abraded. This reflects deposition that occurred on a sea bottom that was above the normal wave base, meaning waves continually swept the bottom sediments; this was a shallow subtidal environment. Grant Lake strata outcrop at Bellevue Hill in Cincinnati, Georgetown, Manchester, Maud, Ripley, Sharonville, West Middletown, along Stonelick Creek, and elsewhere.

Arnheim Formation

The overlying 60 feet of shale with thin wavy beds and nodules of limestone that was deposited on top of the Grant Lake formation suggests continued shallow subtidal deposition. Sometimes the uppermost layers of the Arnheim formation contain abundant dwarf clams and snails, perhaps representing particularly harsh environmental conditions where animals never reached adult size. Geologists first studied this formation on Straight Creek, south of Arnheim in Brown County. Good exposures are in Dent, along Flat Fork near Oregonia, in Caesar Creek Gorge near Waynesville, along Four Mile Creek near Oxford, and in North Fork Eagle Creek near Russellville.

Waynesville Formation

About 100 feet of fossiliferous shale with scattered but prominent beds of limestone compose the next-youngest unit. The Waynesville formation marks a return to offshore water, like the conditions that existed when the Kope formation was deposited earlier. *Flexicalymene*, a trilobite, is common in certain layers of the

Generalized geologic column for Ohio

millions of years ago	Period	southwest Ohio	northwest Ohio	central Ohio	northeast Ohio
2		QUATERNARY	QUATERNARY	QUATERNARY	QUATERNARY
318	MISSISSIPPIAN	unconformity	unconformity	**Logan fm.**: Maxville ls. — Rushville sh.; Vinton sandstone; Allensville conglomerate; Byer sandstone; Berne conglomerate — **Cuyahoga fm.**: Black Hand sandstone; Buena Vista sandstone; Sunbury shale	**Cuyahoga fm.**: Meadville shale; Sharpsville sandstone; Orangeville shale; Sunbury shale
			Bedford shale; Ohio shale (Antrim shale)	Berea sandstone; Bedford shale; **Ohio shale**: Cleveland shale; Chagrin shale; Huron shale	Berea sandstone; Bedford shale; **Ohio shale**: Cleveland shale; Chagrin shale; Huron shale
359	DEVONIAN	unconformity	Tenmile Creek dolomite; Silica formation; Dundee limestone	Prout limestone Olentangy shale; Plum Brook shale; Delaware limestone; Columbus limestone	Olentangy shale; subsurface units
		Tymochtee dolomite; Greenfield dolomite; Cedarville dolomite; Springfield dolomite; Euphemia dolomite; Massie shale; Laurel limestone; Osgood shale; Dayton formation; Brassfield formation	**Detroit R. Grp.**: Lucas dolomite; Amherstburg dolomite; Sylvania sandstone; Holland Quarry shale; undiff. Salina group; Tymochtee dolomite; Greenfield dolomite; Lockport dolomite; subsurface units	Hillsboro sandstone; undifferentiated Salina group; Tymochtee dolomite; Greenfield dolomite; Peebles dolomite; Lilley formation; Bisher formation; Estill shale; Dayton formation; Noland formation; Brassfield formation	subsurface units
416	SILURIAN				
443	ORDOVICIAN	Drakes formation; Whitewater formation; Liberty formation; Waynesville formation; Arnheim formation; Grant Lake formation; Fairview formation; Kope formation; Point Pleasant formation	subsurface units	subsurface units	subsurface units
488	CAMBRIAN	subsurface sandstone and dolomite			
542	PRECAMBRIAN	igneous and metamorphic rocks			
4,600					

Generalized geologic column for Ohio (Pennsylvanian through Tertiary time not represented). This column is nonlinear, meaning the space each chunk of time takes up in the column does not reflect its range of years relative to other chunks of time; for example, Precambrian time spans over 4 billion years of history while the Silurian Period only covers 27 million years of time. Wavy lines represent local and regional unconformities.

lower part of this formation. These rocks, first studied at Waynesville, surround the Cincinnati area, usually forming the lowest visible portions of hillsides.

Liberty Formation

Another 30 feet of limestone beds lie above the slopes of the Waynesville formation. Brachiopods, bryozoans, and an abundance of fossilized animal burrows (trace fossils), which were shallow, indicate a deep subtidal environment that had slightly shallower water than the underlying Waynesville formation. Only during heavy storms was the bottom sediment stirred by waves. The Liberty formation, named after Liberty, Indiana, grades into the Whitewater formation.

Whitewater Formation

Limestone and shale up to 70 feet thick—the Whitewater formation—lie above the Liberty formation in southwestern Ohio. The beds are typically wavy, thickening and thinning in a lateral direction. This was caused by sea animals reworking the sediments, much like earthworms rework soil. Thick-shelled brachiopods and robust bryozoans, commonly broken and worn, are typical fossils of this formation. Near the Indiana border, dolomites of the Saluda formation replace the middle Whitewater strata. The scarcity of fossils and the presence of laminated dolomite and desiccation cracks suggest that the Saluda formation was deposited in a peritidal environment, varying from above-normal tide to intertidal to the shallowest part of the subtidal zone. Bryozoans, brachiopods, ostracods, stromatoporoids (a type of sponge with a calcareous skeleton), and a colonial coral (*Tetradium* species) occur in some parts of the Saluda formation. Eventually, the sea flooded the area of Saluda deposition and more typical Whitewater deposition returned.

Drakes Formation

The youngest Ordovician rock in southwestern Ohio consists of 20 feet of shales and thin dolomite beds. The Drakes formation grades imperceptibly into the rocky soils of the region or is capped by the Brassfield formation of early Silurian age.

A Missing Interval

Sea level dropped about 443 million years ago in response to glacial ice buildup in present-day northern Africa (large volumes of seawater were consumed to make the ice), leaving western Ohio and most of the North American continent high and dry. The exposed seafloor wore away until a rising sea again submerged the area in early Silurian time. The contact between the Drakes and the Silurian strata is the erosional surface, or unconformity, that developed while sea level was low. Marine fossils decline and disappear in the upper portion of the Drakes formation, but the overlying Brassfield formation has a rich assemblage of fossils indicative of open sea.

Silurian Formations

By early and middle Silurian time, the Taconic Highlands to the east were mere stubs of once-tall mountains, so less clay was reaching the sea that once again covered Ohio. The North Africa ice cap had also melted, leading to a rise in sea

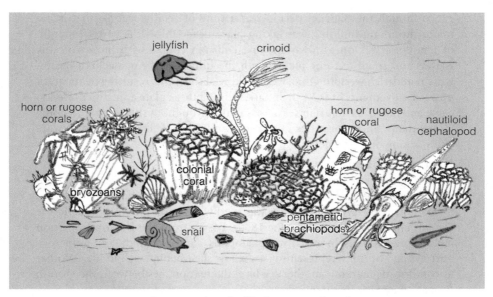

Reconstruction of a Silurian-age patch reef.

level. As a result, limestone was the main rock type that was deposited during this period. Much of it has been altered by magnesium-rich water to dolomite.

Silurian stratigraphy is complicated by unconformities and the lack of marker beds that allow immediate recognition of a point in time, such as the ash beds of Ordovician strata. As a result, geologists use a different set of names for Silurian strata in southern Ohio, northwestern Ohio, and central Ohio. The base of Silurian strata is marked by an unconformity throughout Ohio.

Brassfield Formation

The oldest Silurian rock in Ohio is a 20- to 60-foot-thick fossiliferous limestone that spans from near Fairhaven at the Ohio-Indiana border east to Miami County, then southeast to Rome in the Ohio Valley. The Brassfield formation, which was deposited around 440 million years ago, varies from crystalline limestone to a limestone with thin shale layers. Drillers for oil and gas often called this formation the Clinton sands, strata, or horizon. In Adams, Clinton, and Highland Counties, iron-rich zones typify the upper part of the formation. The iron came from the eroded remnants of the Taconic Highlands and spread into the Silurian sea in Ohio where, because of its density, it was concentrated in layers and nodules in the Brassfield formation. The Brassfield represents deposition that occurred on a shallow, warm, carbonate-rich bank, like that which occurs along present-day southern Florida. A wide assortment of fossils occur in this formation; small segments of stems of crinoids—a type of echinoderm with feathery arms—are very common. A major unconformity marks the top of the Brassfield formation in west-central Ohio, suggesting another drop in sea level that resulted in the exposure of the seafloor to the elements. In places—for example, on the east side of the Cincinnati Arch—the Noland formation, a shale, separates the Brassfield and Dayton units.

Although the Silurian period was a time of relative calm—that is, no major mountain building was taking place near Ohio—the eastern margin of Laurentia was on a slowly moving collision course with more volcanic islands and small tectonic plates. Both the Illinois and Appalachian Basins, which sandwich western Ohio, were sinking at this time, resulting in apparent upward movements of the Cincinnati Arch. All these plate adjustments led to a rise and fall of the Silurian sea across Ohio throughout this interval, leading to the many unconformities and discontinuous rock layers. Geologists are still working out local details.

Dayton Formation

A bluish gray crystalline dolomite and limestone, which averages 8 feet thick in western Ohio but thickens below the surface to the east, lies above the Brassfield formation. The Dayton formation was used widely as a building stone in the Dayton area, especially during the 1800s. Fossils from the Dayton formation include brachiopods, bryozoans, cephalopods, corals, crinoids, and numerous burrows and borings, representing a shallow shelf habitat rich in life. Another unconformity, marking another sea level fluctuation, lies between the Dayton formation and overlying rock strata.

Osgood Shale/Laurel Limestone/Massie Shale and Equivalents

Shale 40 to 90 feet thick covers the Dayton formation. It's called the Estill shale near the Ohio River in central Ohio, but from Greene County north, the shaley deposit is split by 5 to 9 feet of Laurel limestone. The older shale is called Osgood shale and the younger is called Massie shale. The Osgood shale is greenish gray and has thin beds of limestone. Around Yellow Springs, the Osgood shale is 35 to 40 feet thick, but it thins to the north and west and thickens below the surface to the east. The 5- to 6-foot-thick Massie shale, named for exposures along Massies Creek near Cedarville, contains abundant brachiopods and other fossils that easily weather from its calcareous matrix.

Middle Silurian Formations

The middle Silurian (about 435 to 425 million years ago) marks a time of warm, shallow carbonate platforms colonized by a wide assortment of corals, brachiopods, bryozoans, mollusks, crinoids, and stromatoporoids—an extinct group of sponges. Sea life was so plentiful that living mounds of animals formed wherever currents brought food and nutrients. As older animals died, new ones grew upon their skeletal remains, forming reefs. Sedimentation during this period of time was continuous, making it difficult to distinguish one formation from another. Unconformities are absent, indicating a steady or rising sea level. Most of the fossils in these strata are molds created during the dolomitization of limestone, when magnesium-rich groundwater replaced calcite with dolomite.

In south-central Ohio, around Greenfield, Hillsboro, and Peebles, the rocks of middle Silurian time are known as the Bisher and Lilley formations and Peebles dolomite. Farther north toward Lewisburg, Springfield, and Yellow Springs, similar rocks are referred to as the Euphemia, Springfield, and Cedarville dolomites. In northwestern Ohio, the strata are lumped together as the Lockport dolomite.

The gray to yellowish brown Bisher formation, a dolomite with some shale seams, averages 45 feet thick in Adams and Highland Counties and typically forms the lowest prominent cliffs and benches in the region's hills. It contains brachiopods, trilobites, and other fossils in its lower section, and white chert nodules in the upper part. It is named for the former Bisher Dam near Hillsboro.

The Lilley formation was deposited next. The first 20 feet of it is tan and contains thin shale layers in clay-rich dolomite, corals and fossil fragments, and thin calcite veins; the next 20 feet of rock has characteristics of the tan rocks below it and the overlying Peebles dolomite. The Peebles dolomite averages 45 feet thick but thickens to nearly 100 feet around Rocky Fork Lake. It is massive with lots of cavities, especially in the upper part. *Pentamerus* brachiopods occur in the lower part. The Peebles dolomite makes up the lower of two cliffs on the hillcrests of Adams and Highland Counties; above it is another cliff of Devonian-age Berea sandstone.

In west-central Ohio the oldest visible Middle Silurian dolomite is the Euphemia dolomite, named for a small community near Lewisburg. It occurs in thick, homogenous beds of mottled stone with *Pentamerus* brachiopods near the top and molds of brachiopods, crinoids, and other fossils scattered throughout. It averages 6 feet thick. About 10 feet of buff to brown Springfield dolomite occurs above the Euphemia dolomite. This dolomite occurs in 5-inch beds, generally has few fossils, and has been used as a building stone. The uppermost dolomite that is visible in this region averages 70 feet thick and is named after the town of Cedarville. The Cedarville dolomite is generally thick and homogenous with lots of large cavities. Weathering often produces

Silurian-age Cedarville dolomite forms cliffs along Massies Creek in Indian Mound Reserve at Cedarville. Thinner-bedded Springfield and Euphemia dolomites lie below. Massie shale occurs at trail level.

a honeycomb effect. *Pentamerus* brachiopods occur near its base; molds of brachiopods, corals, and crinoids are scattered throughout.

About 50 to 80 feet of crystalline Lockport dolomite from middle Silurian time is exposed in quarries of northwestern Ohio. The rocks are tan to light gray and have many cavities that are lined with crystals of celestite, fluorite, calcite, and other minerals. The Lockport dolomite exhibits many massive and tilted fossiliferous rock layers that mark former reefs. *Megalomoidea*, a large clam, *Trimerella*, a brachiopod, and colonial corals are characteristic fossils of this dolomite.

Greenfield Dolomite
Lying above the Peebles, Cedarville, and Lockport dolomites is a 50-foot-thick unit named for a quarry exposure east of Greenfield. The Greenfield dolomite has both evenly bedded rock and unbedded masses surrounded by tilted beds, which mark where reefs once occurred. Poorly preserved brachiopods and corals and abundant ostracods, small bivalved animals, occur at certain horizons; chert and sphalerite, a zinc mineral, are also present.

The best outcrops of Greenfield dolomite are in southern Ohio, but it extends into northwestern Ohio along the flanks of the Findlay Arch. Sea level lowered during this time, creating areas of shallow, very salty water where few creatures survived, except for algae and the ever-present ostracods. The Greenfield dolomite represents deposition that occurred in shallow subtidal environments.

Tymochtee Dolomite
The 80-foot-thick Tymochtee dolomite, with thin and wavy bedding and thin shale laminae, occurs above the Greenfield dolomite. It roughly parallels U.S. 23 from northern Fayette County to Ottawa County. The formation's type area is along Tymochtee Creek in Wyandot County. This formation has few fossils, but mud cracks and ripple marks are prevalent. In Ottawa County the Tymochtee formation thickens as it alternates between gypsum, anhydrite, and dolomite. Shallow water caused an increase in salinity where water flow became restricted, and these minerals precipitated out of it; tidal flats existed elsewhere, allowing the mud cracks and ripple marks to form.

Undifferentiated Salina Group
The youngest Silurian strata are poorly exposed in Ohio because their deposition preceded a major withdrawal of the late Silurian sea from Ohio, and much of the strata was lost to erosion during that time; they form the Bass Islands and quarry walls near Holland and Gypsum along Lake Erie's southern shore. The undifferentiated Salina group includes dolomite and gypsum and represents deposition that occurred in a salty, landlocked sea that was similar to the present-day Persian Gulf. Under eastern Ohio, Michigan, New York, Pennsylvania, and Lakes Erie and Huron there are deposits of halite, which was precipitated from the ancient sea. These deposits mark the sea's farthest northern and eastern development. Silurian deposition ended about 416 million years ago with the great withdrawal of the sea from ancient North America. As the sea withdrew, most of Ohio was once again above sea level. The sea didn't return to Ohio until middle Devonian time, about 375 million years ago.

Alternating beds of Silurian-age Salina group gypsum and dolomite-anhydrite in a quarry at Gypsum.

Devonian Formations

The undifferentiated Salina rocks of Silurian time unconformably underlie rocks of middle Devonian age in the northwestern corner of Ohio and along a band stretching from Sandusky to the Ohio River west of Portsmouth. There are no visible rocks from early Devonian time in Ohio except for a small patch of shale that workers discovered in a quarry near Toledo. Much of what we know about Ohio in early Devonian time we have interpreted from the sub-surface of eastern Ohio where a Devonian sea periodically covered the area. The strata consist of a number of limestones, dolomites, and sandstones separated by unconformities and reflect tectonic developments taking place along the edge of Laurentia far east of present-day Ohio. Much of Ohio's land surface was at the mercy of ancient streams and other forces of erosion. Much of the rock record of this time was lost.

The seas returned, though, and much of Ohio was underwater by about 375 million years ago. Ohio's exposed Devonian formations comprise limestones, dolomites, organic-rich shales, and sandstone.

Sylvania Sandstone and Hillsboro Sandstone

As the sea returned to western Ohio during Middle Devonian time, sand and other land-derived sediment spread across the shallows. In northwestern Ohio on the west side of the Findlay Arch, these sediments became the Sylvania sandstone, named for exposures west of the city of Sylvania. This white to buff, cross-bedded sandstone averages 50 feet thick in Lucas County, thickens in the subsurface into the Michigan Basin to the north, and disappears in the subsurface south of the Maumee River valley due to erosion. This formation is nearly pure quartz; only a few other minerals have been found as part of its composition. The Sylvania sandstone is the oldest formation within the primarily carbonate unit named the Detroit River group. Far south in southwestern Ohio near Hillsboro, a similar rock—the Hillsboro sandstone—occurs on scattered hilltops. Although the Sylvania sandstone grades upward imperceptibly into Detroit River dolomite, the Hillsboro sandstone is bounded above and below by unconformities. Exact correlations are questionable.

Amherstburg and Lucas Dolomites

Detroit River group dolomite, marking a slow deepening of the Middle Devonian sea, overlies the Sylvania sandstone in northwestern Ohio and the Silurian Salina group in the subsurface from Kelleys Island to around Bloomville in Seneca County. Around 50 to 75 feet of Amherstburg dolomite, named after Amherstburg, Ontario, and 30 to 75 feet of Lucas dolomite, named after Lucas County, occur on both sides of the Findlay Arch. Both of these formations vary from brown to gray and are difficult to distinguish. The major difference involves the more abundant and distinctive fossils of the Amherstburg dolomite.

Columbus Limestone

Buff to gray limestone, 100 feet thick, deposited in shallow water, overlies the Silurian Salina or lower Devonian rocks from Kelleys Island to Pickaway County on the eastern flank of the Findlay and Cincinnati Arches and around Bellefontaine. The lower part of the Columbus limestone is an unfossiliferous, dolomitic limestone; the upper part is a limestone with abundant,

Upper part of Devonian-age Columbus limestone in an abandoned quarry at Marblehead.

well-preserved fossils, including horn and colonial corals, brachiopods, snails, nautiloid cephalopods, and trilobites. Layers and nodules of white chert are common at certain horizons. Two layers, 2 to 16 inches thick, of eroded mineral grains and tiny fossils, including ostracods, conodonts, scolecodonts, and scales, teeth, plates, spines, and bone fragments of early fish, occur within the Columbus limestone (two more similar layers are in the overlying Delaware limestone). The uppermost layer, or bone bed, marks the boundary between the Columbus and the overlying Delaware limestone. The origin of these bone beds is still unclear, and perhaps there is not one explanation. In most cases the fossil material appears to have been transported by wind or water to the site of deposition from another nearby place. The Columbus limestone was an important building stone in the late 1800s to mid-1900s; many Ohio churches were built of this stone.

Delaware Limestone and Dundee Limestone
Roughly 35 feet of dark gray limestone occurs above the Columbus limestone on the east side of the Findlay and Cincinnati Arches. Fossils and chert nodules occur in certain layers within the Delaware limestone. Two more bone beds, similar to those in the underlying Columbus limestone, occur within the formation. A few thin continuous layers of clay appear in the Delaware limestone, representing periods of ashfall from volcanic eruptions in the rising Acadian Mountains far east of Ohio.

The Dundee limestone, which geologists have correlated with the Delaware limestone on the opposite side of the Findlay Arch, forms a band from southeastern Michigan to west of Toledo and then along the southern edge of the Michigan Basin to the Indiana border near Payne. The Dundee limestone, named for Dundee, Michigan, is a fossiliferous, buff to gray crystalline limestone that averages 60 feet in thickness. It contains a diverse assemblage of fossils, including brachiopods, corals, snails, trilobites, and crinoid stems. Unconformities lie above and below both the Delaware and Dundee limestones, marking the temporary existence of shallow Devonian seas on either side of the Findlay Arch.

Olentangy Shale and Equivalents
Strata, 30 feet thick, cap the Delaware limestone along the Olentangy River in central Ohio, marking the arrival of mud eroded from the rising Acadian Mountains. The grayish green Olentangy shale has thin limestone beds throughout; small concretions, especially in the lower part; a subtle unconformity in the middle; and some thin beds of black shale in the upper part. It thickens below the surface of southern Ohio; to the north, it correlates with 30 feet of Plum Brook shale and 9 feet of Prout limestone. The Plum Brook shale is rich in fossils, containing the same fossils as the Silica formation from Lucas County. The Prout limestone contains abundant horn corals, though there are few exposures of it.

On the west side of the Findlay Arch, the Silica formation and overlying Tenmile Creek dolomite were deposited at the same time as the Olentangy shale. The Silica formation ranges between 10 and 54 feet thick and consists of fossiliferous shale and limestone layers. Well-preserved brachiopods,

Typical animals of a middle Devonian sea that are found as fossils in Ohio.

bryozoans, corals, snails, clams, trilobites, crinoids, and fish make this one of the best-known layers for fossil collecting in Ohio. The Silica formation is named after a small community that developed near an early sandstone quarry west of Toledo. An unconformity separates this formation from the 40-foot-thick Tenmile Creek dolomite, which was named for exposures along a nearby creek. This part of Ohio must have been above sea level between the deposition of these two formations.

Ohio Shale

A 250- to 500-foot-thick black shale stretches from Huron on the shore of Lake Erie to the Ohio River, west of Portsmouth. Stagnant bottom waters existed across western Ohio around 365 million years ago, leading to the accumulation of muds rich in organic matter. The fossils in this shale are mainly the remains of animals and plants that floated into the area. Large oblong to spherical concretions occur in the lower and upper parts of this formation. This rock layer thickens dramatically eastward into the Appalachian Basin, where the sea was deeper.

The Pleistocene Ice Age

Only in the Cincinnati area did pre-Illinoian sediments escape reworking and burial; today they form the area's hilly landscape that rests on Ordovician-age bedrock. Illinoian tills still blanket southwestern Ohio from the Indiana border across Hamilton County, up the Little Miami River valley as far as Waynesville, and in a 10-mile-wide band from Hillsboro to Chillicothe. As with most of Ohio, the majority of western Ohio's glacial geology is related to the Wisconsinan glacial stage, which occurred between 117,000 and 10,000 years ago.

The large number of spruce logs found buried in till of the Wisconsinan glaciation tell us that Ohio was an open spruce woodland at that time.

The Vermilion River cuts through Devonian-age Ohio shale in Birmingham.

The majority of Wisconsinan ice moved across Ohio in two large lobes: one traveled down the Scioto Valley on the east side of the Bellefontaine outlier—an area of relatively higher bedrock—and the other down the Great Miami Valley on the west side of the outlier. The Scioto lobe reached its southernmost point at Chillicothe; the Miami lobe stopped in northern Hamilton County. Glacial geologists still debate about how the ice lobes behaved. Some feel that the lobes underwent rapid retreats and readvances—periods when the ice sheets melted and shrank back or grew and moved forward; others feel that the lobes remained relatively stable in southwestern Ohio. Whatever the case, the lobes left their mark on western Ohio in the form of thick, curving piles of till, called *end moraines*, from Lake Erie to the Ohio River, and a thick blanket of till called *ground moraine*. The thickness of these ice lobes varied from hundreds of feet near their margins to thousands of feet in their interiors.

Around 14,000 years ago, the last Wisconsinan ice left Ohio and the current climate and landscape developed. Lower terraces along the Great Miami, Little Miami, Mad, and Scioto Rivers mark Wisconsinan outwash deposited by torrents of meltwater that splashed across Ohio as the ice melted (upper terraces are Illinoian in age). Lakes formed where thick piles of outwash prevented tributaries from entering their main channels. These lakes eventually drained, but they left wide, flat patches of lake sediments near their mouths. Peat and marl deposits dot the western Ohio landscape where other glacial lakes existed. Loess, windblown silt from outwash plains and sand dunes, forms the soils of southwestern Ohio. Loess is negligible in northern Ohio.

An ancient spruce log lies exposed in Wisconsinan till on the south edge of Sidney.

Wisconsinan ice sheets in the lowland where Lake Erie now resides formed an effective barrier to any drainage of meltwater to the north and east—the directions many of Ohio's preglacial rivers flowed. Meltwater flooded the northwestern counties of Ohio, dammed against the ice to the north and lapping against end moraines to the south. As the ice receded or advanced, lake levels changed. The earliest lakes had levels over 200 feet higher than present levels. Silt and mud accumulated in the deeper parts of these proglacial lakes—lakes that formed along an ice sheet—and sand accumulated on their wave-washed shorelines and shoals. The lake silts now characterize the flat landscapes of northwestern Ohio; former sandy beaches, spits, bars, and deltas stand relatively higher, approximating the borders of the various lake levels. These are referred to as *beach ridges.*

Sometimes wind whipped the beach sands into dunes, which are still visible in some areas of northwestern Ohio, like the Oak Openings Region. Because of good drainage, many beach ridges served as the foundation of many of Ohio's highways. Beach ridges marking the final stage of Glacial Lake Maumee are the highest and most prominent beach ridges in Ohio. For examples follow Ohio 12 from Findlay toward Delphos or Ohio 613 from Van Buren toward McComb or Fostoria. Glacial Lake Warren beaches from 13,000 years ago are visible between Bowling Green and Weston along Sand Ridge Road. (For more information about Lake Erie's predecessors, *see* the Lakeshore Ohio chapter.)

The Great Black Swamp

By 9,000 years ago the melting ice sheet lay far to the north of the Lake Erie Basin, but the region's climate was still cool and wet, and tundra plants blanketed northern Ohio. Because of its flatness and clay substrate, the former lake plain was flooded or soggy most of the year. Tundra gave way to spruce-fir for-

Sand Ridge Road follows the Glacial Lake Warren beach ridge west of Bowling Green. The house to the right is built on this ridge.

ests and then temperate vegetation farther away from the glacial influence. Lake Erie assumed its present appearance about 5,000 years ago. A great swamp, dubbed the Great Black Swamp, developed in low-lying areas of northwestern Ohio. It stretched nearly 100 miles from Lake Erie to New Haven, Indiana, and 20 to 30 miles south of the Maumee River.

By the time European settlers arrived, the area consisted of swamp forest, prairies, and marshes. Occasional sand ridges and bumps on the dolomite bedrock provided the only high and dry ground. Ash, basswood, elm, hickory, maple, and oak trees towered 100 feet or more above the wet areas, oak and hickory dominated the sand ridges, prairie grasses graced openings in the forests, and cattails and reeds flourished in marshes along the lakeshore.

The swamp was a major obstacle to travel; travelers forging early routes avoided it. There are many tales of travelers wading knee-deep in the black muck and of wagons buried to their axles. Soldiers hacked the first "road" through the swamp as they marched to retake Detroit during the War of 1812. Hull's Trace, as it became known, was the only north-south passage through Hancock, Wood, and Lucas Counties for many years. An east-west route—the Maumee and Western Reserve Road—surveyed between 1797 and 1811, was a mere blazed trail from rapids in the Sandusky River (now the city of Fremont) to rapids in the Maumee River (now the city of Perrysburg) until 1824 when construction of the actual road began. Between 1824 and 1845 workers struggled to complete canals through the western part of the swamp along the Maumee River from Indiana to Toledo and from near Defiance to Delphos. Communities slowly developed on its edges; larger towns like Defiance, Delphos, Findlay, Fostoria, Fremont, Sandusky, Tiffin, Toledo, and Van Wert were outside the Black Swamp's margins.

In 1852 and 1859 railroad companies completed lines through the swamp and settlement followed the tracks. Early settlers, intent on farming the rich soils, found it nearly impossible to keep their fields drained. Laws, first passed by the Ohio General Assembly in 1859, led to the digging of large, deep ditches across the Black Swamp, which effectively drained the standing water. In Wood County alone, 140 ditches carved the countryside for a total of 495 miles by 1869. The Jackson Cutoff Ditch, a 9-mile-long trench dug between 1878 and 1879 that ran from east of Deshler to west of Weston, eventually drained 30,000 acres and ranked as the largest. Tiling of fields, the laying of drainage tile below the surface, lowered groundwater. What once was a morass of mud, dense vegetation, and relentless hordes of mosquitoes became some of the best cropland in the state by the late 1890s.

Lima-Indiana Oil and Gas Field

When organic matter is buried deep in rock, it can become oil and gas over time. The formation of oil and natural gas is complex, but it involves the accumulation of plant and animal matter in marine sedimentary rocks where the organic matter is protected from extensive oxidation. As the material is buried increasingly deeper over millions of years, the organic matter transforms into the mobile hydrocarbons we call oil and gas. Oil and gas can migrate through

Flowing Oil Well on Eiting Farm, North Baltimore, O.

A gusher near North Baltimore around 1909. Oil came from the Ordovician-age Trenton limestone far below the surface.

porous rocks, or they can become trapped where folds, faults, and lateral changes in rock type bring the fluids into contact with less-permeable rocks. Ordovician-age Trenton limestone lies some 1,100 feet below northwestern Ohio. It's a porous dolomite under Findlay but becomes the less permeable limestone and shale of the Point Pleasant formation to the south along the crest of the Cincinnati Arch. Oil and gas migrated to this porous dolomite from the Point Pleasant formation deeper in the Appalachian Basin and was trapped.

The first wells that tapped this rock, starting in Findlay in 1884, only produced natural gas. Oil appeared the next year and soon the field grew to cover an area up to 20 miles wide, stretching 120 miles from Toledo to Fort Recovery and then far into Indiana. Over sixty fields appear on maps, but there are no distinct boundaries to what is called the Lima-Indiana Oil and Gas Field. This was the first giant oil and gas field in North America (2.5 million acres total, 550,000 acres in the Ohio portion). The most productive portions of the field in Ohio occurred along the trend of the Bowling Green fault in Wood and Hancock Counties. This is a 45-mile-long fault, buried deep in the Precambrian-age basement, that stretches from southeastern Michigan into Hancock County. From 1895 to 1903, Ohio was the leading oil producer in the United States. Production reached 20 million barrels in 1896, but the discovery of vast Texas fields beginning in 1901 shifted attention westward. By 1910 the Lima-Indiana field was largely depleted, and by 1934 annual production was less than 1 million barrels.

Lack of conservation methods and environmental awareness led to great waste. Geologists estimated that between ten and twenty-five times the amount of oil that was sold was lost to poor collection practices—oil coated trees, flowed into nearby streams, and soaked into the surrounding soil. Too many wells and unrestricted venting caused the reservoir's pressure to drop too quickly before more oil could be pumped. The total volume of oil taken from some 76,000 Ohio wells between 1885 and 1939 was estimated at 375 million barrels; gas volume was not estimated. Attempts to remove oil that remains in the Trenton limestone have had limited success. Because the field is too leaky, there is not enough pressure to get the oil to the surface. Too many wells were improperly plugged and many well locations were never properly recorded, so geologists can't locate them today.

The Stone Industry

Silurian- and Devonian-age bedrock, which is the bedrock throughout much of western Ohio, was used in many ways in the past. It was used to make lime, especially if it was free of impurities such as quartz and iron. Lime is used in the steel-making process, as a fertilizer and flux, in sewage treatment, in food processing, and in the production of alkali chemicals, glass, and paper.

Some limestone was pulverized, mixed with clay or shale, heated in a rotary kiln, and eventually mixed with gypsum to make cement. In the 1800s and early 1900s, western Ohio limestones and dolomites that occurred in thick beds and had few fossils were often quarried for building stone. A good building stone was one that could be readily removed in substantial blocks or flat, thinner slabs called *flags*. The more massive rocks were used for building foundations and bridge abutments and piers where attractiveness or color of the stone was not important. Slabs, or flags, found use as wall material, fences, and paving for streets and sidewalks. Stone that could be cut or chiseled into different sizes (dimension stone) was used in many public and private buildings, particularly town halls, courthouses, schools, and churches. Today, Ohio's building stones come from the north-central, northeastern, and south-central regions.

Today the Devonian and Silurian bedrock of western Ohio is a common aggregate material. Quarries in Delaware, Erie, Franklin, Ottawa, and Wyandot Counties produced over 22 million tons of aggregate, around 40 percent of the state's total production, in 2004. Road and building construction and portland cement and lime production are the major uses of aggregate.

Hundreds of quarries have provided jobs and materials to local residents from the mid-1800s to today. Quarries enlarged as long as workers continued to quarry good stone and the quarries did not become surrounded by other land use. Many quarries grew deep, and quarriers extracted different, older strata. A few even tapped adjacent rock by tunneling into quarry walls, which essentially became underground mines. Abandoned quarries became lakes and ponds; others were filled in and reclaimed. Some lost their identity when neighboring pits were combined as one. Western Ohio's landscape is dotted with quarries.

INTERSTATE 70 AND U.S. 40
Buckeye Lake—Indiana Line
143 miles

I-70 and U.S. 40 traverse Wisconsinan sediments of the Scioto lobe from Buckeye Lake to Springfield. End moraines clump together from east of the Madison and Clark County line to Springfield. Tills of the Miami lobe underlie the routes from the Mad River valley to the Indiana line. Devonian- and Mississippian-age rocks underlie Columbus and continue west to just east of West Jefferson. From West Jefferson to the Indiana line the bedrock is of Silurian age.

Buckeye Lake to Springfield

Rocks of Devonian and Mississippian time underlie I-70 and U.S. 40 beginning about 18 miles east of downtown Columbus. A quarry operated in Reynoldsburg pulling Black Hand sandstone, which was used as flagging material, from the Mississippian Cuyahoga formation. Reportedly builders used stone from this quarry in bridges along the National Road. This road parallels U.S. 40 in places and in places U.S. 40 has replaced it; only a couple of the original bridges remain, and the rocks they were built of are inaccessible. The quarry closed before 1910. A sandstone quarry at Blacklick, north of Reynoldsburg, provided stone for the original buildings of the Ohio School for the Blind and old Union Station in Columbus. There were more sandstone quarries along Rocky Fork Creek even farther north in Gahanna. This area marks the western edge of the glaciated Allegheny Plateau; the western Ohio till plains lie to the west.

Workers used till at Brice, just east of Columbus, in the 1890s to make tiles. The clay pit that remained is now a lake. Where I-270 intersects both routes, there

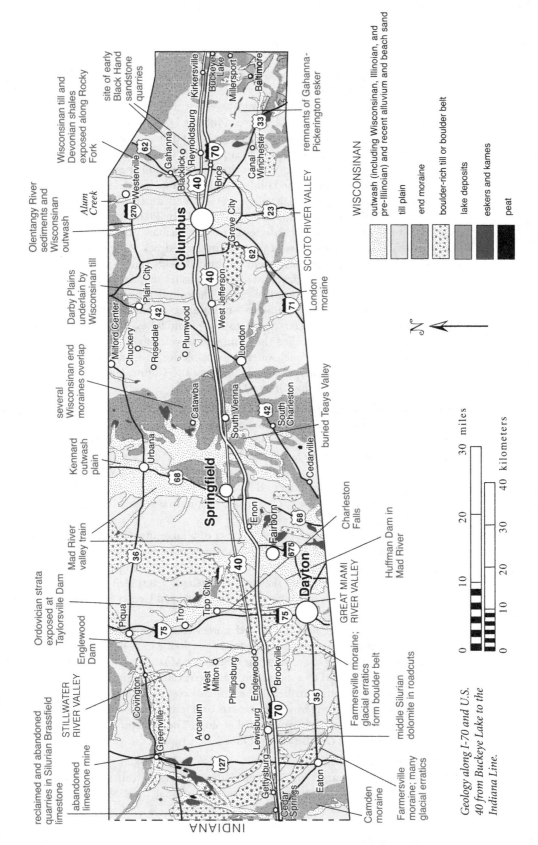

Geology along I-70 and U.S. 40 from Buckeye Lake to the Indiana Line.

WISCONSINAN

- outwash (including Wisconsinan, Illinoian, and pre-Illinoian) and recent alluvium and beach sand
- till plain
- end moraine
- boulder-rich till or boulder belt
- lake deposits
- eskers and kames
- peat

reclaimed and abandoned quarries in Silurian Brassfield limestone

abandoned limestone mine

Camden moraine

Farmersville moraine; many glacial erratics

Ordovician strata exposed at Taylorsville Dam

Englewood Dam

Mad River valley train

Kennard outwash plain

several Wisconsinan end moraines overlap

Darby Plains underlain by Wisconsinan till

Olentangy River sediments and Wisconsinan outwash

Wisconsinan till and Devonian shales exposed along Rocky Fork

site of early Black Hand sandstone quarries

remnants of Gahanna-Pickerington esker

London moraine

buried Teays Valley

Charleston Falls

Huffman Dam in Mad River

Farmersville moraine; glacial erratics form boulder belt

middle Silurian dolomite in roadcuts

SCIOTO RIVER VALLEY

GREAT MIAMI RIVER VALLEY

STILLWATER RIVER VALLEY

INDIANA

N

0   10   20   30 miles

0   10   20   30   40 kilometers

Greenville, Arcanum, Gettysburg, Cedar Springs, Eaton, Covington, West Milton, Phillipsburg, Lewisburg, Brookville, Piqua, Troy, Tipp City, Dayton, Englewood, Urbana, Catawba, Springfield, Enon, Fairborn, South Vienna, Cedarville, South Charleston, Milford Center, Chuckery, Rosedale, Plumwood, Plain City, London, West Jefferson, Grove City, Columbus, Westerville, Alum Creek, Blacklick, Gahanna, Reynoldsburg, Brice, Canal Winchester, Kirkersville, Buckeye Lake, Millersport, Baltimore

127, 36, 68, 75, 70, 35, 675, 40, 42, 62, 270, 23, 71, 33

are remnants of an esker; other remnants of this esker occur at Gahanna, but most of it has been mined away.

A hill of outwash, or kame, appears just west of the Scioto River and to the south of I-70. West of Columbus, between the Big Darby and Little Darby Creeks, I-70 and U.S. 40 traverse the Darby Plains, a flat till plain. It's not noticeable, but both routes rise 100 feet from West Jefferson to the Ohio 56 exit. At this exit, I-70 tops the crest of the Wisconsinan-age London moraine of the Scioto lobe. A London tile plant used till from this moraine in the 1930s.

The Teays Valley—a preglacial valley that drained much of Ohio in a northwesterly direction—lies 10 miles south of the Darby Plains and is buried beneath 200 to 300 feet of glacial sediment. It is not visible at the surface. At about South Vienna, Silurian strata begin to form the bedrock. At the intersection of I-70 and U.S. 40 east of Springfield, the buried Teays Valley turns to the north and passes under both highways. Drilling indicates that the floor of the valley is 600 feet below the surface at this point. The rolling terrain of this area consists of five compressed end moraines—moraines pushed up against one another—of till laid down between the Scioto and Miami lobes. A 2-mile-wide flat deposit of sand and gravel called the Kennard outwash plain cuts through this complex of moraines. The meltwater responsible for this deposit flowed southward across Clark County, eventually carving the Clifton Gorge of the Little Miami River in neighboring Greene County.

Mastodons of Madison County and Elsewhere

Among the common inhabitants of Ohio during Pleistocene time and the most commonly reported Ice Age fossil was the mastodon (*Mammut americanum*), an elephant that stood about 10 feet at the shoulder when fully grown, weighed 4 to 5 tons, and had a heavy coat of hair. Mastodons favored open woodlands, often visiting vernal ponds for water and bathing. They had blunt cone-shaped teeth and were evidently browsers of soft vegetation. Mastodons became extinct around 10,000 years ago.

In 1887 workers unearthed the huge bones of a large male mastodon on the Conway Farm near Catawba. The family that owned the farm reportedly exhibited the bones at county fairs for a year or two before donating them to the Ohio State University in 1894. The skeleton dominated Orton Hall's museum until 1971, when it was moved to the new Ohio Historical Center just east of the university in Columbus.

Workmen in a farm field near Plumwood, northwest of Columbus, were surprised when they unearthed a large bone instead of drain tile in 1949. The owner of the land dutifully notified scientists at the Ohio State Museum, who identified the bone as having belonged to a mastodon. They excavated the site and found the skeleton to be broken and crushed, but essentially complete. The excavation was an early cooperative study involving experts on Pleistocene mammals, pollen, mollusks, wood, archaeology, and glacial geology. The skeleton is housed at the Museum of Biological Diversity at the Ohio State University.

History of the Darby Plains

From 6,000 to 4,000 years ago, warm and dry conditions came to Ohio, allowing an influx of prairie plants, mainly grasses and small trees. More recent

The Mills Quarry at Springfield was a source of Silurian-age Springfield dolomite building stone. Note the blocks of dimension stone ready for shipment in 1909. —J. A. Bownocker photo, Ohio Department of Natural Resources, Division of Geological Survey

climatic changes eliminated most of the prairie vegetation except for resilient patches that maintain a foothold here and there. The flat land between Big Darby and Little Darby Creeks is one of these prairie remnants. Three state nature preserves preserve prairie plants within the Darby Plains: Bigelow Cemetery Prairie near Chuckery; Milford Center Prairie south of Milford Center off Ohio 161; and Smith Cemetery Prairie west of Plain City off Ohio 161. Silurian bedrock remains close to the surface in this area, and workers quarried building stone and lime near Rosedale and east of West Jefferson in the 1870s.

Springfield to the Indiana Line

Quarries in Silurian dolomites scar the walls of the Mad River valley just west of the U.S. 68 junction. Cedarville dolomite forms cliffs along Rock Run, north of U.S. 40, just west of Springfield. On the west edge of Springfield, U.S. 40 crosses the wide Mad River valley, which is filled with hundreds of feet of outwash sand and gravel; I-70 parallels this valley between Enon and Fairborn. Sand and gravel pits dot the Mad River valley train, and huge ones were dug at Fairborn. Quarrying continued downward into the Silurian Brassfield formation, which was crushed and used in the making of portland cement. One of these Fairborn quarries, the Reed North, or Oakes, Quarry is being developed as an educational and research site for natural history studies, particularly paleontology. The Brassfield formation forms low cliffs at the Ohio 202 interchange. The picturesque, 37-foot-high Charleston Falls, one of the highest waterfalls in western Ohio, occurs about 3 miles north of this interchange on the Great Miami River.

The Brassfield rocks form the lip of the falls, while fossiliferous Ordovician limestones and shales form the cliff underneath. Trails in the Charleston Falls Preserve, a Miami County park, lead hikers past a number of rock shelters that formed in the softer Ordovician-age strata.

From where Ohio 201 crosses both I-70 and U.S. 40 to the Stillwater River, a large number of boulders of igneous and metamorphic composition—glacial erratics—appear. Geologists refer to this as a *boulder belt*. The boulder belt is contiguous with the Farmersville moraine to the north and southwest. The boulders may have been left behind as surrounding finer-grained till melted out from around and underneath the boulders. Whatever the exact origin, the boulder belt is a great place to get a look at what forms the bedrock of Canada.

On the southeast side of Dayton, off Ohio 4, is Huffman Dam, an earthen flood control dam that is part of the Miami Conservancy District. Two more similar dams are located on the Great Miami and Stillwater Rivers between U.S. 40 and I-70 on the north side of the Dayton metro area. Ordovician limestone and shale are exposed at the Taylorsville Dam. Boulder-sized glacial erratics are common between these river valleys, marking the Farmersville moraine. Miami lobe ground moraine dominates the landscapes around I-70 and U.S. 40 from Dayton to near Lewisburg.

Workers quarried crushed Silurian dolomite south of Phillipsburg. Silurian-age Springfield dolomite is visible in roadcuts at the Lewisburg exits. Euphemia dolomite outcrops along Twin Creek on U.S. 40 in the village it was named for, which is now part of Lewisburg. Workers quarried Springfield dolomite at Lewisburg for building stone and later quarried deeper units for aggregate. At the Brassfield formation level, inclined drift mines (not visible from the road) extend back into the quarry face. Five species of bats now live in the mine tunnels; other tunnels are used for business storage.

At Lewisburg, both routes cross the bouldery remains of the Farmersville moraine; on either side of Sevenmile Creek near Gettysburg, U.S. 40 crosses this moraine again. The less stony Camden moraine forms the hills at New Paris, where there are more abandoned quarries in Silurian-age dolomites. Around 1900, electric interurban rail lines that ran between Dayton and Richmond, Indiana, stopped at Cedar Springs, where passengers partook of the town's mineral waters, soaking in it and drinking it.

INTERSTATE 71
Columbus—Cincinnati
105 miles

From Columbus to the Warren and Clinton County line, I-71 traverses glacial till of the Wisconsinan-age Scioto lobe. The remaining stretch to Cincinnati skirts the boundary between Wisconsinan and Illinoian till deposits.

The Columbus metro area has spread north along I-71, and from Westerville south not much of the original landscape remains. On the southern edge

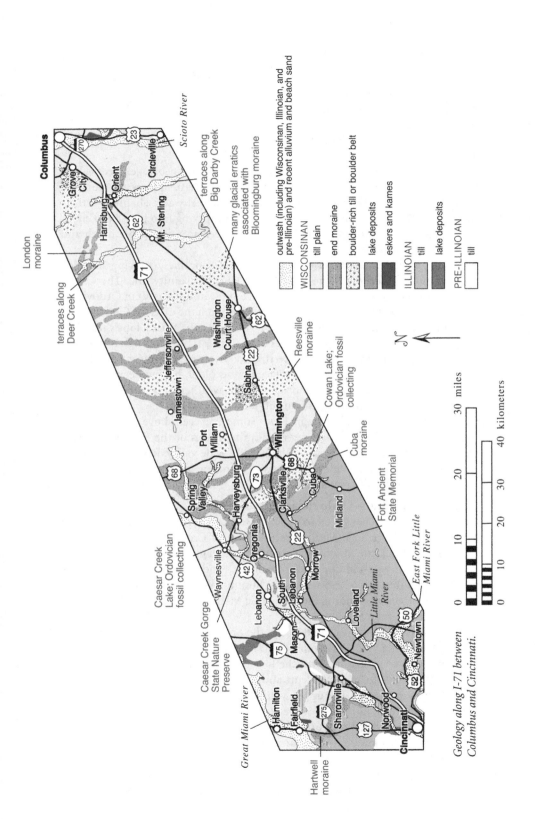

Geology along I-71 between Columbus and Cincinnati.

WISCONSINAN

- outwash (including Wisconsinan, Illinoian, and pre-Illinoian) and recent alluvium and beach sand
- till plain
- end moraine
- boulder-rich till or boulder belt
- lake deposits
- eskers and kames

ILLINOIAN

- till
- lake deposits

PRE-ILLINOIAN

- till

N

0   10   20   30 miles

0   10   20   30   40 kilometers

Columbus

London moraine

terraces along Deer Creek

Grove City

Orient

Harrisburg

Circleville

Mt. Sterling

Scioto River

terraces along Big Darby Creek

many glacial erratics associated with Bloomingburg moraine

Jeffersonville

Washington Court House

Jamestown

Port William

Sabina

Reesville moraine

Wilmington

Cowan Lake; Ordovician fossil collecting

Cuba moraine

Spring Valley

Harveysburg

Clarksville

Cuba

Caesar Creek Lake; Ordovician fossil collecting

Oregonia

Midland

Fort Ancient State Memorial

East Fork Little Miami River

Caesar Creek Gorge State Nature Preserve

Waynesville

Morrow

Lebanon

South Lebanon

Loveland

Little Miami River

Newtown

Mason

Hamilton

Fairfield

Sharonville

Norwood

Cincinnati

Hartwell moraine

Great Miami River

of downtown Columbus, the Olentangy River joins the Scioto River just east of the interstate. From here to I-270, the terrain is scarred by sand and gravel operations mining the glacial outwash and alluvium. At the U.S. 62 interchange, I-71 crosses the London moraine and Big Darby Creek, a terraced meltwater channel. The town of Harrisburg lies on a major terrace, and the town of Orient is above on the margin of the valley. Gravel pits are prevalent along this valley.

Just northeast of the town of Mt. Sterling, the highway bridges Deer Creek. Quarries along Deer Creek once provided good Mississippian-age building stone for local use. Deer Creek also served as a meltwater channel, cutting through the London moraine to the north and emptying into the Scioto channel at Circleville, south of Columbus. Outwash terraces are visible north and south of the highway.

I-71 proceeds across a series of Wisconsinan moraines, covering about 48 miles from Mt. Sterling to the Warren and Clinton County line. The Caesar Creek Lake and Cowan Lake spillways, where you can find fossils in Ordovician rock, lie to the north and south, respectively. Between the exits for Caesar Creek Lake and Fort Ancient State Memorial, I-71 crosses the Scioto lobe's terminal moraine, the Cuba moraine, and passes onto Illinoian till. Although not readily noticeable, the terrain becomes more rugged. The stream bottoms are deeper and the uplands are less flat. The surface here is older, having been deposited 230,000 to 125,000 years ago, and it has been weathering and eroding for a much longer time than Wisconsinan surfaces to the north.

I-71 bridges the Little Miami River some 235 feet above the river—the highest bridge in the state. During the waning stages of the Ice Age, tremendous volumes of meltwater carved the ½-mile-wide valley. Rest areas on both sides of the interstate provide scenic views. From here to Cincinnati, the natural landscape yields to concrete and asphalt.

Caesar Creek Lake—Ordovician Fossils Galore

From the I-71 interchange, Ohio 73 leads west to Caesar Creek State Park and Caesar Creek Gorge State Nature Preserve. Caesar Creek, Flat Fork, and other tributaries of the Little Miami River incise through glacial till into the underlying 445-million-year-old Ordovician bedrock, producing many narrow, rocky gorges in this area. Ohio 73 traverses the hummocky Cuba moraine of Wisconsinan time before descending into the Little Miami River valley at Waynesville and the U.S. 42 interchange.

The oldest exposed rock in the area, the Arnheim formation, is near Caesar Creek Dam along the Clarksville Road. About 12 feet of shale with thin to nodular layers of limestone form the lower part of the dam's bank. About 40 feet of Waynesville formation occurs above this, typically exposed as shale with thicker limestone layers. The upper, younger formations occur along the dam's emergency spillway, just south of the dam. This includes 20 feet of Liberty formation, nearly an equal mixture of limestone and shale, and nearly 30 feet of the base of the Whitewater formation, which is mainly limestone with thin beds of shale. Along Flat Fork, which can be reached by a trail from the emergency spillway, Waynesville to lower Whitewater strata occur. Look in the bed of the creek near the footbridge for ancient ripple marks of the Ordovician seas.

Both the Ordovician-age Liberty and Whitewater formations yield abundant fossils at the spillway for Caesar Creek Lake near Waynesville.

The U.S. Army Corps of Engineers created Caesar Creek Lake between 1971 and 1978 when it dammed Caesar Creek southeast of Waynesville. Normally the water is 115 feet deep near the earth-and-rock-filled dam, making it the deepest lake in the state. The 450-foot-wide emergency spillway stretches from Flat Fork to Roaring Run, about ½ mile west of Clarksville Road, and is where the Ordovician bedrock is exposed. The U.S. Army Corps of Engineers has a visitor center along Clarksville Road with reservoir engineering and fossil information. If you want to collect fossils within the park, you must register at the visitor center. The designated fossil collecting site is the emergency spillway; other rock exposures are for viewing only. You'll find abundant brachiopods and bryozoans; frequent molds of snails, clams, and cephalopods; horn corals, crinoid stems, trilobite fragments, and trace fossils.

At nearby Caesar Creek Gorge State Nature Preserve, off Corwin Road, you can walk through a rock-walled gorge carved by the meltwater of the Wisconsinan ice sheet. Ordovician rock rises 180 feet above Caesar Creek and is capped by glacial till. The gorge occurs where the Miami and Scioto lobes met, what geologists call an *interlobate* area. The gorge cuts through 2 miles of the Hartwell moraine of the Miami lobe. State-owned land adjacent to the Little Miami State and National Scenic River and a segment of the Little Miami Scenic Trail—great places to view the region's geology—are adjacent to the preserve.

Waynesville is where the Waynesville formation was defined. Just north of town on U.S. 42, an exposure yields microscopic, black, toothlike fossils called *scolecodonts*, which are actually pieces of the jaws of marine worms. The best way to find one is to collect lumps of gray mud from this formation and wash it through fine screens that can trap the tiny fossils.

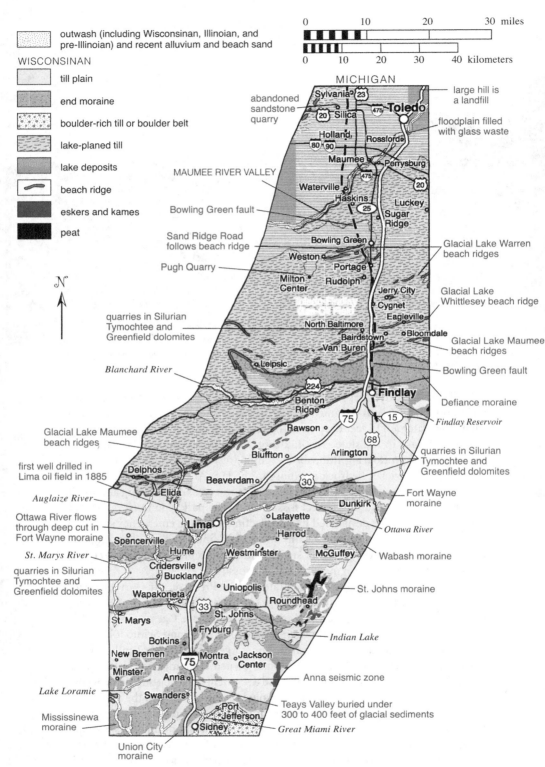

Legend

- outwash (including Wisconsinan, Illinoian, and pre-Illinoian) and recent alluvium and beach sand

WISCONSINAN
- till plain
- end moraine
- boulder-rich till or boulder belt
- lake-planed till
- lake deposits
- beach ridge
- eskers and kames
- peat

0 10 20 30 miles

0 10 20 30 40 kilometers

N

MICHIGAN

large hill is a landfill

abandoned sandstone quarry

Sylvania 23

Silica 20

Toledo 475

floodplain filled with glass waste

Holland

80 90

Rossford

Maumee

Perrysburg

20

MAUMEE RIVER VALLEY

Waterville 475

Haskins

Luckey

Sugar Ridge

25

Bowling Green fault

Sand Ridge Road follows beach ridge

Bowling Green

Glacial Lake Warren beach ridges

Pugh Quarry

Weston

Portage

Milton Center

Rudolph

Jerry City

Cygnet

Glacial Lake Whittlesey beach ridge

quarries in Silurian Tymochtee and Greenfield dolomites

North Baltimore

Eagleville

Bloomdale

Bairdstown

Van Buren

Glacial Lake Maumee beach ridges

Blanchard River

Leipsic

224

Findlay

Bowling Green fault

Benton Ridge

Defiance moraine

Glacial Lake Maumee beach ridges

Rawson

75

15

Findlay Reservoir

first well drilled in Lima oil field in 1885

Bluffton

Arlington

68

quarries in Silurian Tymochtee and Greenfield dolomites

Delphos

Beaverdam

30

Fort Wayne moraine

Auglaize River

Elida

Dunkirk

Lafayette

Ottawa River flows through deep cut in Fort Wayne moraine

Lima

Harrod

Ottawa River

St. Marys River

Spencerville

Hume

Westminster

McGuffey

Wabash moraine

quarries in Silurian Tymochtee and Greenfield dolomites

Cridersville

Buckland

Wapakoneta

Uniopolis

Roundhead

St. Johns moraine

33

St. Johns

Indian Lake

St. Marys

Fryburg

Botkins

Montra

Jackson Center

Anna seismic zone

New Bremen

75

Minster

Anna

Swanders

Teays Valley buried under 300 to 400 feet of glacial sediments

Lake Loramie

Port Jefferson

Mississinewa moraine

Sidney

Great Miami River

Union City moraine

Geology along I-75 between Toledo and Sidney.

INTERSTATE 75

Toledo—Cincinnati

205 miles

I-75 traverses the Huron-Erie lake plain for about 50 miles from the Michigan border to just south of the Van Buren exit. The highway rises between Van Buren and U.S. 224 as it crosses the Defiance moraine, the northernmost end moraine of the Wisconsinan Huron-Erie lobe. Underlying downtown Findlay and extending east toward Vanlue are silts that accumulated in a Wisconsinan-age lake that temporarily formed on the south side of the Defiance moraine. From Findlay to Sidney, the route crosses till plains and more end moraines of Wisconsinan age. Between Sidney and West Carrollton, I-75 travels through the outwash-filled Great Miami River valley. From south of West Carrollton to I-275, I-75 lies mainly on Wisconsinan till, crossing the Camden moraine at the Centerville-Miamisburg exit. From I-275 to the Ohio River, the route follows the Mill Creek valley across Illinoian and older glacial sediments.

Toledo—Glass Capital of Ohio

I-75 passes into Ohio on the industrialized north side of Toledo. Before the days of widespread environmental awareness, all kinds of pollutants from open dumps, industry, and railroads entered streams, soil, and groundwater in this area, all of which drained directly into Lake Erie. Streams with drowned mouths give the shoreline an irregular shape. The entire western and southern shore of Lake Erie has an irregular shape with wide bays. I-75 crosses the wide mouth of the Ottawa River, a seriously polluted stream that flows through a number of old dumps and abandoned industrial sites just south of the Ohio-Michigan border. A man-made hill—a landfill, a modern replacement of earlier surface dumps—lies west of a railroad yard. The highway follows the Ottawa River valley and then heads southeast through the southern edge of downtown Toledo. Swan Creek flows under the highway just before the route passes onto the floodplain of the Maumee River. Once an extensive wetland, the floodplain now sits somewhat higher, underlain by fill material dumped by the glass industry. Rossford, the former location of Libbey-Owens-Ford Glass Company, lies on the river's east bank, just south of the interstate.

In 1887, Edward Libbey of the New England Glass Company of East Cambridge, Massachusetts, contacted towns in western Pennsylvania, northwestern Ohio (including Findlay, Fostoria, Tiffin, and Toledo), and northeastern Indiana looking to relocate his plant in one of them. Findlay had already enticed ten glass factories to relocate there because of free fuel pumped locally from underground deposits, land, and other incentives. Unable to supply energy firsthand, Toledo officials offered to pipe in fuel. The town already had the advantage of lake transportation, and it was the biggest railroad terminal in northwestern Ohio. Toledo officials also promoted its nearby "quartzite" deposits that stretched from Sylvania to the Maumee River as accessible and quality glass-making material. The quartzite was actually a Devonian-age sedimentary rock—the Sylvania sandstone.

Libbey moved his company to Toledo in 1888, occupying a brand-new plant on Ash Street that had free gas service. The company had problems controlling furnace heat at the start since they were not used to using gas fuel, but eventually they were able to produce quality glass. With success in the 1890s, the company officially became the Libbey Glass Company. These beginnings eventually gave rise to the Libbey-Owens-Ford and Owens-Illinois Glass Companies, a couple of Toledo industrial giants.

Toledo to Findlay

I-75 follows glacial lake sediments from the Michigan line to Perrysburg, but few areas of undisturbed ground exist. The interstate is built on lake-planed till from Perrysburg to Van Buren. Bowling Green is built on sandy ridges marking the former shoreline of Glacial Lake Warren, a predecessor of Lake Erie. Except at construction sites, this sandy soil is covered by pavement and buildings. South of North Baltimore the road rises as it crosses the Defiance moraine, also the site of Van Buren State Park. This moraine is the northernmost Wisconsinan-age moraine in Ohio. The Blanchard River flows along its southern edge.

Findlay—Gas and Oil Boomtown

In the 1830s, Findlay was noted for its sulfurous groundwater; locals had a difficult time finding good drinking water. Sulfur seeps lay along the Blanchard River. Even when they were only 8 to 10 feet deep, a number of water wells produced strong odors, which, unbeknownst to most residents, was natural gas rising from the depths. Locals initially discovered gas in Findlay in 1836 on a farm south of town. The story holds that a farmer held a torch to a newly dug well and caused a small explosion, startling all. Later the farmer sank a pipe into the abandoned well and occasionally lit it for curious observers. By 1838 other wells in Findlay tapped natural gas, and reportedly, one enterprising individual piped the gas into his fireplace for heat and to cook with. Early residents didn't really understand natural gas or its potential economic benefits.

Dr. Charles Oesterlen, a German physician and amateur geologist who arrived in Findlay in 1836, possessed the scientific knowledge to shed light on the mysteries of Findlay's explosive gas. Through the years he investigated several gas occurrences around town and spent most of his life trying to convince others of the great wealth under their feet. It wasn't until people opened new gas fields in western Pennsylvania that the citizens of Findlay started to recognize the possibilities. With Oesterlen they formed the Findlay Natural Gas Company. The company drilled a deep well on Oesterlen's farm in 1884, striking gas at several depths, thick oil at a couple of horizons, and eventually saltwater. At 1,092 feet they struck a strong zone of gas. The gas plume shot 20 to 30 feet in the air when it reached the surface. It's estimated that the well produced 250,000 cubic feet of natural gas daily. Finally, after almost fifty years of promoting natural gas, Dr. Oesterlen received his due and the gas boom began, cautiously at first, and then with reckless abandon.

It was the Karg well, started in December 1885 by the Findlay Gas Light Company, an artificial-gas producer, that put Findlay on the map. When workers ignited a stand pipe from this well, roaring flames shot high into the air; reportedly, the flames were visible tens of miles away. This giant torch attracted huge crowds from across northwestern Ohio; railroads even scheduled special excursions to view the spectacle. The well, located on the Blanchard River, burned for four months before employees brought it under control. Initially, 12 million cubic feet of gas poured out of the well daily. Other wells followed: the Pioneer, 2 miles north of the courthouse, and the Kagy well at Van Buren, with a flow that exceeded the Karg's output. Eighteen wells in Findlay were producing natural gas by April 1886, and the gas field spread south of town to Houcktown.

Because of the many successful wells, residents thought that they had tapped an inexhaustible reservoir of gas. The city fathers marketed free fuel, enticing many industries and businesses to locate in town. The population exploded from 4,500 in 1884 to 25,000 by 1889. Extravagance was evident everywhere. Street lights burned through the daylight hours, residents opened windows rather than turning down their gas heat, and wells were ignited solely to impress visitors. A weekend celebration in June 1887 flaunted the abundance of gas as fifty-eight gaslit arches were erected to span Main Street.

A dozen glass factories opened, taking advantage of the cheap power. Sand for the glass came from the Devonian Sylvania sandstone quarried near Sylvania. Brick and lime plants also favored gas as a fuel and operated in Findlay. Beginning in 1888, factories moved their operations elsewhere as gas shortages began. The city responded by piping in gas from more-distant wells, increasing costs. By 1889 it became clear that the gas reservoir was exhaustible, and in 1896 the free supply ended for glass factories and other major consumers because of cost. By 1903 only one glass plant remained; others had moved to places with free or cheaper gas. Findlay's short-lived gas boom was over.

Initially, the deep wells in the Findlay area only tapped natural gas. However, ever since Colonel Edwin Drake—an entrepreneur who believed that drilling was the best way to get oil out of the ground—had struck oil near Titusville, Pennsylvania, in 1859, Findlay men dreamed of striking black gold as well. The intervening Civil War and a number of disappointing drilling adventures delayed the discovery of oil in Findlay. In 1885, however, two wells struck oil; the Mathias Brick Yard well on West Main Cross Street was the most notable of the two, initially flowing at a rate of 300 barrels per day. By September 1886 twenty wells were producing 400 barrels per day, and by December fifty-five wells were producing 2,000 barrels. These figures continued to grow into the early 1900s. The Ohio Oil Company, incorporated in Lima in 1887, moved its headquarters to Findlay in 1905. It became Marathon Oil in 1962. The oil boom provided an energy bridge into the twentieth century just as the gas boom ended.

Both the gas and oil came from the upper part of the Trenton limestone, an Ordovician-age layer some 1,000 feet below the surface. This layer has considerably more and larger openings and fractures than the Silurian-age dolomites above it, allowing gas and oil to readily migrate. Underneath Findlay and extending toward Ottawa County and Lima, there is a slight upward bend, or arch, in the rock layers called the Findlay Arch, which trapped the gas and oil below its crest. Traps also occurred along the buried Bowling Green fault system—a fault system in rocks of Precambrian time—north to Waterville. When drills penetrated these traps, gas and oil, under pressure, raced to the surface.

Vestiges of the gas and oil booms remain today. The Oesterlen well remains on the Hancock County fairgrounds. A plaque marks the wellhead of the Karg well along River Street. Imposing mansions, built in the late 1800s, still line Main Street. The former Ohio Oil office building still dominates the downtown skyline. The Hancock Historical Museum contains informative displays.

Other Wood and Hancock County Gas and Oil Fields

The 1884 discovery of gas in Findlay led many towns in northwestern Ohio to follow suit. Although Findlay was the first community in Ohio to strike gas in the Trenton limestone, important yields came from nearby towns. Just as in Findlay, gas was not properly conserved, and supplies dwindled by 1890.

Bowling Green was another community in Ohio to develop underlying natural gas supplies with the organization of the Bowling Green Natural Gas Company. It drilled a trial well in early 1885. Foul-smelling gas entered the well around 200 to 250 feet down at a horizon called the Clinton formation, a Silurian age limestone-dolomite that correlates with the Brassfield formation where it is exposed in southwestern Ohio. One hundred feet lower in a Silurian sandstone, a pocket of gas exploded from the well. Upon ignition, flames flared 30 feet above the well for two days before dying out. The top of the Ordovician Trenton limestone, at 1,096 feet below the surface, yielded little gas, and drilling deeper didn't help. Only when workers intentionally caused an underground explosion, a drilling technique called *torpedoing*, was the Trenton rock fractured enough to allow gas to flow. Of twelve other wells drilled in this field, the most successful ones were in the Portage area south of Bowling Green. A well drilled by a person independent of the gas company in 1886 was the most productive. The well belonged to the owner of lime kilns in Portage. The availability of gas

Oil wells tap the Ordovician-age Trenton limestone below Cygnet around 1885. —Ohio Department of Natural Resources, Division of Geological Survey photo

in Bowling Green attracted two glass companies and several other industries. By 1890 gas production had slowed, allowing oil to move into the wells. The most productive oil wells also occurred near Portage.

Oil and gas exploration spread south to Bairdstown, Bloomdale, Cygnet, Jerry City, North Baltimore, and Trombley during 1886. Drillers struck oil at 1,400 feet below North Baltimore. The Simons well near Bairdstown tapped gas, but most of it was wasted by burning. Exploration headed north again to Haskins, Weston, and Perrysburg and east into Sandusky County in 1887.

Between Cygnet and North Baltimore, oil tank farms—major storage areas—once dotted the countryside, feeding pipelines that radiated in all directions. They're now gone, but Tank Farm Road still exists. Other tank farms existed between Van Buren and Findlay. Only a few hundred wells still yield oil, and most only produce a few barrels per day.

Traces of this old oil and gas patch lie scattered along I-75 from Toledo to Wapakoneta. Oil field paraphernalia is preserved on the grounds of the Wood County Historical Society in Bowling Green. Pumps and storage tanks still dot some fields, but they no longer produce.

Northwestern Ohio Mineral Industries

Stone quarries, some still in operation, abound along I-75 where Paleozoic bedrock is close to the surface. Undoubtedly, many travelers wonder about the deep holes bridged by the interstate at Lima; they are unused quarries. Many other quarries lie scattered in towns from Toledo to Lima. Silurian and Devonian rocks serve as sources of crushed stone and lime today, but in the past local builders pulled building stone from some of the quarries.

Small quarries, now filled with water, occur near Sugar Ridge and in Bowling Green. Most of this area's quarrying was concentrated between Bowling Green and Portage. Several more occur along Ohio 25, which parallels I-75 in this area; the rock in these quarries is younger Silurian-age dolomite. Pugh Quarry near Milton Center is a famous mineral collecting locality. Doubly terminated crystals—meaning the crystals come to a point at two ends rather than just one—of golden brown calcite and barite occur in numerous cavities of the Devonian Detroit River group. The Quarry Road exit at North Baltimore leads to a former France Stone Company quarry, and another quarry (water-filled) occurs on Main Street north of town. Near Van Buren, between Eagleville and Bloomdale, there are old quarries in Silurian dolomite, mainly the Tymochtee formation. By 1874, at least ten shallow quarries in Findlay supplied stone for foundations and flagging or burned it to make lime; many were located along the Blanchard River. Oakwoods Nature Preserve near Findlay surrounds the old Shank's Quarry. Other quarries flank I-75 and Ohio 15.

Quarries abound in the Silurian-age Salina group dolomites from Rawson to Lima. There are large quarries on either side of I-75 at Bluffton and Lima. Quarries in these dolomites also occur off Main Street and east and west of I-75 at Bluffton, north and south of Buckland, between Delphos and Spencerville, north of Elida, on the northwestern edge of Lafayette, south and southwest of Harrod, and near Westminster, to name a few.

Brick and tile factories used glacial till in this region into the early 1900s. Plants operated in and around Beaverdam, Bluffton, Delphos, Findlay, Fryburg,

Silurian-age Tymochtee and Greenfield dolomite occur in quarries bisected by I-75 on the east side of Lima.

Lima, Minster, New Bremen, Roundhead, Uniopolis, and Wapakoneta. Blue Rock Nature Preserve in Findlay is the site of early clay mining. The bluish till was used to make clay sportsmen's targets in the early 1900s. A Findlay pottery also found a use for local till in the 1800s.

Lima to Sidney

From near Lima to Sidney, I-75 crosses a series of Wisconsinan end moraines. The interstate passes over the rolling terrain of the Fort Wayne moraine from 1 mile south of Beaverdam to the Ohio 81 exit at Lima. The easterly Ottawa River parallels the southern edge of the moraine through Lima and then cuts a 50-foot-deep valley through the moraine southwest of town, flowing north to the Auglaize River. North of Wapakoneta, the Wabash moraine stretches underneath the highway. Similar to the Ottawa River and the Fort Wayne moraine, the Auglaize River flows along the southern border of this moraine before cutting through it northwest of town. A string of sand and gravel pits parallel to I-75 extends south from the Auglaize River along Quaker Run and Pusheta Creek some 6 miles. Botkins sits on the crest of the St. Johns moraine, and Loramie Creek follows its southern edge. These moraines average 50 to 100 feet higher than the surrounding till plain. I-75 crosses over the Mississinewa moraine at Anna, which lies above the ancient Teays Valley that is buried under 300 to 400 feet of glacial fill. From Anna you can take Ohio 119 west to Lake Loramie, a former storage basin of the Miami & Erie Canal, an early waterway that connected Lake Erie and the Ohio River; the lake is part of Lake Loramie State Park. Just north of Sidney, I-75 passes over the Union City moraine.

The Lima Oil Field

In 1885, drillers searching for gas and water for a paper mill along the Ottawa River struck bad-smelling oil instead. Shortly after this discovery, the Citizens

Gas Company of Lima completed a well in Lima that became the first well to supply oil on a regular basis; initially, 40 barrels daily. Since this wasn't the "sweet" oil used in most refineries, its value was questionable. However, testing at a number of plants proved it could be refined into illuminating oil. Drillers hastily drilled more wells, and soon the Lima field became the most important in the state. By the end of 1886, the field had 165 wells; by April 1887 the number rose to 424, yielding over 10,000 barrels of oil per day. Production of this field stretched from Beaverdam southwest to Cridersville and west to Hume. The oil boom was short-lived, just like the region's gas boom, dying out about 1905. Minor production continued until 1935, when the last well of the Lima field was abandoned.

Anna—Earthquake Capital of Ohio

On March 2 and again on March 9, 1937, Anna and its surroundings shook as shockwaves rumbled through thick glacial deposits, rotating cemetery monuments, cracking plaster walls, breaking windows, displacing heavy furniture, and weakening a number of buildings. The shockwaves were the results of the greatest Ohio earthquakes ever recorded, and topped off a period of increased quake activity in the 1930s, in which twenty-three tremors struck this Shelby County area. Anna residents have felt over forty tremors since 1875.

Of all the towns, Anna suffered the greatest damage from the quakes of the 1930s. Its three-story school had to be demolished because of cracks and bulges that developed in the brick walls. Two other brick structures were heavily damaged, and a church steeple tumbled to the street. Most chimneys in Anna fell over or rotated. Merchandise lay strewn on the floors of stores. Damage extended to nearby Botkins, Jackson Center, Montra, Port Jefferson, and Swanders.

Why Anna? Midwest earthquakes are difficult to explain. In most cases they happen when ancient, deeply buried rocks shift, often along an old fault or a failed plate boundary. Anna lies above a 1-billion-year-old Precambrian fault zone known as the Fort Wayne rift. A buried valley below the town served to accentuate the vibrations in the area, like shaking a bowl of gelatin—the bowl being the valley and the glacial material the gelatin. A damage analysis of the Anna area using questionnaires distributed to residents pinpointed the epicenter of the 1937 tremors as having been within a 3-mile radius of town.

Sidney to Dayton

Between Kirkwood and Piqua, large numbers of glacial erratics—mainly granite, gneiss, and schist that likely came from Canada—dot the land, something geologists map as boulder belts. I-75 follows the outwash-filled Great Miami River valley between Troy and Dayton. I-75 passes through another area of boulders, which mark the Farmersville moraine between Vandalia and downtown Dayton.

The Highest Point of the Miami & Erie Canal

West of I-75 between Sidney and Piqua, the Miami & Erie Canal dropped off the "Loramie summit"—a piece of land separating the Loramie Creek and Miami River drainage basins—into the Miami drainage basin at Lockington. At this point, the canal was 395 feet above Lake Erie and 512 feet above the Ohio Riv-

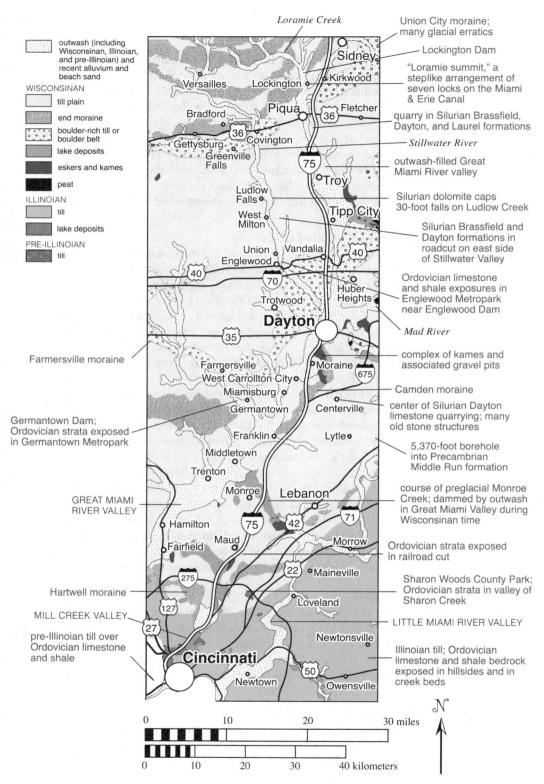

Legend

- outwash (including Wisconsinan, Illinoian, and pre-Illinoian) and recent alluvium and beach sand

WISCONSINAN
- till plain
- end moraine
- boulder-rich till or boulder belt
- lake deposits
- eskers and kames
- peat

ILLINOIAN
- till
- lake deposits

PRE-ILLINOIAN
- till

Loramie Creek

Union City moraine; many glacial erratics

Lockington Dam

"Loramie summit," a steplike arrangement of seven locks on the Miami & Erie Canal

quarry in Silurian Brassfield, Dayton, and Laurel formations

Stillwater River

outwash-filled Great Miami River valley

Silurian dolomite caps 30-foot falls on Ludlow Creek

Silurian Brassfield and Dayton formations in roadcut on east side of Stillwater Valley

Ordovician limestone and shale exposures in Englewood Metropark near Englewood Dam

Mad River

complex of kames and associated gravel pits

Camden moraine

center of Silurian Dayton limestone quarrying; many old stone structures

5,370-foot borehole into Precambrian Middle Run formation

course of preglacial Monroe Creek; dammed by outwash in Great Miami Valley during Wisconsinan time

Ordovician strata exposed in railroad cut

Sharon Woods County Park; Ordovician strata in valley of Sharon Creek

LITTLE MIAMI RIVER VALLEY

Illinoian till; Ordovician limestone and shale bedrock exposed in hillsides and in creek beds

Farmersville moraine

Germantown Dam; Ordovician strata exposed in Germantown Metropark

GREAT MIAMI RIVER VALLEY

Hartwell moraine

MILL CREEK VALLEY

pre-Illinoian till over Ordovician limestone and shale

Sidney
Kirkwood
Versailles
Lockington
Fletcher
Piqua
36
Bradford
36
Covington
Gettysburg
75
Greenville Falls
Troy
Ludlow Falls
West Milton
Tipp City
Union
Vandalia
Englewood
40
40
70
Huber Heights
Trotwood
Dayton
35
Farmersville
Moraine
675
West Carrollton City
Miamisburg
Germantown
Centerville
Franklin
Lytle
Middletown
Trenton
Monroe
Lebanon
Hamilton
75
42
71
Fairfield
Maud
Morrow
275
22
Maineville
127
Loveland
27
Newtonsville
Cincinnati
50
Newtown
Owensville

0   10   20   30 miles

0   10   20   30   40 kilometers

N

Geology along I-75 between Sidney and Cincinnati.

Quarried blocks of Silurian-age Dayton formation form the walls of a canal lock at Lockington Locks State Memorial. Lockington marks the crest of the Miami & Erie Canal, which stretched from Toledo to Cincinnati, more or less paralleling I-75.

er. Through a series of locks, water level dropped over 60 feet in ½ mile as the canal left the Union City moraine and headed south onto the Great Miami River floodplain. Through much of the 1800s, the Miami & Erie Canal system—a conglomeration of rivers, locks, reservoirs, and canals—served as a reliable transportation system that connected the Ohio River and Lake Erie. It was built for economic reasons, so Ohioans could transport goods to eastern markets.

Five of the six locks still survive as a historical site. The lock walls were built of blocks of the Silurian-age Dayton formation that were hauled in by packet and wagon from Dayton-area quarries. Talented stonemasons finished the blocks on site before setting them in the late 1830s.

There is an earthen flood-control dam, which is part of the Miami Conservancy District, just northwest of Lockington. It's used to hold back the water of Loramie Creek during major floods. The Miami Conservancy District was established after the catastrophic flood of the Great Miami River in 1913. The district built levees and five dams in the river's drainage system and straightened the river in several places in order to try to control future flooding.

Quarrying in Piqua

Early Piqua quarries exploited Silurian-age bedrock south of downtown along the Great Miami River. Workers quarried 5 to 16 feet of the Dayton formation that underlies up to 20 feet of outwash and river sediments. Builders used much of the good stone, which was shipped by canal, rail, or wagon, for trim, foundations, and basement walls. Dense 2-foot layers were used for bridges, while large, thin slabs became sidewalks in Piqua and elsewhere. Some variegated layers were polished and marketed as "blue limestone." Surfaces of this rock had a bluish color when fresh, but after exposure to the weather they usually became lighter in color. Quarry workers found highly desired "cutting stone" where the formation was several feet thick; it came out of the quarry in blocks that were large enough to be used as steps. Thinner slabs were sold as flagstone.

Around 1900 the Ohio Marble Company ground the underlying Brassfield formation into fine aggregate that they marketed as "marble" dust. This powder was used to make certain pastes and as an additive in paint and other products. By 1915 the market for Piqua building stone disappeared as it was supplanted by the Berea sandstone from the Cleveland area and Indiana limestone from south-central Indiana. Indiana limestone and the Berea sandstone rapidly were used everywhere because of their beauty and durability, and because they could easily be carved. The quarries continued to dig deeper into the Brassfield formation and sold it as a crushed product. Operating quarries lie west of I-75, south of the U.S. 36 interchange.

A Side Trip to Greenville and Ludlow Falls

Waterfalls of the Stillwater River and its tributaries develop where harder Silurian dolomite meets the softer underlying Silurian and Ordovician shales and limestones. Head west from Piqua on U.S. 36 to 1 mile west of Covington, and then south and west to the parking area of Greenville Falls State Nature Preserve. Here Greenville Creek drops about 20 feet over ledges of resistant dolomite at the head of a 25-foot-deep gorge cut into the Silurian Lockport dolomite. Just downstream of the falls there is a natural arch carved in the same dolomite on the north wall of the gorge. A spring near the top of this cliff has eroded back into the cliff along a layer of softer rock. Surface water flowing down a vertical joint that intersects the roof of this small rock shelter separated the outer part from the cliff face, making the 11-foot-long arch.

The Silurian-age Springfield and Euphemia dolomites form the erosion-resistant cap of Ludlow Falls. To the left, currents carve a rock shelter into underlying soft shale.

From Troy head west on Ohio 55 to the town of Ludlow Falls. The 30-foot namesake of the town is typical of many waterfalls of the Great Miami River and its tributaries that develop where harder Silurian dolomite meets the softer underlying Silurian and Ordovician shales and limestones. At Ludlow Falls, Ludlow Creek flows across a lip of Springfield and Euphemia dolomite, which doesn't erode quickly, and falls some 30 feet to the Dayton and Brassfield formations of the streambed below. Through the years, the creek eroded through 30 feet of shale of the Laurel and Osgood formations, forming the waterfall. The town maintains a small park that offers good views of the falls.

Old quarries occur upstream of the falls where, during the late 1800s, workers quarried the Springfield formation for building stone. By 1915, only one quarry remained open, producing crushed stone.

Dayton to Cincinnati

I-75 crosses a rolling till plain between Dayton and Middletown. Southeast of Middletown there is a 2-mile-wide flat valley bottom. It is a remnant of the larger drainage system of Monroe Creek, which flowed northwest into Indiana before the Wisconsinan glaciers came and reshaped the landscape. During Wisconsinan time, when ice sat in the Great Miami River valley to the west, this small valley flooded, forming a narrow lake. As a result, the valley filled with lake sediment, which remains today, along with a series of short misfit creeks that occupy parts of the valley.

Fossiliferous Ordovician-age bedrock lies in a railroad cut on the south side of I-75 near Maud. This bedrock is exposed at more railroad cuts between Maud and Cincinnati. In the late 1800s, workers quarried Ordovician rocks along the Great Miami River in Hamilton for local building purposes. Older buildings often still have it in their foundations. There is another wide valley bottom—now covered with glacial sediments—south of Maud, that formed when the preglacial Licking River flowed north; after Wisconsinan glaciation and the deposition of various tills, south-flowing Mill Creek developed in the channel. I-75 follows the Mill Creek valley through Illinoian and pre-Illinoian sediments between Maud and downtown Cincinnati.

Dayton Area

I-75 bridges the Great Miami River five times in downtown Dayton. The city developed around the confluence of the Great Miami, Mad, and Stillwater Rivers, and Wolf Creek, and it is built over a thick deposit of alluvium and outwash. The city uses groundwater in this deposit for its water supply and accounts for the largest use of subsurface water in the state (most of Ohio's major cities get their water from Lake Erie or rivers). Just south of the Great Miami River, high hills loom over the city. These hills are kames draped against the higher Camden moraine. The sand and gravel left on these hills is just a remnant of what used to exist. Through the years, aggregate companies have removed a good portion of it. Moraine, formerly Moraine City, sits on the floodplain below the Camden moraine, obviously named for this striking glacial landform. A steep grade marks the northern edge of the Camden moraine where I-75 intersects it, south of West Carrollton.

Outcrops of the characteristic gray limestone and shale of Ordovician-age strata appear in the Dayton area. A number of Dayton metroparks are great places to observe this ancient rock. Englewood MetroPark lies along the Stillwater River northwest of Dayton. Here, the river and its tributaries cut into fossiliferous Ordovician rocks that are well exposed at Martindale and Patty Falls. A large earthen dam at this park—part of the Miami Conservancy District—keeps floodwaters from inundating communities downstream. Starting near Miamisburg, Ohio 725 heads west to Germantown Metropark, where similar Ordovician outcrops are visible. Germantown Dam, another earthen dam, dams Twin Creek at this park. These same rocks are also visible along the Great Miami River at Taylorsville Metropark, which is in north Dayton northeast of the I-75 and U.S. 40 interchange. Wet horizons appear in the rock where groundwater emerges as springs, usually above less permeable shale layers.

A number of communities, now part of the Dayton metropolitan area, have a long history and an early connection to the surrounding geology. Limestone of the Silurian-age Dayton formation, which forms Centerville's bedrock, became a favored building material in this town as early as 1803. The section of the formation that builders used as building stone averages 5 feet thick in this region and breaks naturally into blocks. Some thirty stone houses, many built in the 1830s, remain in the Centerville area. A number of them have shallow quarries on site where workers dug the stone. A larger quarry in Centerville that tapped both the Dayton and underlying Brassfield formations is now a reservoir east of Main Street.

Three quarries off the present-day Wilmington Pike in southeast Dayton once tapped the Dayton formation in nearby Beavertown, which was absorbed by Dayton. These reportedly were the largest quarries in the area, but because of their shallow nature they are difficult to spot. At least nine other quarries in the Dayton area once supplied attractive Ordovician-age building stone to Dayton-area builders. Belmont Park, just northeast of the Smithville Road and Watervliet Avenue intersection, and Cleveland Park, northwest of the Cleveland

Many older buildings in Centerville are composed of Silurian-age Dayton formation. The limestone for this building came from a small quarry, now a backyard pond, 500 yards away. Note the fossil colonial coral.

Avenue and Pershing Boulevard intersection, are thought to be sites of former quarries. This stone was also used in the Dayton area during the late 1830s to extend the Miami & Erie Canal to Toledo.

People stopped quarrying the Dayton formation for building material in the early 1900s. A number of buildings in downtown Dayton attest to its lasting quality, though. The old courthouse (now home of the Montgomery County Historical Society), the Woodland Cemetery buildings, and a number of downtown churches are good examples of this stone. Foundations and retaining walls throughout Dayton may also still exhibit the stone. Not every building with this stone has been documented.

The underlying Silurian Brassfield formation saw less use as a building material in the Dayton area and instead was used as aggregate. Two quarries once operated at the old Soldiers Home (now the Dayton VA Medical Center and National Cemetery), west of I-75 and above the Great Miami River, and the deeper Dayton-formation quarries also tapped into this stone. Workers still quarry the Brassfield formation at large quarries on the east side of Fairborn, northeast of Dayton, for crushed stone and for making portland cement.

Germantown, platted in 1814, was one of the first places people made bricks in this area. Clay came from nearby fields and probably was till. A local mason also fabricated clay roof tiles, the first recorded use of such tiles in the area.

Miami Conservancy District

The 1913 flood wreaked havoc in the Miami Valley, causing $100 million in property damage and killing 360 people. Raging floodwaters inundated Dayton, Hamilton, Piqua, Troy, and other communities. Hoping to prevent future flood disasters, a 1914 act of the Ohio legislature provided for the development of conservancy districts, and in 1915 the Miami Conservancy District was formed. Its mission was simple—build a series of earthen dams on the Great Miami River and its major tributaries that would serve as temporary holding basins during floods, controlling the release of floodwater to downstream communities; and to provide local protection and channel improvements where needed. Between 1918 and 1922, the district built dams on the Great Miami River at Taylorsville, the Mad River in Dayton, the Stillwater River at Englewood, Twin Creek at Germantown, and Loramie Creek at Lockington; it also constructed levees and floodwalls at Dayton, Franklin, Hamilton, Miamisburg, Middletown, Piqua, Tipp City, Troy, and West Carrollton. The district designated over 3,300 acres upstream of the dams as recreation areas, which are now managed and maintained by Five Rivers MetroParks. The Miami Conservancy District continues to serve the Great Miami River valley, providing flood protection and information on the wise use of groundwater and the river valley.

Deep Below Warren County

In October of 1987, in a gravel pit near Lytle east of I-75 and north of Ohio 73, the Ohio Geological Survey began drilling to the Paleozoic-Precambrian boundary, which they projected to be 3,500 feet down. Geologists were hoping to learn more about Ohio's Precambrian past. Since there are no Precambrian rocks exposed in Ohio, it is one of the least-understood time periods of the state's geologic past.

The hole began in Ordovician strata, passing 3,500 feet down into Cambrian-age Mt. Simon sandstone (about 540 million years old), but only more sandstone appeared as the hole probed deeper below the known thickness of the Cambrian sandstone, which is marked by a conglomerate at its base. Monetary restrictions forced them to halt the drilling. Later seismic studies in the area indicated to geologists that they had another 2,000 feet of sandstone to drill through before reaching the basement rocks of Precambrian time. In April 1989, the drilling ended for the second time after the drill rod became stuck deep in the well and couldn't be withdrawn. The well ended in the same reddish sandstone, now 5,370 feet below the surface.

This deep sandstone is now known as the Precambrian Middle Run formation. Geologists have projected that the formation extends, though not as a continuous layer, north to Lima and south to Lexington, Kentucky. It more or less lies under I-75 between these points, mainly just west of the boundary between the Grenville and granite-rhyolite basement rocks. The formation varies in width from about 10 miles in the Cincinnati area to 35 miles between Dayton and Lima.

The Middle Run sandstone is a dense, unfossiliferous rock with a few interbeds of siltstone and lava. It appears to have accumulated in low-lying areas of the supercontinent Rodinia and is associated with folds or faults. Outside Ohio, similar deposits reach thicknesses of 7,000 to 8,000 feet.

Geologic Studies in the Queen City

Since Kentucky's Licking River and Ohio's Mill Creek meet the Ohio River in southwestern Ohio, a considerable area of flat land developed and was available for settlement above annual flood level. The predecessors of Cincinnati began developing in this area in 1788, and the "Queen City," as Cincinnati was known, became a center of commerce and trade by the early 1800s.

Since settlers built the town on an outwash terrace of sand and gravel, the more suitable building stones came from high in the surrounding hills, which rose as much as 500 feet above the river. The hills exposed several Ordovician limestones and shales, but the layer settlers found most useful was what is now

The Ordovician-age Fairview formation was quarried in the uplands of the Cincinnati area. Thicker limestone beds yielded building stone used in many early structures and retaining walls. —Ohio Department of Natural Resources, Division of Geological Survey photo

known as the Fairview formation. Since the usable building stone was located up in the hills, the rocks were called the Hill Quarry beds. A lot of the strata was crumbly shale and practically useless for building. There was better stone to be quarried elsewhere, but in the early days the local stone sufficed. A number of downtown churches and the older part of the art museum still have walls composed of this stone. Cincinnati builders also used the older Point Pleasant formation widely, and it was shipped on the Ohio River to other locales. You'll also find it in retaining walls, foundations, and fences.

When the Miami & Erie Canal opened in 1822, Silurian-age Dayton formation stone became available and Cincinnatians used local stone less and less. Quarries shipped Mississippian-age Buena Vista sandstone from upstream Portsmouth by the 1850s. By 1900 urban sprawl had claimed many of Cincinnati's hilltop building stone quarries, and builders turned away from the use of Fairview formation stone.

The extensive quarrying uncovered fossils—abundant, well-preserved fossils. By the 1870s, Cincinnati became a world-famous fossil-collecting site. A group of local men, often referred to as the "Cincinnati School of Paleontology," began detailed studies of Cincinnati-area fossils around this time. A number of these "amateur fossil collectors" went on to distinguished careers in geology and paleontology; others never became geologists but continued to make great contributions to science's knowledge of Ordovician fossils.

COMMON ORDOVICIAN FOSSILS OF THE CINCINNATI AREA AND HAMILTON COUNTY

Formation	Fossil Type	Some Localities
Grant Lake formation	Brachiopods: *Hebertella, Platystrophia, Rafinesquina* Bryozoans: *Batostomella, Dekayia, Monticulipora, Parvohallopora* Bivalves: *Ambonychia, Caritodens* Snails: *Cyclonema* Straight-shelled nautiloids	Sharon Woods County Park (Sharonville)
Fairview formation	Brachiopods: *Platystrophia, Plectorthis, Rafinesquina, Zygospira* Bryozoans: *Constellaria, Dekayia, Escharopora, Heterotrypa, Homotrypa, Parvohallopora* Trilobites: *Isotelus* Crinoid stems	Cincinnati's Bald Knob, Bellevue Hill, Fairview Park, and Jackson Hill Park; Harrison, North Bend
Kope formation	Brachiopods: *Onniella, Rafinesquina, Zygospira* Bryozoans: *Batostoma, Dekayia, Parvohallopora* Trilobite fragments: *Flexicalymene, Isotelus* Crinoid stems	Cleves, Grand Avenue (Cincinnati), Miamitown

They collected fossils from along the Ohio River in the well-known River Quarry beds (now the Point Pleasant formation) and from the uplands of the Hill Quarry beds (now the Fairview formation). Their work contributed to the designation of Cincinnati as the American Ordovician standard. The Ordovician strata around Cincinnati is loaded with well-preserved fossils of mainly bryozoans and brachiopods, two types of bottom-dwelling sea creatures that had hard calcareous skeletons and shells. Other common fossils include various echinoderms like crinoids and edrioasteroids, trilobites, horn and colonial corals, clams, snails, ostracods, and trace fossils like trails and borings. Sharon Woods Park, a Hamilton County park, is a good place to observe the fossiliferous limestones and shales in the rocky gorge carved by Sharon Creek. Visit the Cincinnati Museum Center in Union Terminal for displays on Cincinnati-area geology.

People manufactured bricks using the Ordovician Kope formation that occurs along the Mill Creek valley, which I-75 parallels through Cincinnati. Brick and tile plants at Mt. Healthy, Saylor Park, and Winton Place used till or Ohio River clays. Cincinnati is also known for the Rookwood pottery, which produced high-quality artistic pieces, and Cincinnati once was the site of many other potteries that used both glacial and river clays.

INTERSTATE 80/90 (OHIO TURNPIKE) AND U.S. 20
Toledo—Indiana Line
61 miles

I-80/90, also called the Ohio Turnpike, and U.S. 20 cross 39 miles of flat Huron-Erie lake plain from the Maumee River to the Fulton and Williams County line and the Fayette area. The remaining stretches to the Indiana line cross the Fort Wayne and Wabash end moraines and intervening ground moraine. Devonian-age bedrock is close to the surface on the west side of Toledo. There are numerous quarries in this area east and west of Centennial Road and north of U.S. 20. Many of them expose the fossil-rich Silica formation.

Cement, Fossils, and Glass Sand—Legacy of the Silica-Area Quarries
Swamplands stretched northwest of the Maumee River in the early years of statehood. Sylvania grew out of the wilderness on bedrock highs and sandy beach ridges above the swamp. By 1863 a company was quarrying a thin patch of underlying sandstone, the Sylvania sandstone of Devonian age, and shipping it to Pittsburgh as glass sand. The quarry operated for a short time, shutting down before 1873. It was reopened by new owners and was connected to the Toledo by the Toledo, Angola & Western Railroad in 1903. The quarry, now filled with water, lies near the intersection of the Sylvania-Metamora and Centennial Roads.

While exploring a mile-wide, low, north-south ridge west of Sylvania, quarriers uncovered Devonian limestone and sandstone in north-south belts. Tenmile Creek dolomite, named after exposures in a nearby creek, is the youngest rock. It caps the quarries west of Centennial Road, lying underneath cobble-rich Wisconsinan till. Nodules of white chert are common in this dolomite, and occasionally a crystal-lined cavity appears. The Silica formation, interbeds

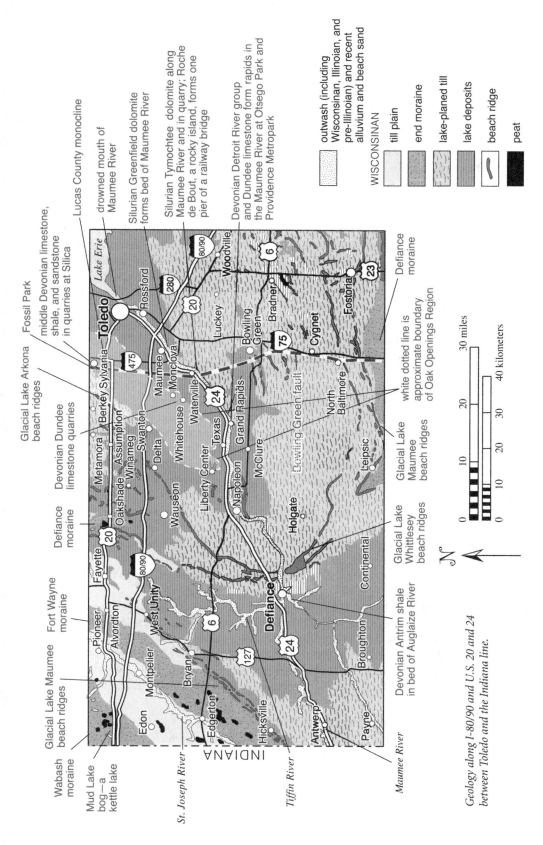

Geology along I-80/90 and U.S. 20 and 24 between Toledo and the Indiana line.

Map labels:

Fossil Park

Glacial Lake Arkona beach ridges

Devonian Dundee limestone quarries

Defiance moraine

Fort Wayne moraine

Glacial Lake Maumee beach ridges

Wabash moraine

Mud Lake bog—a kettle lake

middle Devonian limestone, shale, and sandstone in quarries at Silica

Lucas County monocline

drowned mouth of Maumee River

Silurian Greenfield dolomite forms bed of Maumee River

Silurian Tymochtee dolomite along Maumee River and in quarry; Roche de Bout, a rocky island, forms one pier of a railway bridge

Devonian Detroit River group and Dundee limestone form rapids in the Maumee River at Otsego Park and Providence Metropark

Defiance moraine

white dotted line is approximate boundary of Oak Openings Region

Glacial Lake Maumee beach ridges

Glacial Lake Whittlesey beach ridges

Devonian Antrim shale in bed of Auglaize River

Lake Erie

Toledo

Rossford

Woodville

Luckey

Bowling Green

Bradner

Cygnet

Fostoria

Maumee

Monclova

Berkey

Sylvania

Metamora

Assumption

Winameg

Swanton

Delta

Whitehouse

Waterville

Liberty Center

Texas

Grand Rapids

Napoleon

McClure

North Baltimore

Leipsic

Holgate

Continental

Oakshade

Wauseon

Fayette

Pioneer

Alvordton

West Unity

Montpelier

Bryan

Edon

Egerton

Hicksville

Antwerp

Payne

Broughton

Defiance

Bowling Green fault

INDIANA

St. Joseph River

Tiffin River

Maumee River

Legend:

outwash (including Wisconsinan, Illinoian, and pre-Illinoian) and recent alluvium and beach sand

WISCONSINAN

till plain

end moraine

lake-planed till

lake deposits

beach ridge

peat

0   10   20   30 miles

0   10   20   30   40 kilometers

N

of fossiliferous gray shale and limestone—undoubtedly the most famous rock in northwestern Ohio because of the fossils it contains—lies below the Tenmile Creek dolomite. This formation forms the upper portion of the quarry walls west of Centennial Road. Underneath lies the tan, fossil-rich Dundee limestone and Lucas and Amherstburg dolomites. The older two rocks form the walls of quarries east of Centennial Road, the upper layers having been eroded away before glaciation.

The Lucas County monocline, a steplike bend in the bedrock, brought these formations near the surface along Centennial Road. The monocline is thought to be related to movement along the Bowling Green fault. The monocline dips westward 6 to 7 degrees, extending from near the Maumee River north into Michigan. This monocline, of course, is superimposed on the west flank of the Findlay Arch, which has a westward dip that is more gradual. East and west of the monocline, the dip flattens and the bedrock doesn't change for tens of miles. If not for the monocline, the older Devonian rocks would lie untouched, deep under glacial debris.

People first worked the limestones west of Sylvania in a number of small farm quarries. Dundee limestone was sought after as a flagstone because of its thin bedding; people also burned it to make lime. One Sylvania-area quarry also pulled Detroit River group dolomites overlying the Sylvania sandstone for building stone. By the 1920s, plants making cement and fertilizer operated at Silica. This is when the quarries east and west of Centennial Road opened, south of Sylvania Avenue, before spreading north. The huge Medusa Portland Cement plant, which was located southwest of the Centennial Road and Sylvania Avenue intersection, fell to the wrecking ball in the late 1990s, and the site became another quarry. Companies continue to operate along Centennial Road. The operations are highly mechanized and are designed to efficiently produce aggregate. Two quarries remain open south of the Centennial and Sylvania-Metamora Roads intersection, but not for stone production. One is a swimming quarry, and the other became Fossil Park in 2001, offering fossil collecting in a park atmosphere. This park offers a great selection of Silica formation fossils—trilobites, brachiopods, corals, clams, and snails. The walls of the Fossil Park quarry are Dundee limestone, but park managers have Silica formation collecting piles shipped in from a nearby quarry.

Sandy patches and ridges of vegetation mark the Ohio Turnpike and U.S. 20 from the western reaches of Toledo to Swanton, and in the Silica area. This area is known as the Oak Openings Region—a unique ecosystem that is rich in plants and wildlife. The oak-savanna habitat openings of this region formed on sands of Glacial Lake Warren beach ridges, spits, and deltas, which were redistributed by water and wind. This area has many hiking trails and fire lanes to explore.

Beach Ridges and End Moraines

U.S. 20 crosses a low ridge at the Ohio 295 intersection, which marks the shoreline of Glacial Lake Arkona; the highway crosses another beach ridge just east of Assumption, which marks the shoreline of the earlier Glacial Lake Whittlesey. Beach ridges associated with Glacial Lake Maumee II underlie the routes between Ohio 109 and 108. Just west of Oakshade, U.S. 20 rises on a small

The rise in U.S. 20 marks the shoreline of Glacial Lake Maumee. A depression in the lake plain to the left of the highway once accumulated aquatic plant debris that became peat. Note the piles of excavated peat.

remnant of the Defiance end moraine. This moraine is patchy because portions of it were eroded by Glacial Lake Maumee. Boulder piles outcrop where till plains and end moraines form the landscape. The boulders are mainly igneous and metamorphic glacial erratics from Canada that the ice sheets left behind.

Beach remnants of the highest glacial lake—Glacial Lake Maumee—underlie U.S. 20 at Fayette and the turnpike north of West Unity, but they are not readily recognizable. The turnpike crosses the Fort Wayne moraine at the Ohio 15 interchange and U.S. 20 crosses it just west of Alvordton. Four miles east of the Indiana line both roads cross the Wabash moraine. This area has the hilliest terrain of the routes, and it leads to the Steuben morainal-lake area of northeastern Indiana, a Wisconsinan till and outwash area dotted with many kettle lakes and ice block depressions.

Sand and gravel pits dug in Wisconsinan material dot the St. Joseph River valley around Pioneer and Montpelier. The lake north of U.S. 20 just east of the Ohio 15 interchange is a former pit.

Mastodons, Marl, and Peat

Small wetlands that were once small lakes or ponds are scattered across western Lucas, Fulton, and Williams Counties. Many of these lakes are kettle lakes, which form when ice blocks are left, and consequently melt, after a glacier has retreated. Other lakes are just depressions in till or lake sediment. Sometimes marl, peat, and the fossils of animals and plants living in and around these lakes accumulated. The largest fossils that occur in these sediments are the bones of Ice Age mammals, particularly mastodons (*Mammut americanum*), hairy elephants that frequented the Midwest at that time.

A mastodon vertebra embedded in marl of a former postglacial pond near Winameg.

The earliest reports of Ice Age mammal fossils being found in Lucas County date to the early 1870s: a mastodon skeleton found in an unknown cranberry swamp and a tooth somewhere in Springfield Township. The exact location, completeness of the skeleton, and disposition of the remains are unknown. A mastodon tooth and the remains of a giant beaver (*Castoroides ohioensis*) were found west of Edon in 1959 and 1961, respectively. The first reported mastodon remains from Fulton County were found in 1978. Workers discovered leg bones, ribs, teeth, and vertebrae while excavating a farm pond at Winameg. These mastodon parts are kept at the University of Michigan.

The best remaining kettle lake in northwestern Ohio—most have been destroyed or filled in with sediments and vegetation—is Mud Lake at the Mud Lake Bog State Nature Preserve north of Edon. The most interesting feature of this site is the occurrence of both alkaline and acidic sediments and plants, which are specific to one or the other environment, along the lake margin; most kettles either become bogs with acidic soils or fens with alkaline soils. Some of the unique plant species growing here are lesser bladderwort (*Utricularia minor*), large cranberry (*Vaccinium macrocarpum*), small purple-fringed orchid (*Platanthera psycodes*), swamp birch (*Betula pumila*), and tamarack (*Larix laracina*)—holdovers from an Ice Age climate that now flourish mainly in more northerly latitudes with cooler climates.

U.S. 23
Toledo—Chillicothe
182 miles

U.S. 23 traverses 60 miles of lake sediments and waterworn till of the Huron-Erie lake plain from Toledo to just south of Fostoria. It then passes over a series of end moraines, till plains, and meltwater deposits of the Wisconsinan Scioto lobe to the northern edge of Chillicothe, where Illinoian deposits begin.

Toledo to Upper Sandusky

South of the Monroe Street interchange in Toledo, U.S. 23 crosses the Ottawa River valley, which follows a twisting course to Lake Erie. Between the Michigan line and the U.S. 24 interchange, scattered sand bodies mark remnants of the Oak Openings Region, an oak-savanna habitat that formed on old beach ridges, deltas, and spits of proglacial lakes. Unfortunately, development is fast consuming this unique terrain. The famous Devonian-age fossil locality at Silica is west of U.S. 23 off Central Avenue. The quarries here are known around the world for their fabulous fossils of trilobites, crinoids, and brachiopods. The bike trail just south of the I-475 junction marks the former Toledo, Angola & Western Railroad, which once connected the quarries at Silica with Toledo. Just south of the Ohio Turnpike (I-80/90) junction, where Swan Creek winds its way into downtown Toledo, both directions of Salisbury Road lead to quarries in Silurian bedrock. A quarry at Maumee was well-known for cavities in former reef rock that were lined with calcite and filled with natural asphalt, a thick, sticky, black hydrocarbon derived from ancient animal and plant matter. Look for rocky flats of Silurian Greenfield dolomite, visible at low water, as U.S. 23 bridges the Maumee River. You might see rounded stone knobs that mark fossilized mounds of ancient marine algae.

U.S. 23 follows U.S. 20 from Perrysburg to 3 miles east of Lemoyne, where U.S. 23 turns south. There are numerous quarries along this stretch since Silurian bedrock is near the surface. Several low hills of sand rise just south of the Portage River valley, highlighting former shorelines of Glacial Lake Warren. At the U.S. 6 junction, U.S. 23 crosses a more continuous section of Warren beach ridge. The beach ridge continues southwest and is topped with Ohio 281, which passes through Bradner and Wayne. Many oil wells—some still producing small amounts of oil—that tapped Ordovician-age limestone dot this part of the route. Fostoria sits on sandy, well-drained terrain of the Glacial Lake Maumee beach. Early on the city was a glass-manufacturing center as a result of the cheap fuel that was available during the northwestern Ohio gas boom.

South of town at the U.S. 224 junction, U.S. 23 leaves the Huron-Erie lake plain and crosses 4 miles of the Defiance end moraine. During Wisconsinan time, drainage was poor on the south side of the moraine, so a large glacial lake—separate from Lake Erie's glacial lake predecessors—spread east and west of where the highway is now to 2 miles south of Upper Sandusky. As this lake drained, it became part of the Sandusky River valley, which flows south. A smaller lake developed west of this lake near Findlay. Between the lakes there is an old channel, now filled with peat, probably abandoned when the

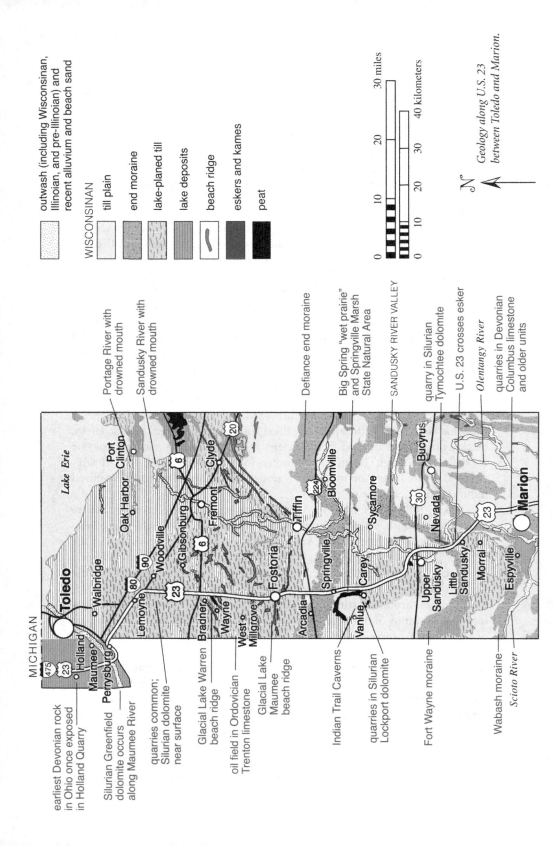

Geology along U.S. 23 between Toledo and Marion.

WISCONSINAN

outwash (including Wisconsinan, Illinoian, and pre-Illinoian) and recent alluvium and beach sand

till plain

end moraine

lake-planed till

lake deposits

beach ridge

eskers and kames

peat

0   10   20   30 miles

0   10   20   30   40 kilometers

N

MICHIGAN

Lake Erie

Toledo

Holland

Maumee

Perrysburg

Walbridge

Port Clinton

Oak Harbor

Woodville

Lemoyne

Wayne

Bradner

West Millgrove

Gibsonburg

Fremont

Clyde

Fostoria

Arcadia

Springville

Vanlue

Carey

Tiffin

Bloomville

Sycamore

Upper Sandusky

Little Sandusky

Nevada

Bucyrus

Morral

Espyville

Marion

earliest Devonian rock in Ohio once exposed in Holland Quarry

Silurian Greenfield dolomite occurs along Maumee River

quarries common; Silurian dolomite near surface

Glacial Lake Warren beach ridge

oil field in Ordovician Trenton limestone

Glacial Lake Maumee beach ridge

Indian Trail Caverns

quarries in Silurian Lockport dolomite

Fort Wayne moraine

Wabash moraine

Scioto River

Portage River with drowned mouth

Sandusky River with drowned mouth

Defiance end moraine

Big Spring "wet prairie" and Springville Marsh State Natural Area

SANDUSKY RIVER VALLEY

quarry in Silurian Tymochtee dolomite

U.S. 23 crosses esker

Olentangy River

quarries in Devonian Columbus limestone and older units

present-day Sandusky River drainage developed in the basin of the larger lake near Upper Sandusky. The Sandusky River developed as the lake drained. Lake sediments cover Silurian dolomite from Carey to south of Upper Sandusky.

Earliest Devonian Rock in Ohio—Holland Quarry

A water-filled quarry along Salisbury Road between Monclova and Holland, now in a housing development, is an important geologic site because in the early days of the operation it was the only spot in the state where rocks of early Devonian age could be seen. The France Stone Company operated the quarry into the 1950s, producing crushed stone. Rocks in the area slope, or dip, to the west at 6 to 8 degrees, a little more steeply than the region's larger-scale dip because they are part of the Lucas County monocline, a local bend in the strata. The oldest rocks in the quarry are Silurian-age Greenfield dolomite; the youngest are Devonian Detroit River group. The most important layer is the one just above the wavy surface that marks an unconformity between the Devonian Sylvania sandstone and Silurian undifferentiated Salina group dolomites. About 10 feet of brownish shale lies in a 30-foot-deep rock channel that was carved into the Silurian dolomite by a river that appeared when the sea had withdrawn from the area. It is exposed in the west wall of the quarry, which is now filled with water. The shale formed in an early Devonian-age sea and is the oldest Devonian rock known in Ohio. People have collected many fossils in the shale, including plates of eurypterids (sea scorpions), ostracoderms (jawless fish), acanthodians and placoderms (early jawed fish), and fragments of land plants. Ostracoderm fossils are not plentiful in Ohio, but two new species were discovered in the Holland Quarry shale in 1960. After it was abandoned, the quarry area became a good place to collect golden brown calcite from the undifferentiated Salina dolomites.

Eastern Wood County Oil Fields

Little did settlers realize that untapped riches of oil and natural gas in the Ordovician Trenton limestone rested under the inhospitable Great Black Swamp. The stretch of U.S. 23 between Woodville and Fostoria passes many towns that once prospered from "oil fever." It began in Bradner in 1889 when a nitro blast brought a gusher from below. Since the drillers had made no preparations for tanking the oil, it coated trees and the ground in the immediate vicinity. Twelve more wells went in by spring 1890, and all produced 70 barrels an hour. A well that brought in 210 barrels in two hours started a stampede. By late November four hundred wells tapped the reservoir, and Bradner built a refinery that year. In 1894 Bradner residents boasted that they had the largest oil well in the world—the Kirkbride Bros. No. 3, northeast of town, which initially yielded 100 barrels an hour. People built derricks everywhere on the north and west sides of town. The field was doomed to early failure, though, for as elsewhere folks gave little consideration to the spacing of wells, which caused the field to lose pressure early on.

Carey and Upper Sandusky—Clay and Quarries

A brick kiln, built in 1859, was one of the earliest mineral industries to develop in Carey. A plant began producing tile in 1877, and by 1895 it employed thirty men. Two resistant 5-mile-long bedrock ridges radiate from Carey north toward Vanlue and Springville. The ridges formed from reef-bearing parts of the upper Lockport dolomite, which tends to resist erosion. What early settlers called Big Spring "Wet Prairie," a poorly drained area of black muck that was a former stream channel, rested between the ridges; however, by 1870 farmers began draining the area, and that combined with peat extraction virtually destroyed this significant wetland. Springville Marsh, a state nature preserve, encompasses part of the surviving wetland. It's 1 mile west of U.S. 23 off Township Road 24, also called Muck Road. A boardwalk trail leads into the marsh and surrounding habitats.

Workers removed dolomite from the ridges in six shallow quarries in 1872. Builders used most of the stone in foundations and walls. Northeast of Carey, a small kiln produced lime from these ridges as well. These meager beginnings led to the incorporation of National Lime and Stone Company in 1903, which is still a major supplier of crushed stone products in Ohio. In a few years the company had two lime plants and one stone crusher in Carey, and by 1926 it had acquired quarries at Arlington, Bucyrus, Findlay, Lewisburg, Lima, Marion, Rimer, and Spore. The company became a major supplier of all kinds of stone products, from building stone and crushed stone to a special dolomite lime for the glass and steel industry. As of 2004, National Lime and Stone still had major quarrying operations in Carey, harvesting the Silurian-age Greenfield and Lockport dolomites.

The first quarries at Upper Sandusky operated along the Sandusky River. Where Ohio 199 crosses U.S. 23 south of town, a large quarry still works the Tymochtee and Greenfield dolomites. Upper Sandusky also had a tile plant in its early years, which mined local glacial lake clays. The clay pits and plant, closed in the late 1990s, occur along the tracks in the southwestern corner of town. Another plant that was located across the tracks is no longer there.

The Lockport dolomite underlying Carey contains evidence of large barrier reefs that flourished in Silurian time. More-massive portions of the quarry face are reef rock; bedded sections represent surrounding sediments.

A Treasure-Bearing Sinkhole

Ohio 568 stretches across the crest of the Silurian-age dolomite ridge between Carey and Vanlue. The earliest settlers knew there were caves under the ridge, but they didn't explore them to any great extent. These caves developed along cracks in the Greenfield dolomite that gradually widened as slightly acidic groundwater dissolved their walls. Funnel-shaped entrances, or sinkholes, formed at the surface first, leading to horizontal passages at depth. Flowstone coated the walls at places, and dripping water formed small stalactites. From 1927 into the late 1930s, Wyandot Indian Caverns offered tours through one of the caves just east of Vanlue. It was reopened in 1973 as Indian Trail Caverns and now offers an interesting excursion 42 feet below the countryside.

In 1990, a crane operator excavating an adjacent sinkhole discovered human and animal bones and charcoal. This sinkhole, like most, had filled with debris over thousands of years. While carefully removing sediment from the sinkhole he exposed another cave, which has since been named Sheriden Cave. Over a ten-year period geologists, archaeologists, other scientists, and numerous skilled volunteers and students have removed the bones of many animals and evidence of early Ohioans, dating from 13,000 to 10,000 years ago. The most exciting were the remains of the extinct Ice Age mammals—giant beaver (*Castoroides ohioensis*), short-faced bear (*Arctodus simus*), long-nosed peccary (*Mylohyus nasutus*), and stag moose (*Cervalces scotti*)—and a stone scraper and stone and bone projectile points. Nomadic Paleo-Indian hunters had used Sheriden Cave while the Great Lakes were still forming and ice sheets still covered Canada. Sheriden Cave is one of less than fifty places in North America where the remains of people and extinct species occur in a thoroughly studied and well-dated sequence of sediment.

Upper Sandusky to Marion

East of Upper Sandusky a long ridge of sand and gravel, an esker, stretches north to south. It is marked by a string of borrow pits. Between Upper Sandusky and Little Sandusky, U.S. 23 crosses the Fort Wayne moraine; however, another

A gravel pit in a Wisconsinan esker south of Upper Sandusky. Cobbles, mainly of Silurian-age dolomite, have been separated from sand and gravel—the main products of this operation.

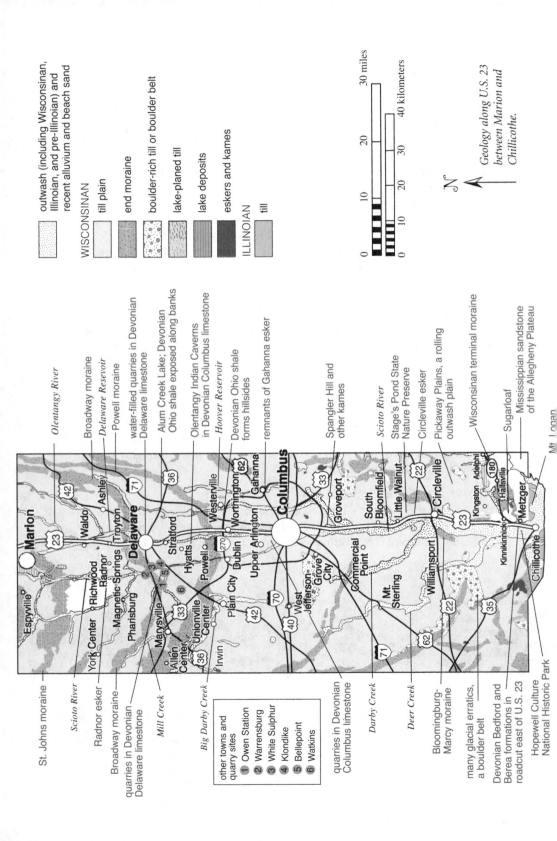

Geology along U.S. 23 between Marion and Chillicothe.

WISCONSINAN

outwash (including Wisconsinan, Illinoian, and pre-Illinoian) and recent alluvium and beach sand

till plain

end moraine

boulder-rich till or boulder belt

lake-planed till

lake deposits

eskers and kames

ILLINOIAN

till

0   10   20   30 miles

0   10   20   30   40 kilometers

N

St. Johns moraine

Scioto River

Radnor esker

Broadway moraine

quarries in Devonian Delaware limestone

Mill Creek

Big Darby Creek

other towns and quarry sites
1 Owen Station
2 Warrensburg
3 White Sulphur
4 Klondike
5 Bellepoint
6 Watkins

quarries in Devonian Columbus limestone

Darby Creek

Deer Creek

Bloomingburg-Marcy moraine

many glacial erratics, a boulder belt

Devonian Bedford and Berea formations in roadcut east of U.S. 23

Hopewell Culture National Historic Park

Olentangy River

Broadway moraine

Delaware Resevoir

Powell moraine

water-filled quarries in Devonian Delaware limestone

Alum Creek Lake; Devonian Ohio shale exposed along banks

Olentangy Indian Caverns in Devonian Columbus limestone

Hoover Reservoir

Devonian Ohio shale forms hillsides

remnants of Gahanna esker

Spangler Hill and other kames

Scioto River

Stage's Pond State Nature Preserve

Circleville esker

Pickaway Plains, a rolling outwash plain

Wisconsinan terminal moraine

Sugarloaf

Mississippian sandstone of the Allegheny Plateau

Mt. Logan

Marion

Espyville

Richwood
Radnor

York Center

Magnetic Springs

Pharisburg

Allen Center

Marysville

Unionville Center

Irwin

Plain City

42

West Jefferson

40

70

Mt. Sterling

71

62

35

22

Williamsport

South Bloomfield

Little Walnut

Groveport

33

Commercial Point

Grove City

Upper Arlington

Dublin

270

Powell

Hyatts

Stratford

Delaware

71

36

Troyton

Ashley

Waldo

23

42

33

36

Worthington

62

Gahanna

Westerville

Columbus

Circleville

23

Kingston

Adelphi

180

Hallsville

Kinnikinnick

Metzger

Chillicothe

Williamsport

section of the esker is more evident than the moraine. The esker cuts across the moraine where the highway passes over it 1 mile north of the turnoff for Little Sandusky. Another glacial lake developed south of this end moraine, and workers quarried its lake clays at Morral. At the Wyandot and Marion County line, U.S. 23 cuts through the Wabash moraine. Low outwash terraces flank the Little Scioto River, which parallels the southern edge of the end moraine. Downstream the river cuts through another area of lake sediments, which were deposited in a small glacial lake. U.S. 23 passes east of Marion, traveling past quarries in Devonian rock. The landscape rises noticeably to the northwest, marking the appearance of more-resistant Devonian strata overlapping the Silurian strata that composes the bedrock of the northern part of this route. Geologists refer to this as the Columbus escarpment, and it can be traced as far as the Lake Erie shore.

Marion to Delaware

Marion, like Carey, is an important quarrying center for Devonian-age limestones. As early as 1850 people were operating lime kilns just south of town. As quarries developed here, the town of Owen Station developed along the Columbus & Hocking Valley Railroad. Besides lime, quarriers extracted Delaware bluestone, a highly desired building stone cut from the Delaware limestone. You can still see it today in many Marion-area buildings, especially churches. Builders used the more massive limestone blocks as bridge piers. Owen Station is now an area filled with newer homes; the water-filled quarries serve as scenic lakes. Small quarries lie along Ohio 4, north of downtown Marion. This limestone ridge is mainly composed of Columbus limestone. Other quarries occurred at Espyville and north of Waldo.

At Marion, U.S. 23 passes through a wide gap in the St. Johns moraine, which underlies part of western Marion, and skirts the Olentangy River across 18 miles of gently rolling till plain. A dam and earthen levee, completed in 1951, backs up waters of the Olentangy River, forming the Delaware Reservoir between Waldo and Delaware. The land around this reservoir is part of the Delaware Reservoir Wildlife Area. From Troyton, a side trip west on Radnor Road will lead you to a long ridge of outwash, which is an 8-mile-long esker that stretches north and south of Radnor.

West of U.S. 23 along this stretch there were numerous brick and tile plants that exploited glacial tills at Allen Center, Irwin, Marysville, New California, Pharisburg, Plain City, Richwood, and Unionville Center in the late 1800s. Around 1874 Richwood also had a pottery. Workers quarried Silurian-age dolomite near Radnor and York Center and Devonian-age Columbus limestone in New Dover, in Unionville Center, and near Watkins.

The hilly area around the dam on Delaware Reservoir marks another end moraine, the Broadway moraine. At many places downstream, the river flows across Devonian bedrock, which is visible at low water stage. A horseshoe-shaped meander of the Olentangy River appears north of Delaware.

Delaware Area

By the 1850s people had begun quarrying in Delaware in a west-side valley. Initially, they produced building stone and lime from the Devonian-age Delaware limestone, which was named for rock outcrops along Delaware Run east of the tracks in western Marion. Once the Columbus & Hocking Valley Railroad was completed, Delaware-area quarries shipped stone from quarries on both sides of the tracks, just north of U.S. 36, to Columbus, Springfield, and Toledo. By 1915, these quarries were exploiting the underlying Columbus limestone and mainly produced crushed stone. The water-filled Campbell Quarry is now part of Blue Limestone Park in Delaware. At least five other quarries operated in this area in the 1870s, and a couple of them occur along the Olentangy River and are visible east of U.S. 23 as you pass through Delaware.

The Olentangy shale, which overlies the Delaware limestone, is named for a cliff exposure along the east bank of the Olentangy River on the southeastern edge of town. It is capped by Ohio shale, which lies under eastern Delaware. People knew that the Columbus limestone was fossil rich and often contained bone beds—layers with dense concentrations of fossils. Most people, however, didn't think that fossils occurred in the thick shales, not until a Delaware minister began poking around in concretions along the river. He found that they often contained petrified wood and bone fragments. Later he made a great discovery—the nearly complete crushed skull of a large Devonian-age fish now called *Dunkleosteus*. This was the first find of this now well-known predatory fish. The reassembled skull ended up at Columbia University.

Quarrying was also important in the early days of Stratford. Old stone buildings with locally quarried limestone abound. Delaware limestone forms

A tributary of the Olentangy River flows across rocky ledges of Devonian-age Delaware limestone near Stratford.

rapids in the Olentangy River in this town, on which water-powered mills were established. Local streams also flow over beds of Delaware limestone, sometimes forming scenic waterfalls as they enter the Olentangy floodplain.

In the 1870s, Delaware had four brick and tile plants using the glacial clays of the Delaware area. Other plants occurred at Ashley, Olive Green, Powell, and Stratford. In 1878, the Denison family began making tile on their farm in northwest Delaware; in 1892, they moved the operation to the east side of town and incorporated as the Delaware Clay Manufacturing Company. The company began producing tiles with ceramic fireproofing, and later hollow bricks. This pioneering company last dug clay at the foot of Hayes Street in the late 1960s.

Springs and Caves

As with many areas of the state where limestone forms the bedrock, Delaware County has its share of caves, sinkholes, and springs. The smell of rotten eggs, which was hydrogen sulfide, led many early settlers to mineral springs. At the time, many people thought that this water had medicinal properties and used it to "cure" various ailments. A Delaware sanatorium, the Mansion House Hotel, opened at one of these springs in 1833. It later became the site of Ohio Wesleyan University in 1842. Sulphur Spring, as it is known today, still seeps, especially in the spring when it fills a small man-made pool near Phillips Hall. At least four other mineral springs exist in Delaware.

In 1879, while drilling a water well, workers tapped artesian groundwater along Bokes Creek in adjacent Union County. Locals felt that the mineral-laden water was a cure-all, which led to the establishment of Magnetic Springs. Reportedly, the mineral water had magnetic properties. Hotels and bathhouses popped up and catered to the health conscious. Not to be outdone, Marysville boasted of its own magnetic springs. The resorts in both towns are now gone and the springs mostly buried by development.

A number of caves dot southern Delaware County on either side of the Olentangy River. Joints widened by acidic water extend downward, sometimes leading to horizontal passageways. Best known, of course, is Olentangy Indian Caverns west of U.S. 23 and Lewis Center on Home Road. The Wyandottes, one of the Indian tribes that lived in Ohio, reportedly used the caverns as late as 1810. The accidental death of an ox that fell into the sinkhole entrance brought it to the attention of white settlers in 1821. The caverns, first opened to the public around 1935, feature four cave levels carved into fossil-rich Devonian-age Columbus limestone. An underground stream connects the caverns with the nearby Olentangy River.

The land surrounding the caverns is typical karst: it has many sinkholes and generally lacks surface streams since the water travels underground through caverns. The Powell moraine lies just to the south, and there are other similar caves nearby, often with sinkholes leading to openings along the Olentangy River's bluffs. Some locals claimed that caves linked the Olentangy River valley with the Scioto River valley to the west, but these claims have never been substantiated. There is still much that geologists don't know about this karst landscape. As late as 1988, leakage in a local reservoir was traced to a large sinkhole underneath it, which nobody knew existed.

Delaware to Columbus

Torrents of summertime meltwater during late Wisconsinan time formed the wide floodplain of the Olentangy River that U.S. 23 passes through from Delaware south to its juncture with the Scioto River in downtown Columbus. The elevated flat shelves that appear on both sides of the river are outwash terraces. Between the east-west roads to Hyatts and Powell, a number of limestone quarries lie hidden along the river valley. The scene is similar 4 miles west in the Scioto River valley.

Highbanks Metro Park, south of Ohio 750 (Powell Road), is aptly named. The crumbly, blackish brown Devonian-age Ohio shale and the more grayish Olentangy shale below it are exposed in 100-foot-high banks along the river. Watch for these rocks at construction sites, because they occur just below the surface in the Columbus metro area. Roadcuts expose these shales at the I-270 junction; however, because they are soft, plants bury them quickly. You should catch glimpses of them if you hike along the tributaries of the Olentangy River.

Columbus—Namesake of the Columbus Limestone

Pioneers seeking both lime and building stone established the first extensive quarries in the capital city along the Scioto River. The Devonian-age Columbus limestone forms bluffs along the river from Warrensburg to just north of downtown Columbus. The pioneers' earliest excavations were just upstream of the Scioto's juncture with the Olentangy River on the west side of downtown, just north of the state hospital. The Sullivant Quarry, which was in Columbus, provided the stone for the State House that workers built between 1839 and 1861. Construction proceeded slowly until steam-powered equipment became available and a railroad spur from the quarry to the building site was built. This classic building stands at High and Broad Streets. The quarry, just west of I-70, has been reclaimed, or filled in.

The rolling landscape overlying Olentangy Indian Caverns is caused by the slow sinking of surface material into dissolved areas of bedrock, or sinkholes. Tours start from the brick building.

Devonian-age Ohio Shale is the bedrock in northern and eastern parts of Columbus. Maryvale Ravine just north of the I-270 interchange and Worthington shows the typical stratigraphy.

Marble Cliff, just upstream of the Sullivant Quarry, was also a quarrying center. (Limestone that undergoes heat and pressure changes, or metamorphism, is called *marble*. In the 1800s, it was standard practice to call limestone that took a nice polish "marble." Even today, people informally refer to some limestone as marble.) Between 1899 and 1901, builders constructed the Judiciary Annex of the State House using Columbus limestone from the Taylor & Bell Quarry across the Scioto River.

The Columbus limestone occurs in massive layers up to 65 feet thick at Columbus, making it a favored cutting and building stone. Quarriers cut blocks that were several hundred square feet in size from certain horizons. Gradually, quarrying spread upstream into Delaware County, eventually to Bellepoint, Klondike, White Sulphur, and Warrensburg. These early quarries also made lime. By 1915, Columbus limestone fell into disfavor as a building stone because it discolored with age and didn't carve well. Around 1910, another market for this rock developed. The iron industry used it as flux to make pig iron. Quarries also sold it as crushed stone. The many small quarries of yesteryear are mostly gone now, having been consumed by the mammoth mechanized quarries of modern aggregate companies or reclaimed by Mother Nature.

East of the Scioto River, the Olentangy and Ohio shales underlie the Columbus limestone. These soft, crumbly rocks appear in new excavations throughout the metropolitan area. They are well exposed in the valleys of streams flowing into the Olentangy River and in the valleys of the tributaries on the east bank of the Scioto River. The Ohio shale was also known as "Ohio slate" in the 1800s because of its dark color and flaking layers. Slate Run in Upper Arlington and Slate Run Metro Park near Lithopolis are reminders of this nickname. Two Columbus companies fashioned sewer pipe from the shale in the 1870s. Ohio shale continued to serve the Columbus tile industry until the 1930s.

The Fish Pressed Brick Company in Columbus made bricks using shales of the Mississippian-age Cuyahoga formation as the raw material before 1910. The brickworks at Taylor Station (now part of Columbus) used Devonian Bedford shale. Brick plants at Groveport and Westerville used alluvial clays from the Scioto and Olentangy Rivers, respectively. The Everal plant near Westerville was the first to make vitreous tile—a tile that resembles glass—in the state. Since 1910, sand and gravel dredged from the Scioto River has been another important resource of the Columbus area.

Urban sprawl has eliminated much of Columbus's surficial geology. Entire glacial moraines and outwash landforms have met earthmovers or lie buried under concrete and asphalt. Thirty-foot Hayden Falls survives as an example of the many waterfalls that once emptied into the Scioto and Olentangy Rivers. It's located in a park between Dublin and Upper Arlington along the Scioto River, where Hayden Run empties into the Julian Griggs Reservoir. Many similar falls are submerged under the water of various reservoirs. Rocky Fork flows through a country club in Gahanna east of Columbus, along which there is an impressive bank of glacial sediments and underlying Devonian-age rocks

Orton Hall on the Ohio State University campus in Columbus displays the important building stones of Ohio in stratigraphic order.

exposed. Remnants of the Gahanna esker are nearby. Battelle Darby Creek Metro Park, southwest of Columbus, has a 70-foot bluff of glacial strata and six burial mounds of the Adena people.

Orton Hall, on the Ohio State University campus, houses the Orton Geological Museum and Memorial Library of Geology and is itself a geological monument to Ohio building stones. From its foundation of the Silurian-age Brassfield formation to the walls of Devonian-age Berea sandstone, five Ohio building stones are arranged in stratigraphic order. Inside the building more than thirty Ohio stones form trim, walls, and twenty-four columns. Perhaps the most significant Ohio fossil on display in the Orton Geological Museum is a mounted skeleton of a ground sloth (*Megalonyx jeffersoni*). The bones of this large Ice Age beast came from the bog deposit of a marl lake near Berlin in Holmes County. It was unearthed in 1890.

Ohio's Geological Survey—the Early Years

After Caleb Atwater, a minister and a lawyer, moved his family to Circleville in 1815, he became interested in the earthworks and artifacts of the Scioto River valley. He was among the first residents of Ohio to note its geological features. Beginning in the 1820s, Samuel P. Hildreth, a Marietta physician and astute observer, began writing a series of scientific papers on Ohio's natural features and potential mineral resources. Inspired by the findings of these two men, scientists across the state lobbied for a formal government survey dedicated to studying the state's natural features. In 1835 the governor appointed Hildreth and others to a committee to design such a survey.

The first Geological Survey of Ohio began in 1837 with William W. Mather at the helm as Ohio's first state geologist. Although the survey focused on establishing the value of Ohio's mineral resources, other accomplishments included the description of new fossils, correlation of strata, recording of archaeologic sites, compilation of lists of living animals and plants, and relating soil type to agriculture. Unfortunately the Panic of 1837, an economic depression, caused the state to not fund the survey in 1838. The survey completed fieldwork in 1838 using leftover funds from the first year. Attempts were made to reestablish the survey work in the 1840s and 1850s, but the survey's hiatus lasted until 1869.

The second Geological Survey began under the direction of John Strong Newberry. By this time, great advances had been made in the geological sciences, but the survey continued to gather information about plants, animals, and agriculture as well. Due to internal bickering and financial problems, the survey ended in 1874.

In 1889 the third Geological Survey was organized with Edward Orton at the helm. The survey fell on meager times once again, and Orton mostly served as an honorary state geologist, since no funds were available, until his death in 1899.

Edward Orton Jr. followed as the next state geologist as the fourth manifestation of the survey began in 1900. The early surveys lacked a formal permanent office, but in 1900 a permanent headquarters was established at the Ohio State University. Great strides had again been made in geology since the previous survey had worked, and a new series of bulletins served to update previous volumes.

Wisconsinan kames rise above U.S. 23 just south of the I-270 interchange on the south edge of Columbus.

Ohio's Geological Survey, now formally known as the Ohio Department of Natural Resources (ODNR) Division of Geological Survey has come a long way. It now has modern research and office facilities in the ODNR complex on Morse Road in Columbus, a large storage facility in Alum Creek State Park north of Columbus, an ever-increasing inventory of scientific and educational publications and maps, and a knowledgeable staff that continues to gather and disseminate information about Ohio's geology.

Columbus to Chillicothe

South of Columbus the combined courses of the Olentangy and Scioto Rivers form the major meltwater channel of the Scioto Valley, which is characterized by a wide floodplain, steplike terraces, kames, kettles, and a deep fill of outwash sand and gravel. At 230.8 miles, the Scioto River is the longest river in Ohio—its headwaters are in Auglaize County. One mile south of the I-270 interchange, U.S. 23 encounters Spangler Hill, which marks the northern edge of a 4-mile stretch of kames. As the Scioto lobe melted, accumulations of sediment in ice depressions and crevasses were lowered to the ground, creating the kames. There is a smaller, yet prominent kame in St. Joseph Cemetery just southeast of Shadeville and the junction with Ohio 317. There is a major sand and gravel mining operation across the road from the cemetery.

The Circleville esker is another prominent glacial landform. It runs north to south, starting 2 miles north of South Bloomfield and ending in Circleville. Disconnected sandy ridges between the highway and the Scioto River mark its location, and a number of historic brick farmhouses dot its crest. A tree-covered rise north of Little Walnut marks the intersection of the esker and U.S. 23. The highway then crosses the Bloomingburg-Marcy moraine at Circleville. South of this moraine there is a wide spot in the outwash plain called Pickaway Plains. Once an undeveloped prairie, Pickaway Plains features more prominent kames. Remnants of another esker lie between the highway and Kingston.

Sand and gravel spread by raging meltwater accumulated in a tunnel or channel of the Wisconsinan Scioto ice lobe. After the ice melted in the area a long, sinuous esker remained, marking the course. Recent drainage has cut through this ridge in places, such as here, south of South Bloomfield.

Off U.S. 22 on the southern edge of Williamsport, Deer Creek cuts through the Devonian Ohio shale. At low water, boulder-sized concretions composed of siderite, an iron carbonate, are visible in the streambed. If you have the time, they're worth the side trip.

Illinoian-age outwash forms terraces along the bedrock escarpment to the east along U.S. 23 south of Circleville. East of U.S. 23 between Circleville and Chillicothe, rounded bedrock mounds loom 200 to 400 feet above the Scioto River valley, marking the Allegheny Escarpment, a rise in land that separates the Allegheny Plateau from the western till plains. Sulphur Spring Road north of Sugarloaf, a high knob east of Delano, suggests that early settlers were aware of the sulfur-enriched groundwater that seeps from the Mississippian-age Sunbury shale and Cuyahoga formation where they come into contact with the base of the mounds. Ohio 180 follows the escarpment's rise onto the Allegheny Plateau from Kinnikinnick northeast to Laurelville and the Hocking Hills region. The area is rich in early Ohio history with many Native American village sites, ceremonial and burial mounds, Logan Elm State Memorial, and Mt. Logan.

Mastodon remains were uncovered in kettles near Hallsville in the 1960s and 1970s. Prominent Wisconsinan kames are also visible between Hallsville and Adelphi. You can view them well from Adelphi Cemetery. A complex of kames occurs east of U.S. 23 between Kinnikinnick and Metzger. Sand and gravel operations continue to strip away this resource. North and south of Chillicothe there are roadcuts exposing Devonian-age Bedford shale and Berea sandstone. U.S. 23 crosses the Wisconsinan terminal moraine at Chillicothe, and the route continues south onto the Allegheny Plateau.

An Ice Block Lake

If you turn east on Hagerty Road at Little Walnut, you will hit Stage's Pond State Nature Preserve. The centerpiece of this preserve is a kettle lake west of the Circleville esker. The lake formed when ice blocks broke off the receding Wisconsinan ice sheet some 15,000 years ago. Outwash eventually, at least partially, covered the ice, insulating it. Once the blocks finally melted, they left a water-filled depression. Trails wind through this reclaimed farmland, which has been a preserve since the mid-1970s.

Chillicothe—Edge of the Allegheny Plateau

Chillicothe's location at the edge of the plateau and along the Scioto River led to its selection as territorial capital in 1800 and later state capital from 1803 to 1810 and 1812 to 1816. The Scioto flows through a 2-mile-wide valley along the edge of the plateau at Chillicothe. Floods plagued the town from its 1796 founding until 1979, when the U.S. Army Corps of Engineers completed 2 miles of levee through town. Sometime before 1820, floodwaters cut through the neck of a tight meander on the north side of downtown. An island formed from the old point bar and the river abandoned the old channel. The old channel, still filled with water, and island are now part of Yoctangee Park. It's west of U.S. 23 as the road crosses the river. Mt. Logan, a 690-foot knob and the inspiration for the Great Seal of Ohio, is visible to the east from this bridge. Mt. Logan has about 120 feet of Devonian Ohio and Bedford shales at its base, 8 feet of Berea sandstone, 25 feet of Mississippian Sunbury shale, 342 feet of Cuyahoga formation, and 180 feet of Logan formation, and is capped by about 15 feet of Pennsylvanian Pottsville group sandstones. Paint Creek, a major tributary from the west, joins the Scioto River on the south edge of town. Just west of Chillicothe, Paint Creek flows through a 350-foot-deep gorge cut into Devonian shales.

Hopewell Culture National Historic Park

One of the best areas to observe mounds, earthworks, and artifacts of the Native American Hopewell culture (2,200 to 500 years ago) is just north of Chillicothe along the Scioto River. Two amateur archaeologists, Ephraim G. Squier and Edwin H. Davis, first explored the mounds in 1846. This area was named Mound City. Since then farming practices flattened some of the earthworks, while the construction of Camp Sherman during World War I led to the destruction of others. Camp Sherman no longer exists; in its place stands the

The Allegheny escarpment southeast of the Kinnikinnick Creek crossing marks the edge of Mississippian- and Pennsylvanian-age rocks. Ice only penetrated the edges of this more resistant upland. The prominent knob to the left is Sugarloaf.

Chillicothe Correctional Institute. Today the National Park Service manages the Hopewell Culture National Historical Park, which encompasses the Mound City Group National Monument, Hopeton Earthworks, Hopewell Mound Group, High Bank Works, and Seip Mound State Memorial.

The Hopewell culture traded widely for geologic materials, including native copper from Upper Michigan, mica from the Great Smoky Mountains, Cenozoic-age shark teeth from Maryland, and obsidian—a black, glassy volcanic glass—from the Yellowstone National Park area. The Hopewell culture built earthworks shaped like circles, squares, and octagons.

U.S. 24 (See map for I-80/90 and U.S. 20 west on page 79)
Toledo—Indiana Line
82 miles

U.S. 24 crosses 82 miles of Huron-Erie lake plain, paralleling the Maumee River, from south Toledo to the Indiana line just west of Antwerp. The ancestral Maumee River flowed southwest while serving as an outlet for the proglacial lakes that preceded Lake Erie; however, until Glacial Lake Warren developed sometime between 13,000 and 12,000 years ago, the river only existed in Indiana. Drainage switched to a northeasterly direction as the edge of the ice sheet moved north of the St. Lawrence Valley, freeing a northeast outlet for the Maumee River and Glacial Lake Warren. Between Toledo and Napoleon, U.S. 24 lies on 45 miles of sand and clay, once the bottom and shoreline of these Wisconsinan-age glacial lakes.

Toledo to Maumee

From Toledo to Maumee, U.S. 24 lies on the filled bed of the Miami & Erie Canal; this stretch of the highway is also called the Anthony Wayne Trail. Works Progress Administration workers salvaged stone from the locks and aqueducts of the canal and used them in building projects around Toledo in the 1930s. Many of the older buildings at the Toledo Zoo benefited from this recycling. Much of the original canal stone reportedly came from Marblehead quarries near Lake Erie's shore and the bed of the Maumee River. Ewing Island, a large bar of river sand and gravel just downstream of Maumee, is the largest natural island along the Maumee River. This is where the drowned mouth of the river ends. The upstream channel alternates from bedrock rapids to deeper sediment-bottomed pools all the way into Indiana.

Ancient Tidal Flats at Maumee and Waterville

Late Silurian-age Greenfield, Tymochtee, and undifferentiated Salina group dolomites form the river bed between Maumee and Waterville and occur in local quarries. Sidecut Metropark, located along River Road just south of U.S. 24, preserves a canal lock where a "sidecut" canal connected the Maumee River to the Miami & Erie Canal. Bumpy bedrock flats of Greenfield dolomite occur upstream of the lock along the river's floodplain. If you walk along these

Bedrock flats along the Maumee River at Sidecut Metropark in Maumee. Swirls in Silurian-age Greenfield dolomite mark mounds of fossilized algae.

flats, you will see all kinds of rock evidence indicating that this was a tidal flat 420 million years ago, when much of Ohio was inundated by Silurian seas. These formations are similar to the rocks currently forming in the Bahamas.

Hemispherical rocky mounds dot the flats. They have fewer layers than surrounding rocks, which seem to lap up onto the flanks of these mounds, forming a distinctive, rocky scrollwork. Although these mounds seem to lack fossils, the entire mound actually is a fossil—a growth of algae and other simple organisms that formed a mossy mat in shallow seawater and was subjected to the daily fluctuation of tides. If you were to view a cross section of one of these mounds with a microscope, you would see many fine laminations that represent alternating layers of cells and sediment. Through time the organisms grew upward through the sediment that covered them, maintaining a mat of growth on the seafloor. Some mounds have small scour channels that seawater carved into the tops or along the margins, indicating submergence. Ripple marks in the Greenfield dolomite indicate areas of shallow moving water, and mud cracks suggest that areas were periodically exposed at low tide. A company quarries the Greenfield dolomite north of U.S. 24 in Maumee; one of the oldest quarries that pulled rock from this strata in the area was near Monclova, northwest of Sidecut Metropark.

Upstream at Farnsworth Metropark, just southwest of Waterville, slightly younger Silurian dolomite indicates that the tidal flat environment lasted a long time in northwestern Ohio, perhaps 10 to 20 million years. The Tymochtee dolomite forms high bluffs along the river at the north end of this park. This was an important point in the Maumee River valley because of the isolated 30-foot rocky landmark, Roche de Bout, in the middle of the channel, which served as a landmark to people traveling the river, and because of the undercut

Roche de Bout in the Maumee River at Farnsworth Metropark in Waterville. This prominent island of Silurian-age Tymochtee dolomite is just north of the trace of the Bowling Green fault.

river bluffs that served as shelter and a good camping spot. Roche de Bout, an erosional remnant of Tymochtee dolomite, became a pillar of the concrete arch electric railway bridge that was completed in 1912. The dolomite in Roche de Bout and the north bank reveals clues that it was deposited in an intertidal to above-tide environment some 425 million years ago. Look for ripple marks, cross-bedding, and tidal channels in layers and occasional ostracod, brachiopod, and trace fossils in the 35-foot bluff; small faults and broad dips are also evidenced in these layers. To see a fault, look for a place where layers look like they have been cut vertically and don't match up on either side of the line. The small faults may be related to the differential settling of sediments or may have been caused by earthquakes along the Bowling Green fault zone. The layers of the Tymochtee formation thin and thicken across the cliff face, and sometimes they wedge out.

The daily rise and fall of sea level on a typical tidal flat causes seawater to flow back and forth across it, seeking out the lowest spots on the surface. These preferred flow paths grow deeper because of erosion, making channels, while the higher spots in between often become encrusted with algal growth. Storms may lead to the cutting of new channels and the filling of old ones. Mud may accumulate in channels that are not being continually flushed by tidal currents or waves, often forming lens-shaped deposits that mark the former channels or depressions. Newly exposed areas may dry out, forming mud cracks. Tidal currents and waves may also produce ripple marks in the bottom sediment. Ripple marks on the surface mean there are cross-beds within the sediments. As these features are buried, compacted, and cemented, they become limestones or dolomites. Few fossils, except for laminated algal mats and ostracods, are found where these harsh environments of exposure and submergence existed.

The Bowling Green Fault

Farnsworth Metropark has another geological attraction. The Bowling Green fault trace crosses the river just upstream of Roche de Bout. It can be recognized by a riffle zone during low water. The fault originates far below northwestern Ohio in the Precambrian basement, a good mile below the surface. It is difficult to interpret, but evidence suggests it is a complex fault zone, perhaps consisting of a series of separate faults, with the upthrown block to the east; a component of lateral movement is also indicated. The complex stretches for at least 45 miles, from southeastern Michigan to south of Findlay in Hancock County. At Findlay several deep faults trend to the southeast to Bucyrus and Marion. The maximum offset between the blocks on either side of the Bowling Green fault is 500 vertical feet, though this isn't readily apparent at the surface because of overlying rock strata and glacial sediments. Reactivation along the fault displaced Silurian-age rocks at Waterville, but it's not possible to say when this happened since overlying younger rocks are missing.

At one time a cross section of the fault showed in the Tymochtee dolomite of Waterville Quarry in Waterville, but quarrying erased it. This quarry is across U.S. 24 from the metropark. North of Waterville, though, the fault fails to reach the surface, but a monocline, a local steepening in the large-scale tilt of strata, suggests that it exists here. The fault could have created the monocline without fracturing the overlying rocks. Are future earthquakes in northwest Ohio possible? We can't rule them out.

Quarrying at Whitehouse

By the late 1800s several quarries operated in the Whitehouse area west of Waterville. Some quarried Devonian-age Sylvania sandstone, but most sold Dundee limestone. Building stone and flagging were the major products.

Wisconsinan ice gouged the surface of the Devonian-age Dundee limestone near Whitehouse, leaving grooves that show the direction ice flowed.

Quarries shipped out many carloads of these rocks on the Toledo Wabash & Western Railroad. When the county built a welfare farm—a prison for minor offenses—west of town, a quarry became part of its operation. Inmates worked a shallow quarry in Dundee limestone into the 1960s. The stone quarried in this area came out in thin slabs and was perfect for flagging. During the Works Progress Administration years, workers used loads of these slabs in Toledo city parks and at the Toledo Zoological Gardens. It also became popular as garden stone. The Whitehouse quarries now lie abandoned, and a couple of them are ponds in the downtown area. One pond is part of a developing park of the Toledo Metroparks system.

Waterville to the Indiana Line

Six miles southwest of Waterville, opposite Wood County's Otsego Park, Devonian rocks of the Detroit River group form rapids in the river. Upstream and across from Grand Rapids, Devonian Dundee limestone forms rapids in the river at Providence Metropark. A restored section of the Miami & Erie Canal, including canal boat rides and an operating lock, is a major attraction of this park. Devonian Columbus limestone from Marblehead; Sylvania sandstone, probably quarried downstream of Grand Rapids; and possibly local Devonian dolomite compose the lock's stonework. Dams on either side of Buttonwood Island, which were initially built during the canal era, create a pool upstream and effectively cover rocks exposed at river level.

Between Texas and Napoleon, U.S. 24 crosses the valley of meandering North Turkeyfoot Creek, which cuts through glacial lake deposits and river alluvium. Just north of U.S. 24 at Liberty Center, sandy soils mark the beach ridge of Glacial Lake Warren; remnants of this shoreline are also south of the Maumee River between McClure and Napoleon.

U.S. 24 leaves the Maumee River valley to bypass Napoleon and Defiance. To continue along the river and the abandoned Miami & Erie Canal, follow Ohio 424. Another canal-era dam crosses the Maumee downstream of Defiance at Independence Dam State Park. Sandy patches here mark the shoreline of Glacial Lake Whittlesey. U.S. 24 crosses lake sediments from Defiance to 3 miles east of Antwerp. Just west of the Ohio 18 junction, U.S. 24 bridges the Tiffin River and then crosses to the south side of the Maumee River valley. Devonian-age Antrim shale forms the banks of the Auglaize River not far upstream from its convergence with the Maumee River in downtown Defiance. Limestone quarries were common along this stretch of the Auglaize before the Defiance Power Dam was constructed and the river inundated the quarries. Limestone for the dam's concrete was quarried just downstream of it as it was being constructed. Four miles upstream of Defiance, quarriers used to cut Devonian-age limestone from the Maumee River bed. Dikes built in the river kept the quarrying area relatively dry. Brick factories in Defiance, Farmer, and Napoleon utilized the clayey glacial lake sediments.

Between Defiance and the Indiana line, the Maumee follows a twisting course through flat glacial lake plain. There is a large meander about 7½ miles west of Defiance where the river approaches the highway. A town developed here, not surprisingly named The Bend. Even larger meanders occur at Antwerp.

At the Indiana line, U.S. 24 lies on wave-planed till. The till was deposited by the same ice lobe that deposited the Fort Wayne moraine and was later planed by Glacial Lake Maumee.

U.S. 30 AND U.S. 224
Crestline and Willard—Indiana Line
120 and 122 miles

Between Crestline and Williamstown and Willard and Findlay, U.S. 30 and U.S. 224 cross Scioto lobe sediments. Miami lobe tills form the landscape from Williamstown and Findlay to the Indiana line. Beginning near Gomer, U.S. 30 plays tag with the southernmost shorelines of Glacial Lake Maumee, a predecessor of Lake Erie from Wisconsinan time, all the way to Indiana—sometimes crossing over it, and sometimes not; U.S. 224 first touches this ancient shoreline around Tiffin, and then lies well within the lakebed to the Indiana line. Silurian- and Devonian-age strata underlie both routes.

Crestline to Bucyrus and Willard to Tiffin

West of the Sandusky River crossing north of Leesville, U.S. 30 crosses onto Devonian-age shales; this change is not detectable at the surface. To the north is Sulphur Springs, once the site of a small health resort that entrepreneurs built around an odiferous spring rising from the contact of Devonian shale with underlying limestone. The St. Johns end moraine causes a slight rise in the road at Bucyrus, and after crossing the Sandusky River U.S. 30 cuts through the Wabash moraine.

U.S. 224 follows the Defiance moraine from Willard to east of Tiffin. Small proglacial lakes formed on the south edge of this moraine, stretching from New Haven, just east of Willard, to Attica. Honey Creek, a tributary of the Sandusky River, parallels the southern edge of the moraine before cutting through it just south of Tiffin and dumping into the Sandusky River.

Leesville Building Stone

Just south of U.S. 30 on Ohio 598, there is an area known for its "hog's back" and building stone. The hog's back is a ridge of sand and gravel, several kames and crevasse fillings, that runs between Leesville and Galion, just west of Ohio 598. Pittsburgh, Fort Wayne & Chicago Railroad workers used this gravel to build many miles of roadbed for their tracks. The Sandusky River has incised deeply into the Broadway moraine at Leesville, exposing underlying Devonian-age Berea sandstone. At places the relief is nearly 70 feet.

People began quarrying Berea sandstone along the river in the 1830s; five quarries were in business in the 1870s, exposing 12 to 35 feet of good-quality stone. It was readily shipped on the nearby railroad and brought prosperity to Leesville until increased competition, and less demand for building stone, caused the quarries to close around 1907. Lowe-Volk Park, along Ohio 598, has trails that lead into old quarries along the Sandusky River and a nature center with geological displays.

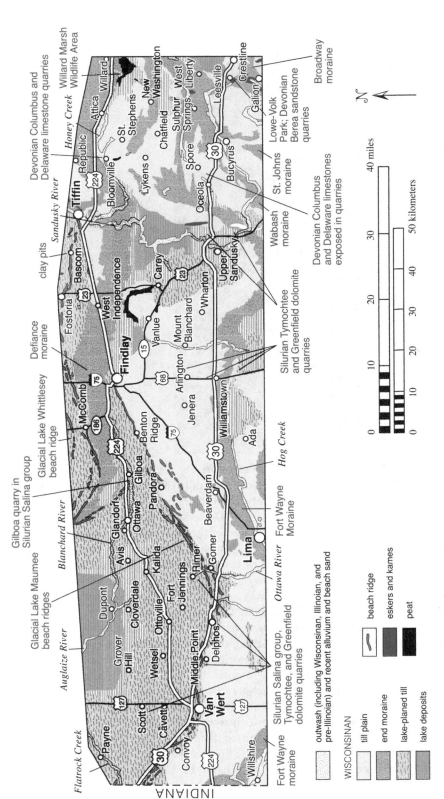

Geology along U.S. 30 and U.S. 224 between Willard and the Indiana line and Crestline and the Indiana line.

WISCONSINAN

- till plain
- end moraine
- lake-planed till
- lake deposits
- beach ridge
- eskers and kames
- peat
- outwash (including Wisconsinan, Illinoian, and pre-Illinoian) and recent alluvium and beach sand

Fort Wayne moraine

Flatrock Creek

Glacial Lake Maumee beach ridges

Auglaize River

Gilboa quarry in Silurian Salina group

Glacial Lake Whittlesey beach ridge

Defiance moraine

Blanchard River

clay pits

Sandusky River

Devonian Columbus and Delaware limestone quarries

Honey Creek

Willard Marsh Wildlife Area

Silurian Salina group, Tymochtee, and Greenfield dolomite quarries

Ottawa River

Hog Creek

Fort Wayne Moraine

Silurian Tymochtee and Greenfield dolomite quarries

Devonian Columbus and Delaware limestones exposed in quarries

Wabash moraine

St. Johns moraine

Lowe-Volk Park; Devonian Berea sandstone quarries

Broadway moraine

INDIANA

0 — 10 — 20 — 30 — 40 miles
0 — 10 — 20 — 30 — 40 — 50 kilometers

N

Devonian-age Berea sandstone lies in the Sandusky River valley at Leesville, just south of U.S. 30. There are abandoned building stone quarries in nearby Lowe-Volk Park.

Bucyrus-Area Quarries and a Mastodon

The shallowness of glacial till and close proximity of limestone bedrock to the surface in southern Crawford County led to a flurry of quarrying activity in the mid-1800s. Initially, people quarried the Devonian-age Delaware and Columbus limestones for building stone or crushed them to make lime. Both products found ready markets, especially to the east where limestone was scarce. Oceola became the hub of the quarrying district, and a water-filled quarry, Quarry Lake, exists just north of U.S. 30 in this town; a larger operation continues to quarry northeast at Spore. Spore was a station on the Toledo & Ohio Central Railroad. Developers established it in 1888 to serve the flourishing quarries nearby, but an official town never developed. Other water-filled quarries lie west of Lykens, north of Bucyrus and Oceola. Entrepreneurs also used glacial tills of this area at brick plants, tile factories, and a pottery in the late 1800s.

Workmen digging a millrace—a canal that water flows in to and from a mill wheel—in what is now southeastern Bucyrus uncovered the nearly complete skeleton of a mastodon in 1838. The bones rested 5 to 7 feet below the surface in a swampy area and reportedly were in an excellent state of preservation. This was one of the earliest reported discoveries of a preserved mastodon skull in Ohio. The miller displayed the bones for a time and then sold them. Eventually, some of the bones fell in the hands of P. T. Barnum and reportedly were lost in the spectacular 1865 fire that destroyed his American Museum in lower Manhattan. Some reports indicate that other bones ended up at the Ohio Agricultural and Mechanical College, which later became Ohio State University in the late 1870s, but nobody knows where the bones are today.

Willard Marsh Wildlife Area

West of Willard, U.S. 224 passes just north of what was once an extensive 13,000-acre marsh that formed at the site of a glacial lake, which had formed along the Defiance moraine some 25,000 years ago. Eventually, over 15 feet of peat and muck accumulated in the area as the lake lowered. The marsh was impassable until settlers built a road through it in 1884. The southern part of the marsh was known as New Haven, or Cranberry, Marsh. Although cranberries were a profitable crop there, other residents felt that a drained marsh might be a good place to grow celery. By the late 1890s, an influx of Dutch farmers from Wisconsin capably managed twenty-seven vegetable farms on drained portions of the marsh. The community became known as Celeryville. Deep, rich soils continue to provide quality vegetables. In 1942 the state began acquiring land and turning it into a hunting area. This area became the Willard Marsh Wildlife Area, which is 3 miles west of Willard and south on Section Line Road.

Seneca County Quarrying— Birthplace of the France Stone Company

In 1881 Enoch H. France began quarrying stone just north of Bloomville from the old Koller Quarry and used it to improve the railroad that was being constructed through Bloomville. In 1883 he purchased 59 acres of farmland on the northeastern edge of Bloomville and opened a larger quarry to supply railroad ballast. This purchase was the birth of the France Stone Company, which eventually operated numerous quarries and related firms across the Midwest.

The upper 35 feet of this second quarry—the France Quarry—consists of Devonian Delaware limestone; about 2 feet of it was suitable for building stone. The lower part of the quarry is Columbus limestone with a number of thin horizons that were suitable for building stone. Builders at Tiffin used this "blue limestone" to construct buildings at Heidelberg College. The France Stone Company also marketed it to builders in Alliance, Bucyrus, Columbus, Mansfield, and Toledo, where they used it as trim and storefront facing and in stone churches. Other quarries in Devonian-age strata operated around Flat Rock, Frank, Republic, and St. Stephens. Except for the quarry at Flat Rock, they are all abandoned and filled with water.

A few miles east of Tiffin, U.S. 224 passes onto Silurian dolomite of the Salina group, which people quarried widely in this area. A water-filled quarry lies east of Bascom on Ohio 18. The huge quarry at Maple Grove, about 6 miles north of Tiffin, opened in 1903 and supplied crushed stone for the construction industry. After a change in ownership in 1908, the quarry began supplying high-purity dolomite as flux and refractories, materials used in the production of metals. By the 1920s the company was a national leader in the production of refractories for open-hearth steel furnaces, and it expanded its Maple Grove quarry and plant. In its peak years, workers removed 2 million tons of rock from the quarry each year, and 500 to 700 full railroad cars left the plant weekly. France Stone was just one of several companies with quarries in the area. A quarry at Bettsville, north of Tiffin, now serves as a community recreation area. Tiffin and Fostoria quarries produced lime, crushed stone, and stone for foundations and bridge piers.

The water-filled clay pits—dug in glacial till—and plant of the former J. A. Miller Tile Company, in operation by 1889, are on the northern edge of Bascom. Many brick and tile plants once served the area, including ones at Attica, Iler, and St. Stephens, but most of them closed long ago and left little trace of their existence.

Bucyrus and Tiffin to the Indiana Line

U.S. 30 crosses two more end moraines, barely distinguishable from intervening till plain, between Bucyrus and Upper Sandusky. Around Upper Sandusky, the fine silt and clay that were deposited in an ancient lake, which formed along the Defiance moraine, are also indistinguishable from the till plain when viewed from the highway. The lake disappeared thousands of years ago during the final stages of the Ice Age. U.S. 30 crosses 16 miles of till plain between Upper Sandusky and Williamstown. The highway then parallels the Wabash moraine between Williamstown and the I-75 interchange. Abandoned quarries in Silurian Salina group dolomites occur along Hog Creek south of U.S. 30 and north of Ada. At Gomer, U.S. 30 rests on sandy ridges marking the shores of Glacial Lake Maumee. The road angles northwest toward Fort Wayne, Indiana, following the ridges. If you look to the north from either route, you can view the flatness of what was once lake bottom.

Numerous quarries in Silurian dolomites are scattered along both routes. U.S. 30 meets younger Silurian-age dolomites at Upper Sandusky; these rocks compose the bedrock to the Indiana line. Most towns in this relatively flat region had a quarry nearby; most of them were short-lived, but those that were favorably located and blessed with abundant reserves of rock continue to operate.

By the 1870s, workers were prying dolomite from the bed of Jennings Creek at Delphos; the Tymochtee dolomite still feeds the crushers on the southwestern side of town. Quarrying at Middle Point began around the same time, and a large water-filled quarry lies about 1 mile south of U.S. 30 and west of town. Between Middle Point and Van Wert, there is another active quarry producing crushed Tymochtee dolomite. This quarry is unique in that it belongs to the township and was opened in 1915 for the sole purpose of providing stone for public road construction and repair.

U.S. 224 skirts what was the edge of Glacial Lake Maumee from Tiffin to the U.S. 23 junction and cuts through the Defiance moraine from West Independence to Findlay. U.S. 224 leaves the moraine briefly at Findlay, and returns to follow it to east of Gilboa. Litzenberg Memorial Woods, a Hancock County park between Findlay and Gilboa, is a pleasant place to hike the end moraine. The Blanchard River closely parallels the south side of the highway from Findlay to Ottawa. Several oxbow lakes show how the river has shifted its course, especially between the junction with Ohio 186, south of McComb, and Ottawa. Ohio 12 to the south follows the Maumee beach ridges before intersecting U.S. 30 near Delphos. Silurian-age bedrock is at the surface at Gilboa where the highway passes a closed quarry in the Salina group, which instructors now use to train scuba divers. The quarry is on the south side of the road. Quarriers began much of their activity in the beds and bluffs of the meandering Auglaize, Blanchard, Little Auglaize, and Ottawa Rivers and their tributaries.

Flat lake plain composes the roadbed from Gilboa to Van Wert. Lake clays served as raw material for tile and brick manufacturing plants at a number of locations. The ruins of a tile plant lie on the southern edge of Dupont. The

OTHER QUARRIES AND MINERAL INDUSTRIES ALONG U.S. 224 AND U.S. 30

Town	Quarry and/or Mineral Industry
Avis	Tile and brick plant; used glacial lake clays.
Bucyrus	Brick plant and tile factory (near town) that used glacial till.
Cavett	Quarry; Salina group.
Chatfield	Brick plant that used glacial till.
Cloverdale	Quarry; Salina group.
Convoy	Quarry; Salina group.
Crestline	Brick plant and tile factory (near town) that used glacial till.
Dupont	Tile and brick plant; glacial lake clays.
Fort Jennings	Quarry; Salina group.
Galion	Brick plant that used glacial till.
Glandorf	Tile and brick plant; glacial lake clays.
Kalida	Two water-filled quarries on Township Road 85 west of town; Salina group.
New Washington	Brick plant and a pottery that used glacial till.
Ottawa	Quarry; Salina group. Tile and brick plant and roofing tile plant; glacial lake clays.
Ottoville	Quarry; Salina group.
Pandora	Active quarry in Greenfield dolomite.
Rimer	Active quarry in Greenfield dolomite.
Scott	Active quarry in Salina group southwest of town and north of U.S. 30.
Van Wert	The Palmer Quarry near town was the first commercial stone quarry in Van Wert County; opened in 1875 and operated into the early 1900s. A tile plant used surface clays west of town.
West Liberty	Tile factory that used glacial till.
Wetsel	Quarry; Salina group.
Willshire	Quarry; Salina group.

Ruins of a tile plant at Dupont that used local glacial till as the raw material for a number of ceramic products. Stacks remain but the beehive kilns are gone.

Glandorf Tile Company operated into the 1980s, and the abandoned pits of its quarrying activity are just north of Dupont. Slight ridges of sand at Van Wert mark the shorelines of Glacial Lake Maumee. U.S. 224 then crosses till plain southwest of Van Wert and enters Indiana on the Fort Wayne end moraine.

U.S. 50
Chillicothe—Indiana Line
119 miles

West of the Scioto River at Chillicothe, U.S. 50 follows Paint Creek valley, crossing over Devonian-age Ohio shale and an Illinoian till plain for about 25 miles. Silurian-age bedrock begins west of Bainbridge, with excellent exposures of the Peebles dolomite along Rocky Fork. A flatter Illinoian till plain marks the route for some 50 miles to Cincinnati. Between Cincinnati and the Indiana line, some 22 miles, U.S. 50 lies within the Ohio Valley and passes over river sediments and outwash.

Chillicothe to Hillsboro

Chillicothe is where Wisconsinan and Illinoian deposits come to an end at the edge of the Allegheny Plateau. Illinoian deposits occur south of town, while Wisconsinan deposits are to the north. West of town, Devonian Ohio shale is spectacularly exposed in Paint Creek Gorge. West of Chillicothe, U.S. 50 follows the northwest-trending valley of the North Fork until it crosses the stream at Slate Mills ("slate" being a reference to black shale exposed in the valley). The highway bends southwest and follows a streamless section of Paint Creek valley for 2 miles. Paint Creek used to flow through this section of valley before the glaciers came and the creek cut a deep rock-walled gorge—a new channel—through the uplands of the plateau. Wisconsinan ice blocked the present-day North Fork valley, which had been the channel of Paint Creek. When the ice began to melt, Paint Creek valley filled with a lake, and the ponded water sought a new outlet. A low spot in the upland to the east became the new lower channel of Paint Creek. The strong current of the ponded water had little trouble cutting a deep gorge into the soft Ohio shale. This channel is bounded by 250-foot-high shale cliffs called Alum Cliffs. After the glaciers melted back into northern Ohio, the North Fork assumed the old lower valley of Paint Creek and flowed into Paint Creek southwest of Chillicothe.

Natural springs occur at the contact between Mississippian-age Sunbury shale and the overlying Cuyahoga formation, which is sandstone. Groundwater leaches sulfur ions from the pyrite in the shale, resulting in place names like Sulphur Lick in the North Fork valley west of Chillicothe, Alum Cliffs (named for white, hydrous iron sulfate minerals, incorrectly called "alum") in Paint Creek Gorge, and Sulphur Lick Flat near Spargursville. Other springs flowed at Kincaid Springs and Sinking Spring south of Bainbridge.

Silurian-age Peebles dolomite forms rapids in Paint Creek on the western edge of Bainbridge. Paint Creek turns abruptly north at the Ross-Highland County line, where U.S. 50 climbs into hills developed on Silurian limestones and dolomites, passing the turnoff for Seven Caves and the Highlands Nature Sanctuary. Rocky Fork has helped carve these landscapes in Silurian Peebles dolomite. The well-jointed and soluble Peebles dolomite forms a landscape of rounded knobs. Groundwater solution has created areas of reddish brown soil, or terra rossa. South of the route, Illinoian outwash and kames are exposed along the valley of the Rocky Fork. Paint Creek Lake, a 1,200-acre reservoir completed in 1974, inundates an upstream portion of Paint Creek. Rocky Fork also has a reservoir, but it was built in the 1950s.

West of Rainsboro, which was built on Illinoian outwash, you can follow Ohio 753 south to Fort Hill State Memorial for more Silurian-age scenery and Hopewell culture mounds. Quarries tapped Silurian dolomites around Byington and Hillsboro, and tapped Devonian-age Berea sandstone from the hills above Cynthiana.

Copperas Mountain

At 550 feet high, Copperas Mountain, along the south side of Paint Creek Valley, is a landmark in south-central Ohio. Paint Creek slices into the northwestern edge of this ridge between Bourneville and Bainbridge, exposing 285 feet of Devonian-age Ohio shale, 86 feet of Bedford shale, and a little Berea sandstone. It is clearly visible south of U.S. 50. If you want to observe this mountain up close, travel south on Jones Levee Road, which is just west of Bourneville. Then follow Spargursville Road southwest to Storms Road and turn west on Storms Road, north on Storm Station Road, and finally sharply west on a narrow township road to the foot of Copperas Mountain.

The Huron shale, with two zones of concretions, is at road level; one zone occurs at eye level and the other is about 50 feet above the first zone. The concretions are spherical to slightly discoidal and have soft centers. They formed as dolomite, calcite, or other carbonate minerals grew around a mineral grain or

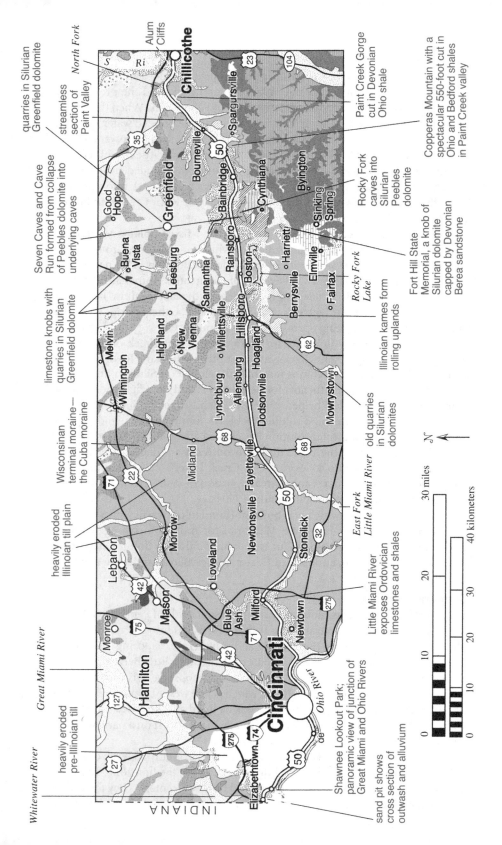

The view north across Paint Creek valley southwest of Bourneville. The uplands are Illinoian till on top of Ohio shale. The valley bottom contains recent stream sediments with Illinoian outwash terraces at the margins; Illinoian lake silts and clays mark where small lakes occurred where tributaries enter the valley.

fossil fragment while the clay sediment hardened into well-bedded shale. The larger concretions usually have encircling horizontal ribs that match up with layers of shale. The shale layers also bend around the concretions. The concretions at Copperas Mountain often have hollow centers where the soft centers weathered away. There are still unanswered questions concerning the origin of these concretions.

The 20-foot-thick Three Lick bed, which is easily recognized by its three greenish gray layers separated by brownish black shale, lies above the Huron shale. The Three Lick bed was deposited at the same time the Chagrin shale was being deposited in northern Ohio—about 365 million years ago. The Cleveland shale comes next, which differs from the Huron shale mainly because it lacks concretions. The upper portion of the exposed cliff is Bedford shale, which includes interbeds of shale, siltstone, and sandstone with ripple marks. Berea sandstone caps the ridge.

The mountain is named for the flower-shaped chemical growths, including iron sulfate and melanterite (a hydrous iron sulfate mineral), that coat the

Geology along U.S. 50 between Chillicothe and the Indiana line.

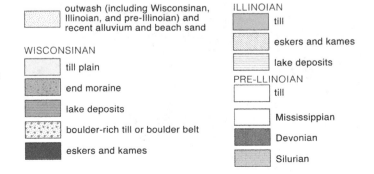

outwash (including Wisconsinan, Illinoian, and pre-Illinoian) and recent alluvium and beach sand

WISCONSINAN

till plain

end moraine

lake deposits

boulder-rich till or boulder belt

eskers and kames

ILLINOIAN

till

eskers and kames

lake deposits

PRE-LLINOIAN

till

Mississippian

Devonian

Silurian

Huron shale where pyrite is common. *Copperas* is a synonym of "iron sulfate." Look for sandstone with ripple marks along the road at the foot of the Copperas Mountain. It erodes from high on the cliff. These ripple marks were made by the waves of Devonian seas before the sandstone hardened.

Rocky Fork and Its Caves

South of U.S. 50, west of Bainbridge, Rocky Fork gouges deeply into Silurian-age bedrock before joining Paint Creek. Some 25,000 years ago, an advance of the Scioto lobe blocked preglacial Rocky Fork from flowing east. Meltwater poured into the valley, reversing Rocky Fork's flow and cutting the spectacular 75-foot-deep gorge it now flows through. During the retreat of the Wisconsinan ice sheet, Rocky Fork began flowing east again. Groundwater seeping downward widened joints in the Peebles dolomite that composes the walls of this gorge. The joints became caves near the water table. Over time the ceiling of one cave collapsed, allowing daylight into the main cave passageway. This was the beginning of present-day Cave Run. Cave Run continued to cut into the Peebles dolomite, forming a 25-foot-deep notch, which is high on the eastern wall of this gorge. Other tributary passages cut other caves, similar to the above, seven of which were managed as a tourist attraction called Seven Caves for many years.

Interest in the caves began over a hundred years ago when visitors, some staying at the nearby Rocky Fork Hotel in Paint, would hike or ride into the gorge, explore mud-filled caves with lanterns, and perhaps spread a blanket

Copperas Mountain southwest of Bourneville provides a spectacular view of the Devonian-age Ohio shale where Paint Creek cuts into its base.

Dancing Cave near Bainbridge circa 1908. This cave, carved in Silurian-age Peebles dolomite along a tributary of Rocky Fork, is now part of Seven Caves.

along the dolomite cliffs and enjoy a picnic lunch. In 1890 entrepreneurs began charging fees to various caves; caves were dug out and trails and steps were built to make the caves more accessible. Seven Caves, just about 13 acres of land, became a magnet to tourists by the 1950s. It also became an Ohio Natural Landmark. The longest cave in the park was 315 feet long. A number of separate nature preserves eventually surrounded the park and many people hoped that they could all be united one day as a large preserve. That dream took one more step toward reality when Seven Caves was purchased by neighboring Highlands Nature Sanctuary in 2006. The preserved lands of the Rocky Fork Gorge now total over 1,700 acres; the preserve contains one of the highest concentrations of caves in the state.

Fort Hill State Memorial

A 1,300-foot-high erosional knob, upon which a Hopewell culture earthwork is preserved, stands south of Bainbridge off Ohio 41. Devonian-age Berea sandstone caps the knob, and Baker Fork has carved a deep gorge in underlying Silurian-age Peebles dolomite around the north and west sides of it. This knob is one of many that the elements have carved into the edge of the Allegheny Plateau; they are preserved at Fort Hill State Memorial.

Baker Fork formed when an Illinoian ice lobe blocked north-flowing streams in this region, creating a lake that spread out around present-day Cynthiana. The lake eventually drained along a low spot, cutting through Ohio shale and the underlying Peebles dolomite to form the gorge of south-flowing Baker Fork. Tributaries seeking out cracks or joints in the rock perhaps formed the natural rock arches in this area, although geologists do not fully understand their origin.

Some twelve miles of hiking trails allow people access to the geologic features in the memorial. Seven natural arches occur along the gorge. You can easily see Spring Creek Arch in the Peebles dolomite along the Gorge Trail; its opening measures 18 feet long by 6 feet high. The Keyhole, another arch, is 77 feet long and 35 feet high; it narrows to 2 feet wide at points. You can also see the Baker Fork Bridge from this trail. All of these features were carved in Peebles dolomite. An overlook at the earthwork offers a great view across Highland County.

Greenfield-Area Quarrying

North of Rainsboro, quarriers have been extracting Silurian-age bedrock since 1810. The Greenfield dolomite is 40 to 50 feet thick at quarries along Paint Creek on the southeast edge of Greenfield. Although this stone is widespread across much of western Ohio, Greenfield is the only location where quarriers were able to find strata suitable for building material. Workers quarried the best building material from 22 feet of thick, even beds of gray stone. Most of it was used for general building purposes, flagging, and sidewalks; some of it was shipped as far as Cincinnati, Marietta, and Toledo. Quarriers produced lime from the more fractured layers that were not suitable for building.

In later years, the Rucker Quarry, on the site of the pioneering Lang Quarry, produced crushed stone from the Greenfield dolomite and used an elaborate system of steam-driven cable cars to shuttle stone from the quarry to crushers and waiting railroad cars. Between its quarries in Greenfield and Hillsboro, the Rucker Company employed over one hundred men around 1905. The Greenfield quarries south of town and in Hillsboro only sporadically operated after 1920, and quarrying shifted to north of Greenfield in the 1950s.

The first Greenfield structure built of local stone was Traveler's Rest, an inn, built in 1811 and still standing on Main Street. The Greenfield Seminary

Traveler's Rest in Greenfield—the first major structure built of Greenfield stone.

and Presbyterian Church are other good testimonials to the Greenfield dolomite's quality. Irish stonemasons, who settled in this area between 1840 and 1850, applied their talents at the local quarries. Their legacy lives on in cemetery monuments and the trim of larger buildings.

West of Greenfield, erosion on the flanks of the Cincinnati Arch eliminated the Greenfield dolomite, exposing older Silurian dolomites. Isolated eroded remnants, or outliers, of Greenfield dolomite are marked on the landscape by quarries, such as those at Highland and Leesburg.

Brick and tile factories used local floodplain clays and Illinoian till. The largest tile factory extracted its clay from a pit that is now disguised as a pond north of Jefferson Street, on the west side of Greenfield. The tile industry prospered into the early 1900s. Another tile plant worked the tills around Good Hope into the mid-1950s.

Hillsboro to the Indiana Line

From Hillsboro to Cincinnati, U.S. 50 crosses a relatively flat Illinoian till plain. Just south of Hillsboro, flanking the headwaters of Rocky Fork, are hills of sand and gravel—Illinoian-age kames. Extensive quarrying in the Hillsboro area exploited Silurian-age rocks and Illinoian tills. Quarries in the area, now abandoned, once exposed the Peebles, Lilley, Bisher, and Brassfield formations. Two quarries in Lilley dolomite, on the east side of Hillsboro, date to around 1850. About 2½ feet of the 40-foot-section of this dolomite was of building stone quality; quarriers crushed most of it, though, or turned it into lime. The first quarry in the area opened at Lilley Hill, now part of easternmost Hillsboro along Ohio 124, where workers abandoned a railroad tunnel they had been excavating. Entrepreneurs marketed a porous Lilley dolomite filled with asphalt, which occurred between Hillsboro and Willettsville, as asphaltic rock in the mid-1900s.

A pottery opened near Hillsboro in 1806, making it one of the first opened in Ohio. Brick making was prevalent by 1850, and numerous kilns dotted Hillsboro. A tile factory operated for twenty-five years on Ohio 138, southwest of town. Most of these plants had closed by the early 1900s, but a factory at Mowrystown on U.S. 62 south of Hillsboro continued operating until 1940. The plants used Illinoian tills.

Ordovician-age limestones and shales form the bedrock under U.S. 50 from Allensburg to the Indiana line. Dodson Creek and its tributaries provide some of the easternmost exposures of Ordovician strata near Allensburg and Dodsonville. U.S. 50 bridges the East Fork Little Miami River at Fayetteville, offering more views of Ordovician rocks. Dissected Illinoian till plains characterize the route near Cincinnati. Between Stonelick and Milford, U.S. 50 traverses the East Fork Little Miami Valley. Milford developed around the base of an outlier composed of Ordovician strata where the East Fork enters the Little Miami River. U.S. 50 follows the Little Miami to the Ohio River and then parallels the riverbank through Cincinnati and on to North Bend.

On the west side of downtown Cincinnati and Mill Creek, the highway passes onto pre-Illinoian till, the oldest glacial material in the state; however, you can't see it because of pavement and buildings. Hills rise on either side of downtown, underlain by Ordovician-age Kope, Fairview, and Grant Lake strata. From Riverside to North Bend, U.S. 50 lies on pre-Illinoian outwash on a terrace above recent river alluvium and the current floodplain of the Ohio River. Little of these sediments is visible because of industrial and commercial development.

North of the U.S. 50 viaduct over the Mill Creek valley and on the west side of the valley is a flat-topped area, called Bald Knob, that is considerably lower in elevation than the surrounding hills. From 1930 to 1932, when construction of the Central Union Terminal—Cincinnati's main railroad depot—was underway, workers stripped Ordovician-age Grant Lake and Fairview shales and limestones from the top of this hill for use as fill in the Mill Creek valley in order to locate the railroad facilities above flood level. A tremendous volume of Wisconsinan outwash sand and gravel from the Great Miami River valley was also used as fill. Bald Knob was further leveled about 80 to 100 feet in the 1960s for fill when road crews constructed I-75 through downtown Cincinnati.

Near North Bend a few scattered exposures of Kope formation occur on the north side of the highway. U.S. 50 parallels the boundary between Illinoian and pre-Illinoian outwash between North Bend and Cleves and follows the Great Miami River valley, filled with recent alluvium and slightly higher terraces of Wisconsinan outwash from Cleves to the Indiana line. The highway loops onto the wide floodplain of the Great Miami and Whitewater Rivers between North Bend and the state line. Sand and gravel operations dot the lower Great Miami River valley.

OTHER QUARRIES ALONG U.S. 50

Town	Strata
Berrysville	Lilley formation
Buena Vista	Greenfield dolomite
Elmville	Peebles and Lilley formations
Fairfax	Peebles and Lilley formations
Good Hope	Greenfield dolomite
Harriett	Peebles dolomite
Hoagland	Bisher formation (?)
New Vienna	Peebles dolomite
Samantha	Greenfield dolomite
Sinking Spring	Peebles dolomite
Washington Court House	Greenfield dolomite
Willettsville	Peebles dolomite

The Confluence of the Ohio and Great Miami Rivers

Shawnee Lookout County Park of the Hamilton County Park District is located south of U.S. 50 just east of the Indiana line. The park contains what is known as Fort Hill, which contains Hopewell culture earthworks on a promontory high above the juncture of the Great Miami River and Ohio River. This promontory is composed of Ordovician Fairview and Grant Lake formations and is capped by pre-Illinoian till. At the far western end of this hill there are terrace sands of pre-Illinoian age. At the base of the promontory are slopes developed on the Ordovician Kope formation; however, exposures are few because they are covered with Wisconsinan outwash and recent alluvium deposited in the Great Miami Valley. This outwash is clearly visible in large sand pits just south of Elizabethtown.

Hike the Miami Fort Trail to the nose of the promontory for spectacular views through the trees of both the Great Miami and Ohio Valleys. Directly west and below the promontory is the present channel of the Great Miami River, and beyond this there is a broad C-shaped depression filled with marshes and arcuate ponds that was an earlier path of the river when it swung much closer to what is now Lawrenceburg, Indiana. This is a meander scar, and the arcuate ponds are oxbow lakes. This was still the main channel of the river in 1847. Sometime before 1900 the meander was cut off and it became a meander scar. The most recent development occurred when the Ohio River backed up behind the newly constructed Markland Dam (built between 1959 and 1964) about 40 miles downstream near Markland, Indiana. The mouth of the Great Miami River was flooded a few miles upriver and water reentered the meander scar, partially flooding it. The drainage history, however is much more complex.

Geologists are still trying to work out the preglacial drainage in southwestern Ohio. Several interpretations involving ancient rivers in the present-day Great Miami River valley have been offered since the 1890s. At least two versions put forth the idea that a preglacial river flowed north through the valley, eventually connecting with the Teays River north of Dayton. Another version has the preglacial Hamilton River flowing south through Hamilton to a connection with the preglacial drainage of southern Indiana. The big question is whether or not a drainage divide existed between present-day Dayton and the

Shawnee Lookout Park. The flooding Great Miami River, to the right, enters the Ohio River in the background in May 2003.

Teays Valley near Columbus. Whichever is correct, the Ohio River to the south of Shawnee Lookout probably didn't exist during the pre-Illinoian glacial stage. The first ice advances of pre-Illinioan time led to disruptions and redirections of the Teays River and its tributaries and most likely led to the development of a south-flowing river in the present Great Miami River valley.

Through hundreds of thousands of years the drainage pattern of Ohio was completely changed. Wisconsinan ice of the Miami Lobe got as far as northern Hamilton and southern Butler Counties, forming the Hartwell moraine. The Ohio River began as a stream flowing along the southern edge of the Wisconsinan ice sheet. When the final glaciers began to leave southwestern Ohio, the Shawnee Lookout landscape of today began to develop. At this time the fledgling Ohio River began cutting into its south side, narrowing the once-wide ridge; the new southward-flowing Great Miami River was busy cutting away at the ridge on the other side.

Looking south across the Ohio River to Kentucky reveals uplands of a basically unglaciated terrain (Illinoian and pre-Illinoian sediments cap some of the hills). The hillsides are well eroded by many small streams, and slopes are generally steeper than those north of the Ohio River because streams and landslides have been the dominant slope-forming forces in this area since the glaciers left at least 125,00 years ago. The Shawnee Lookout surroundings, in comparison, have streams only near the edges of wide valleys, like those of the Great Miami and Ohio. Steep, almost vertical, slopes occur here and there around the base of hills where strong river currents erode into the soft Ordovician Kope formation, undercutting the overlying, harder limestone beds. These are called *cutbanks*.

U.S. 68
Findlay—Aberdeen
184 miles

U.S. 68 begins on the edge of the flat Huron-Erie lake plain southeast of Findlay, stretches across a myriad of Wisconsinan deposits, surmounts Ohio's highest terrain at Bellefontaine, enters older Illinoian sediments 1½ miles south of Cuba, and crosses the unglaciated hills of the Allegheny Plateau between Ripley and Aberdeen before crossing the Ohio River.

Findlay to Kenton

Flat, gently rolling land, typical of till plain, characterizes U.S. 68 between Findlay and Williamstown. Meandering Eagle Creek parallels the road to the west, and numerous impoundments modify its flow. Gas and oil wells, holdovers from the great oil boom of the late 1800s, still dot the fields. A farmer digging a water well in 1836, 2 miles south of Findlay, is credited with having discovered natural gas in the area.

The glacial deposits vary in thickness from a few feet to several tens of feet along this route but are thin enough in areas to allow quarriers to dig Silurian-age bedrock. Between Williamstown and Dunkirk the road crosses the Fort Wayne moraine. It's imperceptible to the traveler, but the till thickens,

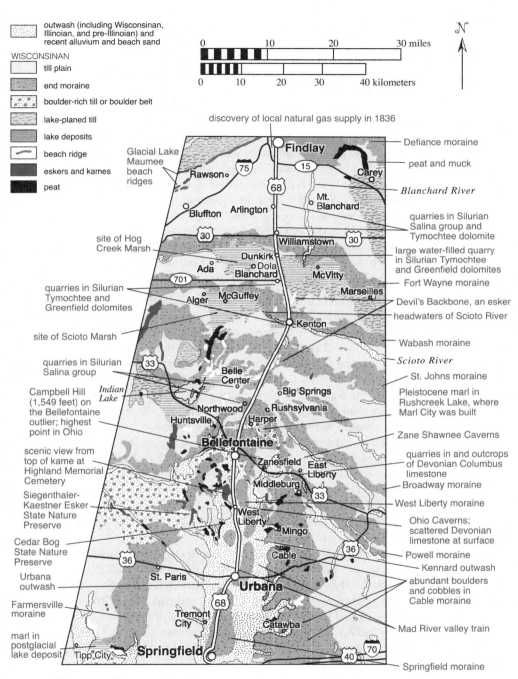

Geology along U.S. 68 between Findlay and Springfield.

causing a rise in elevation over the next 2 miles. The tills of this moraine have a less compacted structure since they formed along the edge of an ice sheet rather than underneath it. A series of glacial lakes existed on the south side of the end moraine. Dammed against piles of till, the lakes flooded the lowlands to the south. Fields southwest and west of Dunkirk yield dark, rich soils and underlying gray calcareous mud called *marl*, which was deposited in these lakes. U.S. 68 lies on waterworn till that marks the edge of one of these lakes for 6 miles south of Dunkirk. A short, sandy ridge intersects the highway about 3½ miles south of the Ohio 701 junction. This is a remnant of a long, winding esker that stretches southwest toward Bellefontaine.

U.S. 68 crosses the Wabash moraine in downtown Kenton and dips gently into the Scioto River valley on the southern edge of downtown. The Scioto's source is just a few miles southwest of town near what was the Scioto Marsh, which farmers drained in the 1800s. Glacial debris is thin in places, and workers still quarry Silurian bedrock west of town; older quarries occur along Tymochtee Creek northeast of town. Water also ponded along the southern edge of the Wabash moraine, forming small lakes west of Kenton.

Quarrying in Southern Hancock and Hardin Counties

Silurian dolomites, mainly of the Salina group and particularly the Tymochtee formation, form the bedrock of Hancock and Hardin Counties. Settlers pried this stone from creek- and riverbeds and used it for foundations, bridge abutments, and flagging, and as a source of lime. At least five small quarries operated around Arlington in the 1870s. One of these, about 2 miles east of town, prospered until 1941; the water-filled pit remains as the center of a private club. U.S. 68 passes an abandoned quarry north of Williamstown, now called Sulphur Lake, and another water-filled quarry lies east of town. Several quarries operated around Dunkirk in the 1870s. One quarry lies tucked away behind vegetation and buildings in the southeastern corner of the village. Opened in 1877, this quarry supplied the Pittsburgh, Fort Wayne & Chicago Railroad with ballast. Initially, two hundred men worked the pit, shoveling stone by hand. It became one of the larger quarries in the state by 1910, with a depth of 35 feet and an area of 40 acres. Crushed stone continued to be its major product until the quarry closed in the late 1930s.

West of Dunkirk, quarries once operated along Hog Creek, north of Ada. Builders in Ada used stone from these quarries in the 1870s in buildings of the school that became Ohio Northern University. Lime was the major product of a quarry along the now-abandoned Big Four Railroad at McVitty, east of Dunkirk. An abandoned Lockport dolomite quarry lies along Ohio 67 in Marseilles. Quarries still operate at Blanchard and on the west side of Kenton; an abandoned quarry, now reclaimed as France Lake–Saulisberry Park, lies southwest of Kenton.

Brick and tile manufacturing plants at Alger, Arlington, Kenton, and Mt. Blanchard made use of local glacial clays. In the 1850s and 1860s, a small pottery operated on Arlington's Main Street, and a cement-tile factory on the east side of town closed in the 1930s. Four brick factories and two tile plants operated in Kenton around 1900.

Abandoned quarry in Silurian-age dolomite north of Williamstown. Undifferentiated Salina group rocks form the upper walls.

Hog Creek and Scioto Marshes

After the ice withdrew from northern Ohio about 16,000 years ago, small lakes were left in low-lying areas, particularly along the southern edges of end moraines. A number of these failed to drain completely because of underlying impervious clay sediments, so marshes formed. West of Dunkirk, Ohio 81 passes through what was once Hog Creek Marsh, which encompassed around 8,000 acres between Dola and Ada. Settlers drained it around 1879 to make fertile farmland. When the fields are bare, black soils and marl provide an idea of the former extent of this feature. Other than that, though, a historical marker on the south side of the road is all that makes this former marsh visible. Tall grasses and bushes covered the area during summer, and water flooded it during winter. In a few places, farmers harvested large quantities of cranberries. Workers encountered difficulties building the Pittsburgh, Fort Wayne & Chicago Railroad across the marsh. For around ten years after it was built, sinkholes caused washouts under the rails. After ditches drained the marsh, farmers grew potatoes and onions until poor conservation methods allowed most of the organic topsoil to burn or blow away.

Ohio 195 bisects a similar but larger wetland—the Scioto Marsh—south of McGuffey and about 9 miles northwest of Kenton. Numerous ditches run through the former wetland, including the channelized path of the Scioto River. Workers unearthed several mastodon bones while draining the marsh, and certainly more lie waiting to be discovered in the rich soils.

Kenton to Bellefontaine

The highway crosses ground moraine for 4 miles south of the Scioto River. South of Kenton, what locals call the Devil's Backbone, the remains of an undulating esker resting on top of ground moraine, parallels U.S. 68. County Route 155

The Devil's Backbone, an esker, stretches south of Kenton.

and Taylor Creek follow this feature. You can trace the esker's path by noting the numerous sand and gravel pits that are dug into it. For 15 miles between Kenton and the junction with Ohio 274, U.S. 68 cuts through a complex of four end moraines and intervening till plain. An upland called the Bellefontaine outlier split the Wisconsinan continental ice sheet into two well-defined lobes—the Miami to the west and the Scioto to the east. The space between them and south of the outlier became the valley of the Mad River, a major glacial sluiceway that extended into southern Ohio. Quarries in the Silurian-age Salina group dot the east side of Belle Center, and another is in Northwood just west of U.S. 68. The highway rises onto the Devonian-age bedrock of the outlier south of Ohio 274, which leads to Huntsville and Rushsylvania.

Bellefontaine Outlier—the Highest Point in the State

Just south of the junction with Ohio 273, U.S. 68 rises onto the northernmost edge of the Bellefontaine outlier, which is capped by Devonian-age strata of the undifferentiated Columbus and Lucas formations and Ohio shale. An outlier is an area of relatively higher, younger rock that is surrounded by older rock, in this case the Silurian-age Greenfield and Tymochtee formations. The Bellefontaine outlier is an isolated segment of Devonian rock, roughly 160 square miles in size, far from other Devonian-age rock. East of the outlier a wide Devonian rock band stretches from the shore of Lake Erie to the Ohio River, and a similar belt of Devonian rock to the west roughly parallels the Maumee River valley from Toledo to the Indiana border. From the outlier, one must travel 25 miles to the east and 130 miles to the west to reach the nearest correlative Devonian bedrock.

The elements have carved the northern part of the outlier into three small mounds that are scattered between U.S. 68 and Big Springs. Closer to Bellefontaine the Devonian cover of the outlier expands, reaching a maximum width of about 10½ miles. The highland continues south of town to near Urbana, a distance of nearly 25 miles from Big Springs. The rise onto the younger strata is not very noticeable when approaching from the north or south, but the view is quite striking when heading east or west from Bellefontaine because meltwater streams that flowed to either side of the outlier dissected it. It may be surprising, but the outlier marks the highest elevation in Ohio. At 1,549 feet, Campbell Hill, on Ohio 540 east of downtown Bellefontaine, is the highest point in the state.

The history of the area goes back 375 million years to the midst of Devonian time, when a broad, shallow sea covered the state. The Columbus and Lucas formations formed in this sea, only later to be covered by Ohio shale. This land likely was above sea level after the Columbus and Lucas formations were deposited because two rock units, which elsewhere lie between the Lucas dolomite and Ohio shale, are missing. After Devonian time, Mississippian and Pennsylvanian seas may have covered the area, but if so, all traces of this time period have disappeared. The other scenario could be that the outlier remained above sea level after the Columbus and Lucas formations were deposited, and later Paleozoic sediments never accumulated.

During glacial times, ice piled up against the upland and gradually covered it, depositing glacial sediments up to 150 feet thick in places. It is the till that makes it the highest point in Ohio. Roadcuts along U.S. 33 between Bellefontaine and East Liberty are good places to observe the bedrock. Just west of East Liberty a quarry still produces aggregate.

The Bellefontaine outlier is an area of Devonian-age bedrock that is completely surrounded by Silurian-age dolomite. Columbus limestone is quarried at this site near East Liberty.

Devonian strata in the Bellefontaine area proved as useful as the surrounding Silurian-age rocks. The limestones and dolomites served as building stone, aggregate, and as a source of lime. Quarries operated at Bellefontaine, Big Springs, East Liberty, Harper, Middleburg, Rushsylvania, West Liberty, and Zanesfield. People found little use for the Ohio shale because of its thin beds and crumbly nature.

Workers excavated marl, a calcareous clay, from a small glacial lake deposit northeast of Bellefontaine. George Bartholomew, a local entrepreneur, purchased the marl pits around 1888. Bartholomew experimented with various mixtures of the marl, local limestone, and other ingredients in the back room of a downtown drugstore while developing a suitable cement. He struck a deal with the city and paved an 8-foot stretch of Bellefontaine's Main Street in 1891, the first concrete street in America. In 1893, the other streets surrounding the courthouse were surfaced with the same mixture, and a sample of the successful Main Street pavement was an attraction at the 1893 Chicago International Exposition. Workers have since resurfaced the streets with the exception of Court Avenue, where a historical plaque and statue commemorate the paving. The old marl pits became the Buckeye Portland Cement Company, and a small community called Marl City, which still appears on some maps, developed around this company.

Bellefontaine to West Liberty

The terrain between Bellefontaine and West Liberty is a particularly complex landscape that developed between the Miami and Scioto lobes. Thick deposits of outwash sand and gravel, ground and end moraine sediments, glacial boulders, lake sediments, and peat are all scattered on either side of the road. North-south-trending end moraines form the horizon to the east and west of U.S. 68. Numerous sand and gravel operations expose the thick sedimentary filling of the valley, which provides the building and construction industry with a seemingly endless supply of raw material. A number of lakes in southern Logan County, for example Newell and Silver Lakes, contain calcareous clay, or marl. These lakes formed as ice block depressions, or kettles, in the interlobate terrain. Kames, the sedimentary deposits of ice crevasses and depressions, also dot the landscape as hills. The flat upland areas paralleling the highway and resting against adjacent end moraines, particularly to the east, are remnants of outwash terraces that developed when tremendous discharges of meltwater, heavily laden with glacial sediments, washed down the valley. When there weren't torrents of meltwater, streams cut deep channels in the valley fill. This alternating pattern of cut and fill formed the present-day wide valley. The current Mad River seems minuscule in comparison to the raging glacial waters that blasted through this area in the past. West of West Liberty the Siegenthaler-Kaestner Esker State Nature Preserve exhibits a sinuous ridge of sand and gravel. You can hike along the crest of this esker at this preserve.

Caverns of the Bellefontaine and West Liberty Area

Limestone close to the surface and adequate rainfall often lead to a karst landscape as groundwater dissolves the limestone. Caves form underground; and on the surface, depressions, or sinkholes, appear as water removes the limestone below the soil and glacial sediments. So why aren't karst features prevalent in the Bellefontaine outlier? Because glaciers covered the outlier at times, sediment from the melting ice sheets clogged the many subsurface passageways in the limestone and filled in the sinkholes in recent time. Only in areas with a thin till cover is the karst landscape still visible, such as around the Ohio and Zane Shawnee Caverns. As the glacial cover of this outlier disappears, large-scale dissolution of the limestone will resume.

In downtown Bellefontaine you can follow Ohio 540 east for 7½ miles, past Campbell Hill to the entrance of Zane Shawnee Caverns and Southwind Park. The caverns follow a system of joints through Devonian-age Columbus limestone. The tour covers about 0.4 mile, although the caverns are much more extensive and there are passageways at three distinct levels. The caverns are wet, and they continue to grow. Numerous soda straws, stalactites, stalagmites, flowstone, and draperies make this one of the better-decorated caverns in Ohio. Cave pearls, which are rare in caves, occur at three sites along the tour. They are small pebbles of aragonite (a form of calcium carbonate) that accumulate in small depressions on the tops of stalagmites and jostle like jumping beans each time a drop of water disturbs their nests. Zane Shawnee Caverns was discovered when a boy's dog fell into a hole in the woods. Around 1892, the sinkhole was widened and served as the main entrance to the caverns. Visitors rode a basket to the bottom of the sinkhole armed with candles and lanterns. Later, owners excavated the present entrance.

The Ohio Caverns—the largest in Ohio—are just southeast of West Liberty. As you drive to the caverns, look for glacial erratics used as fences and the exposed limestone bedrock; the till is thin here, too. The caverns were

Stalactites, stalagmites, and columns in Ohio Caverns near West Liberty are often a milky white color.

discovered in 1897 when locals noticed that surface water in a low area disappeared. A small section of the caverns, originally called Mt. Tabor Cave, opened for tours later that year. New passages discovered in the early 1920s led to a revision of the tour route and the present 1-mile route opened. At the deepest point, visitors are over 100 feet below the surface. The caverns have at least two levels, but the tour route follows only the upper level. The most impressive features of Ohio Caverns are the milky white stalactites and stalagmites that have red iron oxide and manganese oxide growths and stains. They are the dominant feature of the tour route through the Columbus limestone, although in places soda straws, draperies, and helictites—twisted calcite growths—hang from the ceiling. Certainly more caverns remain to be discovered under this region.

West Liberty to Springfield

Early quarries northeast of Urbana along Stone Quarry Road and south of St. Paris focused on dolomites of the Silurian-age Salina group. Workers pulled Devonian-age Columbus limestone from small quarries near Cable and Mingo east of U.S. 68. Much of this stone went to Urbana. Large sand and gravel pits, tapping the valley glacial fill, lie southwest of Urbana, east of the Mad River. The Mad River valley narrows south of Urbana, and U.S. 68 hugs the east side of the river along the edge of the Springfield moraine.

North of I-70 a roadcut exposes the cavity-riddled Cedarville dolomite. A rusty brown soil fills surface cavities where the dolomite dissolved; these are sinkholes and are typical of the Cedarville dolomite, which is a common cliff-forming rock in this part of the state. U.S. 68 crosses a rock-walled gorge that was carved by meltwater at the southern edge of these roadcuts.

Mad River Outwash Plain

The last Wisconsinan glaciers entered northern Ohio around 22,000 years ago. This ice flowed around, and eventually over, the Bellefontaine outlier. The Miami lobe, to the west, reached its maximum extent just north of Cincinnati,

U.S. 68 cuts through an erosion-resistant ridge of Silurian-age Cedarville dolomite near the Mad River crossing in Springfield. Leaching of the dolomite along vertical joints resulted in a number of sinkholes.

while the Scioto lobe, to the east, slowed to a halt at Cuba. The lobes faced one another south of the outlier, forming a narrow north-south drainage trough through which U.S. 68 now passes. As the ice fronts fluctuated between periods of growth and melting, torrents of meltwater gushed through this trough, alternately eroding and depositing outwash. The northernmost deposit of sand and gravel is called the Kennard outwash.

The lobes deposited chunks of ice in the sediment, which later melted and formed kettles. At one point the Miami lobe readvanced south and east over the Kennard outwash, coming to a halt and depositing the Springfield moraine. The fitful ice then retreated to the north and west, leaving a second gravel and sand layer, the Urbana outwash. Another advance formed the West Liberty moraine just west and south of the Springfield moraine, leaving Urbana outwash sandwiched between the two. Both lobes then readvanced: the Scioto lobe to the Reesville moraine about 6 miles east of U.S. 68, and the Miami lobe to the Farmersville moraine 6 miles west of U.S. 68. Then the ice began melting again, forming the third and lowest outwash of the region. These sediments form the most extensive outwash plain in the state. Between West Liberty and Springfield, the Mad River outwash plain reaches widths of as much as 3 miles; the outwash continues southwest to Dayton. When the ice sheets left this area 17,000 years ago, meltwater subsided and the current drainage patterns developed.

Cedar Bog Nature Preserve

Cedar Bog State Nature Preserve, with its holdover of plants and animals that flourished in Pleistocene time, lies between Urbana and Springfield in the middle of the Mad River outwash plain. To get there, follow Woodburn Road west from U.S. 68 for 1 mile to the preserve's entrance. A boardwalk trail leads into a grove of northern white cedars and other plants that are uncommon this far south. Cedar Bog is a misnomer; the area actually is a fen characterized by flowing water, underlying marl, and sedge mats. Bogs are generally ponds that are underlain by acidic organic debris and covered with sphagnum moss.

Early settlers knew the area as an impassable swamp, but naturalists from nearby Urbana noted its unique flora as early as the 1830s. In 1942 the state purchased part of the site for a nature preserve. Originally it covered over 7,000 acres; today the preserve protects 427 acres. Cedar Bog is a haven for boreal plants and animals that once existed along the borders of ice sheets throughout Ohio but now are only common in more northerly areas. Northern plants found in the fen include bog twayblade, shrubby cinquefoil, queen lady's slipper, and swamp lousewort.

Cedar Bog should have disappeared with the ice sheets, but local hydrologic conditions allowed it to survive. Geologists have different views on the fen's source of water, but they all agree that underground springs are the reason it still exists. The local water table is at the level of the Mad River west of the fen, and thus is not high enough to create a wetland. Some geologists believe that cool groundwater seeps into the fen from east of the preserve, originating in the Urbana outwash. Others suggest that it comes from a buried channel of the Mad River near the preserve. Either way, the lime-enriched water elevates the fen's local water table and leads to the precipitation of marl.

Springfield to Xenia

U.S. 68 is on Wisconsinan ground moraine between Springfield and Goes, crosses terrace deposits between Goes and Xenia, crosses the Little Miami River and Massies Creek, and ascends onto Xenia moraine sediments at Xenia. In the 1870s the railroad maintained a spring-fed water tank at Goes at a railroad station. The water seeped from the boundary between Ordovician and Silurian units. West and southwest of Xenia, meltwater in the Mad River valley carved two rocky gorges near Alpha and Bellbrook when an ice dam at Dayton diverted the water to the south. Today, the Little Miami River follows this course. Southwest of Xenia the Little and Great Miami Rivers meander through eroded end moraines and Miami lobe and Illinoian tills toward the Ohio Valley. U.S. 68 veers southeast away from the Little Miami Valley at Xenia and reenters a landscape of deposits laid down by the Scioto lobe.

Springfield Quarrying

The quarries along the Mad River just west of Springfield were major producers of quality building stone from the 1830s until 1900. Builders used much of this stone, the Springfield dolomite, locally. The Springfield dolomite formed even beds of suitable building block thickness in this area, and three beds in the lower part of the formation were suitable for cutting. Some beds that contained chert ended up in bridge piers. You can see the Springfield dolomite at the old Masonic home and in several downtown churches.

Since it was generally unsuitable for building purposes, people burned the Cedarville dolomite, which occurs on top of the Springfield dolomite, for lime or crushed it for aggregate. Limestone City, platted 1886, was mainly established to house quarry workers. Three companies employed around seventy-five men by about 1900. What's left of Limestone City lies north of Ohio 4 just west of its interchange with U.S. 68.

The Yellow Springs–Cedarville Area

Northern Greene County contains some of the best exposures of Silurian rocks in the state in rocky gorges in Glen Helen Preserve, Clifton Gorge State Nature Preserve, John Bryan State Park, Indian Mound Reserve, and along Clark Run. Meltwater flowing from the northwest across the edge of Silurian-age bedrock—the Niagaran escarpment—onto softer, younger Ordovician-age rocks carved these gorges. The Niagaran escarpment in Ohio is not as spectacular as, though it is quite similar to, the one in New York made famous by Niagara Falls. Before the glaciers came and scraped off the top of the escarpment, it may have been 100 feet high. Close inspection of the outwash north of Yellow Springs reveals that the water that deposited it was funneled into these gorges. The outwash deposit narrows in the direction of town and lines up with the gorges. Each gorge displays various Silurian rocks and has old quarries, slump blocks (rock masses that have slipped downslope after breaking away from a cliff face), waterfalls, and sometimes potholes.

Glen Helen Preserve is a nature sanctuary managed by Antioch University on the east side of Yellow Springs. If you decide to visit, be certain to visit the Trailside Museum and Visitor Center to learn of the preserve's nature activities and examine the large glacial erratic near the parking lot before you descend

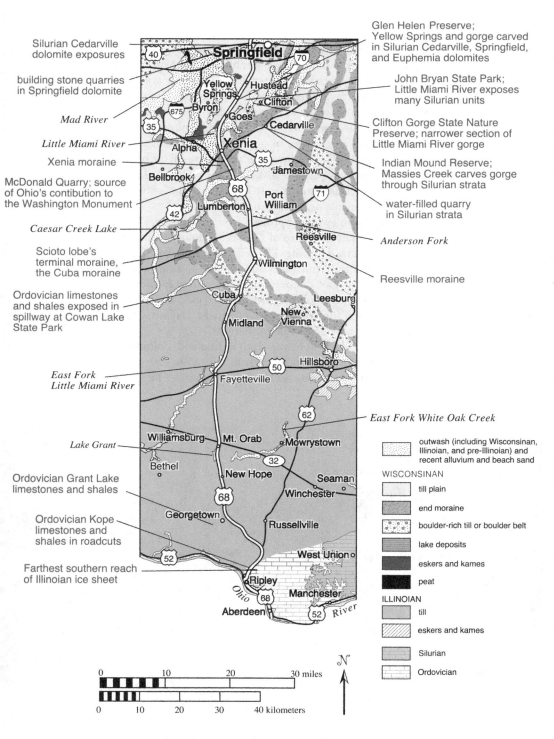

Silurian Cedarville dolomite exposures

building stone quarries in Springfield dolomite

Mad River

Little Miami River

Xenia moraine

McDonald Quarry; source of Ohio's contibution to the Washington Monument

Caesar Creek Lake

Scioto lobe's terminal moraine, the Cuba moraine

Ordovician limestones and shales exposed in spillway at Cowan Lake State Park

East Fork Little Miami River

Lake Grant

Ordovician Grant Lake limestones and shales

Ordovician Kope limestones and shales in roadcuts

Farthest southern reach of Illinoian ice sheet

Glen Helen Preserve; Yellow Springs and gorge carved in Silurian Cedarville, Springfield, and Euphemia dolomites

John Bryan State Park; Little Miami River exposes many Silurian units

Clifton Gorge State Nature Preserve; narrower section of Little Miami River gorge

Indian Mound Reserve; Massies Creek carves gorge through Silurian strata

water-filled quarry in Silurian strata

Anderson Fork

Reesville moraine

East Fork White Oak Creek

Springfield

Yellow Springs
Byron
Goes
Hustead
Clifton
Cedarville
Xenia
Alpha
Jamestown
Bellbrook
Port William
Lumberton
Reesville
Wilmington
Cuba
Leesburg
Midland
New Vienna
Hillsboro
Fayetteville
Williamsburg
Mt. Orab
Mowrystown
Bethel
New Hope
Seaman
Winchester
Georgetown
Russellville
West Union
Ripley
Manchester
Aberdeen

outwash (including Wisconsinan, Illinoian, and pre-Illinoian) and recent alluvium and beach sand

WISCONSINAN

till plain

end moraine

boulder-rich till or boulder belt

lake deposits

eskers and kames

peat

ILLINOIAN

till

eskers and kames

Silurian

Ordovician

0   10   20   30 miles

0   10   20   30   40 kilometers

N

Geology along U.S. 68 between Springfield and Aberdeen.

into the preserve. The preserve's outstanding geologic feature is a flowing spring in the Euphemia dolomite, which has been present since at least the last glacial retreat. The calcium-charged water flows out of the gorge's rock layers, forming a large mass of travertine and tufa, a porous rock encrusted with plant debris. The spring has a flow rate of 68 to 80 gallons per minute, and early on it supplied Antioch College with water. The spring is colored orange, not "yellow," because the water precipitates iron from the rocks. Because of its color, many people in the 1850s thought the water was a cure-all for numerous maladies and a health resort was developed. A dam was built across Yellow Springs Creek and a hotel flourished until the 1890s; the pond eventually filled with sediment and only remnants of the dam remain.

Near the spring, Pompey's Pillar projects from the hillside. This mushroom-shaped erosional remnant has a cap of Cedarville dolomite and a pedestal of Springfield and Euphemia dolomite. The pillar is slowly creeping into the valley, having broken loose from the wall it was attached to along a joint. In nearby Birch Run you can see 6-foot-diameter potholes. These formed when turbulent meltwater created whirlpools that swirled hard glacial cobbles that eroded the softer bedrock. Birch Run exposes the Brassfield, Dayton, Osgood, Laurel, Euphemia, Springfield, and Cedarville formations.

Quarriers worked the Euphemia and Springfield dolomites at Yellow Springs for building stone and lime as early as the 1860s. Abandoned quarries lie north of town and west of Hustead, and large quarries lie north of Byron, west of Ohio 235.

At the namesake of Yellow Springs in Glen Helen Preserve, groundwater flows from joints in the Silurian-age Euphemia dolomite. The groundwater's movement downward is repelled by the less permeable Massie shale.

John Bryan State Park lies east of Yellow Springs on Ohio 370. The Little Miami River flows in a valley bordered by cliffs of Silurian-age rocks through this park, dropping more than 130 feet. Earlier meltwater and not the current river carved most of the gorge. The Little Miami River found the gorge an easy course to follow as modern drainage patterns developed. Most of the strata deposited during Silurian time in this part of Ohio is visible in the park. About 36 feet of the Brassfield and Dayton formations form benches on the slopes of the gorge; above these strata is 24 feet of Osgood shale, 8 feet of Laurel limestone, 5 feet of Massie shale, and about a 33-foot cliff of Euphemia, Springfield, and Cedarville dolomites. The tendency of the Massie shale to wear away, removing support for overlying rock layers, and the presence of vertical joints or cracks in the cliff-forming dolomites, leads to slump blocks of dolomite sliding into and widening the gorge. If you examine the layers of these blocks, you can match them with rocks in the walls above. During wet weather and during spring, waterfalls cascade into the gorge. Groundwater also seeps into it from springs along the Massie shale and overlying Euphemia dolomite contact. Travertine deposits drape the rocks here and there below these springs.

Clifton Gorge State Nature Preserve lies upstream of John Bryan State Park, farther east along Ohio 343. Until 1973, it was part of the park. The gorge grows narrower along this stretch on east into Clifton. The same rocks that compose the gorge in the park are visible here. Geological highlights of the preserve include the 40-foot falls of the Little Miami River; Steamboat Rock, a large slump block of Cedarville dolomite sitting in the river; and potholes up to 8 feet wide that occur along the walls and edges of the gorge, unrelated to current river action.

Pompey's Pillar is a chunk of Cedarville, Springfield, and Euphemia dolomite that broke loose from the valley wall and is slowly sliding into the valley of Yellow Springs Creek in Glen Helen Preserve.

Clifton Gorge was a logical place for harnessing water power. As early as 1830 eight mills operated in the gorge, and by the 1850s the population of Clifton grew to three hundred. Only one gristmill survives, but if you look closely at the gorge walls you can see stone foundations and borings—the sites of other mills.

From Xenia, a side trip to Cedarville on U.S. 42 offers more Silurian scenery. The Cedarville formation was named for exposures on the west side of town in a quarry along the south side of U.S. 42 in 1871. This quarry supplied building stone from the underlying Springfield dolomite; later it produced crushed stone. Several quarries along Massies Creek produced lime.

Indian Mound Reserve, a Greene County park, is just west of Cedarville. Here Massies Creek flows through a rock-walled gorge similar to those near Yellow Springs. To the north and south are segments of the Xenia moraine laid down by the Scioto lobe. The waterfall at the trailhead formed when the creek was diverted by humans, but others are in their natural positions. Trails lead along the rim of the gorge, and at one point a trail descends into the gorge through a joint in the Cedarville dolomite. The trail leading down to creek level passes under an overhanging cliff of Silurian dolomite where the Massie shale has been cut back. Once out of the gorge, a trail leads to a mound of the Hopewell culture.

A Piece of the Washington Monument

Stone Road off U.S. 68, southeast of Xenia, runs past a historic farm and a quarry that dates back to the 1820s. What is now a pond at the McDonald Farm yielded Silurian-age Dayton formation rock from 1820 to 1896. People used the attractive stone widely throughout the region as a durable building material. At one time twenty men worked this shallow pit. In most of Ohio, the Dayton formation averages 4 feet thick, but it measured up to 8 feet thick in the McDonald Quarry. The McDonald Quarry is much like many other building stone quarries in southwestern Ohio in that today, few would even recognize it as a former stone quarry. However, it has a unique claim to fame.

In 1848, the Washington National Monument Society requested that each state supply it with a representative building stone to place in the Washington Monument. Quarries across Ohio competed for the honor, and the McDonald sample was chosen. It now resides on the 90-foot level and is recorded as "marble." Stoneworkers often referred to limestone as "marble" if it could be readily polished. A historical marker along Stone Road recounts this story.

Xenia to Aberdeen

U.S. 68 is on the Xenia moraine as it passes through Xenia and until it crosses Caesar Creek, south of town. Scioto lobe till plain underlies the route from here to Lumberton. U.S. 68 crosses Anderson Fork just south of Lumberton. This stream has cut a deep meandering valley into Wisconsinan-age till, complete with oxbows and meander scars. Quarriers pulled Silurian-age Brassfield limestone from Lumberton quarries into the early 1900s. In the mid-1800s, iron-rich layers northwest of Wilmington were smelted at a furnace along Todd Fork. The iron-rich zones formed when rivers eroded iron-bearing mineral zones in the remnants of the Taconic Highlands east of Ohio. The iron was

Cedarville dolomite

Springfield
dolomite

*The Cedarville dolomite
forms rock shelters along
Massies Creek in Indian
Mound Reserve at Cedarville.
Springfield and Euphemia
dolomites occur underneath
the overhanging cliff and
Massie shale is at path level.*

Euphemia
dolomite

Massie
shale

then reprecipitated in the Silurian sea covering Ohio. The highway crosses the Xenia moraine once again south of I-71 and enters Wilmington, which was built upon boulder-rich till. Quarries west of town also quarried Brassfield and Dayton limestones. Southwest of Wilmington, U.S. 68 cuts through the Cuba end moraine. At the crest you can visit Cowan Lake State Park for an opportunity to collect fossils from the Ordovician-age Whitewater formation along the spillway of Cowan Lake Dam. From here U.S. 68 traverses the oldest bedrock of the state, Ordovician strata, all the way to the Ohio River. Cuba also marks the southernmost reach of Wisconsinan glacial ice. To the south, the landscape is flat to gently rolling Illinoian till; end moraines are no longer distinguishable. Illinoian ice sheets last eroded this area over 125,000 years ago.

About halfway between Cuba and Midland, U.S. 68 bridges the valley of Todd Fork. Wisconsinan terrace deposits fill the channel just to the west, but only modern alluvium lies below the bridge. The East Fork Little Miami River passes under the route just south of Ohio 123, north of Fayetteville. Because the surface sediments along this stretch have been here longer, the streams have dissected the landscape much deeper than to the north.

Lake Grant, a reservoir that formed when Sterling Run was dammed, lies just southwest of Mt. Orab. U.S. 68 bridges White Oak Creek at New Hope. More-rugged Illinoian terrain appears southeast of Georgetown as streams flow directly into the Ohio River. Fossil-rich Ordovician-age Grant Lake limestones and shales form the valley margins and are particularly well exposed just west of Georgetown on Ohio 125. Southeast of Georgetown, U.S. 68 crosses the valley of Straight Creek. The highway follows the narrow valley of Redoak Creek onto the Ohio River floodplain at Ripley. Ordovician-age Kope shales and limestones border the highway as it enters Ripley. From Ripley to Aberdeen, U.S. 68 lies on unglaciated landscape along the Ohio River.

U.S. 127
Michigan Line—Cincinnati
195 miles

U.S. 127 passes into Ohio just north of Fayette on Wisconsinan-age till and skirts along the boundary between till plain and Huron-Erie lake plain 20 miles to Bryan. From Bryan to Van Wert, about 40 miles, it crosses the flat lake plain. The road traverses till plains, end moraines, and occasional outwash deposits of the Miami lobe to just north of Cincinnati. The only topographical relief along this stretch comes from abraded end moraines and stream valleys. The entire stretch through Cincinnati ranges over Illinoian-age or older sediments.

Fayette to Celina

Low rises to the east and west of U.S. 127 near Fayette mark the Defiance and Fort Wayne moraines, the last moraines formed by Wisconsinan ice before it left Ohio. The route follows the western edge of where Wisconsinan-age Glacial Lake Maumee rested from West Unity to Bryan. The lake formed as the ice retreated about 14,000 years ago. Sand in the fields marks former beaches and other shoreline features. Northeast of Bryan and west of the highway, Beaver Creek meanders across a wide floodplain it cut into flat till plain. U.S. 127 bridges the Maumee River south of Sherwood. The Maumee meanders across a ½-mile-wide floodplain carved into what once was the floor of Glacial Lake Maumee. Once out of the river valley, U.S. 127 crosses some of the flattest terrain in the state—lake-planed till and lake deposits—as far as Van Wert.

Thick (20 to 50 feet) sand, silt, and mud underlie the countryside along this route, and the bedrock lies buried. There were few mineral industries on the lake plain. A number of sand pits operated between Bryan and Ney; a brick company worked glacial sediments near Paulding; a tile plant at Haviland used lake clays; and quarries at Charloe, Grover Hill, Junction, Paulding, and Scott in Middle Devonian- and Silurian-age strata produced crushed stone and bridge stone. The most prominent quarry of this region lies west of U.S. 127 between Cecil and Paulding; you can see the plant from the road but not the actual quarry.

The southern shoreline of Glacial Lake Maumee coincides with Van Wert; sandy beach ridges run east and west of town. These correlate with sand ridges between West Unity and Pulaski on the northern shores of the former lake. Just south of Van Wert, the landscape becomes a more rolling till plain.

After crossing more till south of Van Wert, U.S. 127 crosses the Fort Wayne moraine between the Van Wert and Mercer County line and the St. Marys River. This is the first in a series of closely spaced end moraines the highway passes

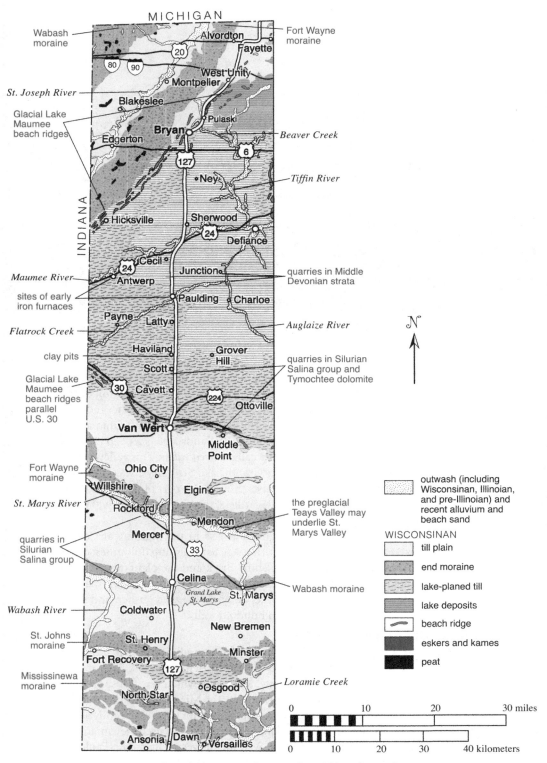

Geology along U.S. 127 between the Michigan line and Dawn.

Haviland Clay Works used glacial lake clay near the plant to fashion drainage tile, an important product in these former Great Black Swamp lands.

over on its way to Greenville. The St. Marys River hugs the southern edge of the Fort Wayne moraine, its course determined by the east-west orientation of the moraine. The river parallels the moraine all the way to Fort Wayne, Indiana, where it joins the St. Joseph River and eventually flows back into Ohio as the Maumee River. Hidden 170 to 300 feet below the landscape is a deep channel cut into Silurian bedrock, what geologists call a *buried valley*. This valley, which extends northwest and southeast from Grand Lake St. Marys, was part of a preglacial drainage system; some geologists think it was part of the Teays River drainage network but others disagree. Regardless, there was a major river flowing across this area long before the glaciers came.

The Wabash moraine marks a significant rise in the highway at Celina. Three miles west there is a quarry in Silurian-age Salina group dolomites; other quarries operated at Rockford and Willshire. Brick and tile plants worked the till at Celina and St. Henry. An oil well drilled at Ohio City in 1902 sparked further exploration. The third well workers dug there was a gusher, and soon after the area was riddled with over two hundred wells.

Iron Smelting on the Lake Plain

The Civil War was raging when iron furnaces appeared at Antwerp and Cecil around 1865 as the Great Black Swamp was slowly being drained. The smelter operators shipped ore in from Marquette, Michigan, which was transported to Toledo on Lake Erie and then to the Furnaces by the Miami & Erie and Wabash & Erie Canals. The furnaces burned charcoal made from cut trees, readily available as the Black Swamp was cleared. The Antwerp Furnace used Devonian-age Dundee limestone from the bed of the Maumee River as flux. The Paulding Furnace at Cecil received limestone from the Devonian strata workers quarried along the Auglaize River east of Junction. Initially, the war effort used pig iron, crude iron pulled directly out of the blast furnace, and then locals began using it for iron implements and horseshoes. By 1870, each furnace produced 3,000 tons annually. These isolated furnaces, destined to fail for economic reasons, had their fires extinguished in the mid-1880s.

Grand Lake St. Marys

Three storage basins, or reservoirs, maintained water level in the Miami & Erie Canal while it operated. U.S. 127 parallels the west shore of the largest of the three—Grand Lake St. Marys—south of Celina. Between 1837 and 1841, what was once a wet prairie between Celina and St. Marys was flooded after earthen dams were constructed, forming the reservoir. In the 1890s the oil boom reached Mercer County and oil rigs sprouted from the lake; at the time, drillers were oblivious to any environmental damage threat they might pose. Exploratory drilling in 1886 had released a small flow of oil and gas near St. Marys. By the next year, wells on the east side of the lake showed good gas yields and a natural gas plant was built in St. Marys. The gas ran out and the plant closed in 1893. Grand Lake St. Marys and the other two reservoirs became Ohio's first state parks in 1949.

Celina to Greenville

From Grand Lake St. Marys, U.S. 127 traverses a Wisconsinan-age till plain deposited by the Miami lobe. At the Ohio 119 interchange, U.S. 127 crosses the St. Johns moraine. Waterworn till lies south of this moraine, marking the edge of a former proglacial lake that formed south of the lake. The Mississinewa moraine, about 5 miles wide, lies just south of North Star. Dawn sits atop the narrow Bloomer moraine.

Quarries once worked Silurian-age strata at Fort Recovery, Gettysburg, and Greenville. Brick and tile plants at Celina, St. Johns, and Versailles exploited the overlying till. U.S. 127 crosses 8 miles of ground moraine and the Stillwater River, a State Scenic River, between Dawn and the Union City moraine on the northern edge of Greenville. Glacial boulders and cobbles are particularly common on the south side of this end moraine, suggesting that some of the finer

An offshore oil well in Grand Lake St. Marys circa 1905. Oil came from the Ordovician-age Trenton limestone.

Glacial erratics cleared from a farm field near Castine. West of here there is a boulder-rich end moraine.

particles were removed while larger particles were concentrated. Flat shelves in the valley near the junction with Ohio 49 are terraces left by meltwater during the waning stages of the Ice Age. West of U.S. 127, and south of the Union City moraine, valley trains with kames and outwash terraces extend as much as 12 miles down the valleys of Bridge, Mud, and West Branch Greenville Creeks. The outwash plain south of the Union City moraine probably contained a number of large ice chunks that trapped outwash between them, protecting it from the rushing meltwater. As these blocks eventually melted, hills of outwash, or kames, were left on the outwash plain. The partially channelized creeks flow past numerous sand and gravel pits. West of Fort Jefferson, old gravel pits are masked as a subdivision.

Darke County Ice Age Mammals

The Greenville area has been the center of several important discoveries of Ice Age mammals. Perhaps this is related to the fertility of the soils and highly desirable farmland surrounding the Darke County seat. Farmers tried to alleviate standing water problems by digging drainage ditches and laying drain tiles in their fields. In the process, bones of beasts that frequented this area 10,000 or more years ago were accidentally discovered.

In 1883, farmers discovered mastodon bones along Kraut Creek west of Greenville. Mammoth discoveries in this area followed: a large jaw section from the shores of Mud Creek west of New Madison, and bones from the Stillwater River and Carter Bog, both near Ansonia. The most complete mastodon was found along Bridge Creek, south of Greenville. This animal was reassembled and displayed for many years in the Dayton Museum of Natural History, now the Boonshoft Museum of Discovery. Mastodon bones are also on display in Greenville's Garst Museum on North Broadway.

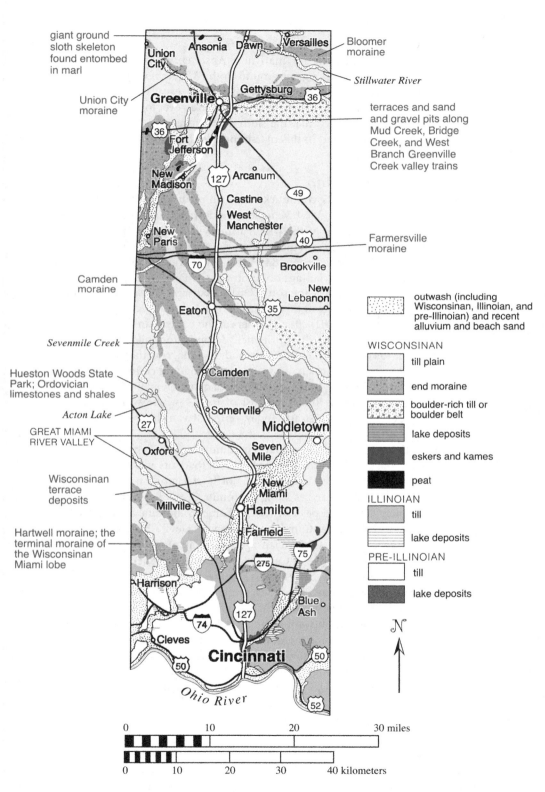

Geology along U.S. 127 between Greenville and Cincinnati.

In 1966, while excavating for marl near Ansonia, workers struck large bones again. This time it was a giant ground sloth (*Megalonyx jeffersoni*). The Dayton Museum of Natural History completed the excavation, eventually assembling the largest and most complete sloth skeleton ever found in the state. Radiocarbon dating indicated that the sloth died between 14,810 and 11,720 years ago. Other bones found in this marsh include those of giant beaver, elk, mammoth, and mastodon.

Greenville to Cincinnati

South of Greenville, U.S. 127 continues across till plain for another 12 miles, passing through Castine and West Manchester. A low rise south of West Manchester marks the beginning of the wide Farmersville moraine of the Miami lobe. The highway leaves this end moraine on the northern outskirts of Eaton. The glacial covering is thinner, about 8 feet thick, in this area; look for bedrock in stream bottoms. The Silurian-age dolomites of the Eaton area proved to be a better building material than the softer Ordovician-age rocks, and since the town was at the southern edge of these Silurian rock layers, it became a quarrying center in the mid- to late 1800s. Quarriers at the Christman Quarry, along Bantas Fork northeast of town, dug Springfield dolomite for building purposes. Builders used it to construct the old Richmond, Indiana, courthouse. Builders used Brassfield formation rocks in bridge piers and to build chimneys because of its fire resistance. By 1909, only one Eaton-area quarry remained open—the Kautz Quarry along Rocky Run, southeast of town—where workers pulled rocks from older Silurian-age strata. Today, only fast-vanishing scars of these operations are left.

Just south of Eaton, strata of Silurian time disappear and Ordovician strata compose the bedrock; watch for gray rock in roadcuts and stream banks. U.S. 127 parallels Sevenmile Creek to just north of Hamilton. Segments of the outwash fill remain as terraces along Sevenmile Creek, flat areas above the present floodplain. The road passes over the Camden moraine at Camden. If you turn west on Ohio 725 at Camden you can visit Devil's Backbone, where Paint Creek cuts through the Ordovician-age Whitewater formation, creating a high cliff. On the southern edge of Camden, ponds mark former gravel pits in the valley fill. Hueston Woods State Park is southwest of town.

Seven Mile marks the juncture of Sevenmile and Four Mile Creeks shortly before they enter the 2-mile-wide Great Miami River valley. U.S. 127 passes a number of gravel pits in outwash as it enters Hamilton. It's easy to see why floods plagued Hamilton since the downtown is on the floodplain. Note the high flood walls. A number of stone quarries operated on the west side of the river on the southwestern edge of town into the early 1900s. They quarried Ordovician-age limestones. U.S. 127 reaches the south edge of the Great Miami Valley south of Fairfield. From Fairfield to Cincinnati the highway crosses dissected till plain; the Hartwell moraine, the terminal moraine of the Miami lobe; and Illinoian till.

Hueston Woods State Park

Usually state-managed areas are off-limits to fossil collecting, but Hueston Woods State Park, southwest of Camden, allows it because of the abundance

Fossils may be collected at designated sites within Hueston Woods State Park. Slabs of Ordovician-age limestone cover the banks of Four Mile Creek.

of Ordovician-age fossils in the park's bedrock. The best collecting areas are along the dam spillway at the southeast end of Acton Lake, a reservoir created in 1956, and along small creek valleys on the north side of the park. The Ordovician Whitewater formation—the main rock exposed in the park—consists of many limestone layers separated by thin shale zones; it is some of the oldest bedrock in Ohio. The most common fossils people find include brachiopods, bryozoans, clams, snails, and horn corals. People also find straight-shelled nautiloids—squidlike animals with shells shaped like ice-cream cones—and trilobites.

OHIO 32
Peebles—Cincinnati
50 miles

Between Locust Grove and Lawshe, the bedrock underlying Ohio 32 is Devonian and Silurian in age and was never glaciated. From Lawshe to the I-275 junction just east of Cincinnati, Ohio 32 traverses 45 miles of Illinoian till and underlying Ordovician limestones and shales, which outcrop in roadcuts between Lawshe and Seaman and at Batavia.

Just northeast of Peebles, Ohio 32 passes north of Plum Run Quarry near Bacon Flat. This large water-filled quarry once exposed Silurian Tymochtee, Greenfield, and Peebles dolomites. The older Lilley dolomite forms the base of the quarry, and younger Devonian-age Ohio shale is at the surface of the quarry.

Quarries in the Silurian-age Bisher formation lie on the west side of Peebles. Between Peebles and Seaman, where Ohio 32 crosses the Spoon River, the Ordovician-Silurian boundary is visible in a roadcut on the west side of the

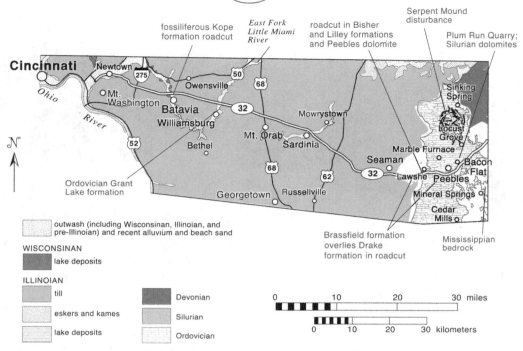

Geology along Ohio 32 between Peebles and Cincinnati.

valley. At the crossing of Ohio Brush Creek, a 25-foot-high roadcut of middle Silurian Bisher formation, Lilley formation, and Peebles dolomite occurs on the northeast side of the road. Illinoian terraces are scattered along the drainage at Lawshe and east of Seaman, and Illinoian tills stretch from Lawshe to Cincinnati. The fossil-rich Ordovician-age Grant Lake formation is exposed in the East Fork Little Miami River valley at Williamsburg, and at Batavia this river exposes fossil-rich limestones and shales of the Kope formation. Kope formation also occurs in a roadcut at the Ohio 222 junction in Batavia.

Adams County Mineral Springs

A number of sulfur and iron springs along Scioto Brush Creek and tributaries southeast of Peebles mark the contact of Silurian-age Bisher dolomite and Estill shale. Groundwater enriched in calcium sulfate, calcium carbonate, calcium chloride, and magnesium chloride seeps from the contact since it can't penetrate the shale. Between 1875 and 1915, the community of Mineral Springs, off Ohio 781, had two large hotels, guest houses, and sporting facilities to accommodate crowds of visitors seeking what they felt were healing waters. A historical plaque marks the site today. A historical marker on Ohio 247 marks the site of Rock Spring, another health resort that developed around similar springs.

Serpent Mound Disturbance

Serpent Mound State Memorial, north of Peebles on Ohio 73, marks not only a site of great archaeologic importance, but one of equal geologic significance. It is the site of one of the best-preserved effigy mounds in the United States. Radiocarbon dating indicates that it was probably built by the Fort Ancient people (AD 1000 to 1550) and not the Adena people (800 BC to AD 100) as originally thought. The mound is nearly ¼ mile long and overlooks Ohio Brush Creek. Archaeologists believe its shape represents an uncoiling serpent. Less obvious, though, is the Serpent Mound disturbance, a circular feature in the bedrock. This 5-mile-diameter enigma stretches east and northeast of the park.

Geologists first noted there was something strange about this area in 1838. John Locke, of the Ohio Geological Survey, noticed that many different sedimentary rocks, more than would normally be expected, outcropped in a small area near the boundaries of Adams, Highland, and Pike Counties. Instead of being essentially flat layers as was typical in this part of Ohio, the layers were tilted at various angles and at places broken and shattered. The first geologic map of the area was penned by August Foerste of Dayton in 1918. In the 1930s the area was described as being of cryptovolcanic origin, meaning that gases deep within the subsurface had exploded, disheveling the overlying rock strata and creating areas of both upwarp and collapse. An opposing view suggested it was an astrobleme, an ancient erosional scar caused by the impact of an extraterrestrial object.

Meteorite impact sites around the world have many of the same characteristics that the Serpent Mound disturbance does, and geologists recently confirmed that it is an impact site marked by areas of intensely broken rock, or shatter breccia; closely spaced faults; younger Ordovician-age through younger Mississippian-age strata instead of just Ordovician or Silurian rock; conelike masses of carbonate rock, called *shatter cones*, in the center; and a trace of iridium, an element associated with high pressure and impact. The characteristics are difficult to see since rock exposures are few in this rolling farm- and pastureland.

The meteorite is thought to have hit during Permian time, 299 to 251 million years ago. Although decreasing in size rapidly as it passed through the earth's atmosphere, it still was a considerable mass when it buried itself in the Permian countryside west of the relatively new Appalachian Mountains. The force of impact would have melted and vaporized the surficial sediments and rock layers, at first pushing them downward and then exploding outward with

Underneath this rolling landscape of northern Adams County, just east of the maximum extent of Illinoian ice, lies evidence of a meteorite impact. Contorted and faulted rock lies under this pastureland. Hills in the distance are composed of Silurian-age dolomites.

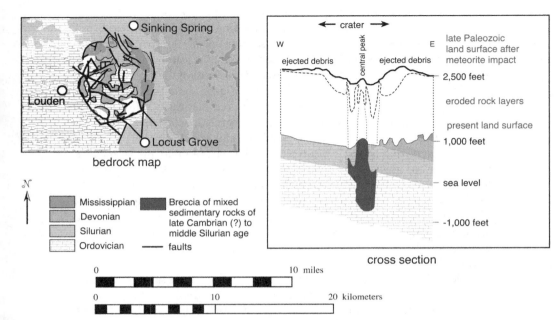

bedrock map

Sinking Spring
Louden
Locust Grove

N

Mississippian
Devonian
Silurian
Ordovician

Breccia of mixed sedimentary rocks of late Cambrian (?) to middle Silurian age
— faults

0 _____ 10 miles

0 _____ 10 _____ 20 kilometers

cross section

← crater →

W   E

ejected debris   central peak   ejected debris

late Paleozoic land surface after meteorite impact

— 2,500 feet

eroded rock layers

present land surface

— 1,000 feet

— sea level

— -1,000 feet

The Serpent Mound disturbance. The faulted Precambrian basement (Grenville province), perhaps an old rift zone, lies 2,200 feet below northern Adams County. These ancient folded and faulted rocks were covered by a mile-thick sequence of Cambrian to Permian sediments as first seas, and later rivers and streams, covered them. Sometime in late Paleozoic time, probably during the Permian Period, a meteorite crashed into the countryside on top of this ancient fault zone. The force of the impact depressed the Paleozoic strata, causing a crater to develop, and reactivated the basement faults; however, now they slipped downward instead of upward. Strata of late Ordovician time or younger rebounded after the impact, causing a domelike uplift in the center of the crater that was underlain by shattered rock, or a breccia. Since Permian time, surface erosion, mainly running water, has removed over 1,000 feet of rock, producing the present landscape.

great force (large impact structures typically have an upraised central portion). This would account for the arcuate faults surrounding the Serpent Mound Disturbance and the Ordovician-age breccia. The great pressure of the impact allowed coesite, a high-pressure form of quartz commonly associated with known meteorites, to form. The surrounding Ordovician limestone has shatter cones, which are cone-shaped series of fractures associated with the impact of rock bodies. This part of south-central Ohio is on the eastern edge of the Cincinnati Arch, so strata incline slightly to the east and grow younger in that direction. This isn't the case at the disturbance. Older rocks, the Ordovician strata, were brought to the surface, and overlying younger layers, Silurian to Mississippian in age, were steeply tilted to the side.

The disturbance consists of a central portion of Ordovician and Silurian strata uplifted as much as 1,000 feet into seven radiating anticlines. A transitional zone of folded and faulted rock, which shows little uplift or subsidence, surrounds this. A series of downfaulted synclines forms the outer part of the disturbance—about 70 percent of the structure. The impact penetrated a fault zone that already existed deep in the subsurface at the Precambrian and Paleozoic boundary; perhaps this was an earlier rift zone, an area where the continent was pulling apart hundreds of millions of years ago. Crystals of barite, calcite, dolomite, fluorite, and sphalerite, among others, formed in the surrounding surficial faults and fractures.

Pioneer Iron Smelters

In the early 1800s, smelter operators used the Silurian-age Brassfield formation along Ohio Brush Creek to make pig iron. A zone in the uppermost layers is often cemented with hematite, an iron oxide, the source of the ore. In other places along the creek, weathered stretches of bedrock contained depressions filled with limonite, up to 10 feet thick in places. Geologists believe that this brown ore formed when pyrite nodules, which are densely concentrated at this horizon, oxidized.

In 1811, workers constructed the Brush Creek Furnace at what is now Cedar Mills. In 1814 and 1816, two more charcoal-fired furnaces followed; the last, the Marble Furnace, was built just north of Ohio 32 and northwest of Peebles. The furnaces lay abandoned by 1839, put out of business by the richer ores in the Hanging Rock Iron Region of southeastern Ohio.

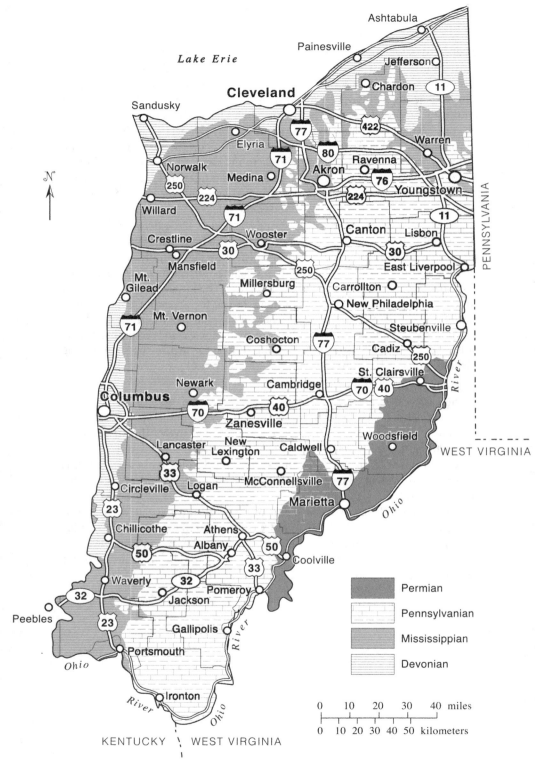

Highways and bedrock geology of the Allegheny Plateau. —Modified from Ohio Department of Natural Resources, Division of Geological Survey publication

OHIO UPLANDS—THE ALLEGHENY PLATEAU

Hilly eastern Ohio lies in striking contrast to the flat plains of the western part of the state. Mountain building that took place far from the Buckeye State in late Paleozoic time accounts for this difference. This region, known as the Allegheny Plateau, was built with the eroded sediments from the third and final Paleozoic mountain-building event, which resulted in the Appalachians.

It may seem contradictory that the Allegheny Plateau lies over the Appalachian Basin, but it is a matter of the erosional resistance of the strata. Paleozoic-age strata thicken into southeastern Ohio and adjacent parts of Pennsylvania and West Virginia, indicating that the region was slowly sinking throughout Paleozoic time, allowing more sediment to build up; the basin was a trap for sediment coming off the mountains to the east. The coarser sediment, sand and pebbles, settled out in the basin; only the mud got farther west into western Ohio. After compacting and cementing into conglomerates, sandstones, and some shales during late Paleozoic time, 350 to 251 million years

Typical Allegheny Plateau scenery; view is from Mt. Pleasant at Lancaster. This knob, composed of Mississippian-age Black Hand sandstone, and several surrounding ones escaped burial by Pleistocene ice sheets.

149

ago, the strata have been under erosional attack. Due to their general hardness and resistance to erosion and the continuing gradual isostatic rise of the folded rocks of the Appalachian Mountains to the east, the plateau has remained an upland area. The general ruggedness of the terrain is due to the downcutting of present streams and rivers. The boundary between the younger Paleozoic strata of the Allegheny Plateau and the older strata of the till plains is an escarpment, a sudden increase in elevation sometimes marked by a cliff or series of rounded knobs, like the one seen at Chillicothe.

Through time the depositional environments on the plateau alternated between river channels, floodplains, coastal swamps, uplands, brackish estuaries, and shallow sea. These different environments are reflected in the Mississippian-, Pennsylvanian-, and Permian-age rocks of southeastern Ohio. This region marks the transition from the flat till plains of western Ohio to the deformed rocks of the Appalachian Mountains in Pennsylvania and West Virginia.

Devonian and Mississippian Formations

Some 370 million years ago, a shallow equatorial sea called the Kaskaskia Sea, home to many brachiopods, bryozoans, crinoids, and fish, covered a sandy platform in eastern Ohio; deeper basins lay in central Ohio. The Devonian and Mississippian rocks record a time in which the sea grew shallower. Basin shales and delta-edge siltstones and sandstones were the dominant rocks that were deposited until the close of the Mississippian Subperiod 318 million years ago.

Devonian-age Bedford shale along Big Walnut Creek in Gahanna. Note the thin bedding and tendency to break into small plates.

Devonian- and Mississippian-age sandstone, siltstone, and shale form the bedrock along the north and west margins of the Allegheny Plateau, stretching in a northeast-to-southwest-trending, 30-to-50-mile-wide band from the Pennsylvania border in Trumbull County west to Huron County and south to the Ohio Valley in Scioto County.

Bedford Shale

About 95 feet of gray or reddish shale, siltstone, and sandstone marks the oldest visible Devonian-age bedrock of the plateau—the Bedford shale. The Bedford shale was formed on a shallow shelf and represents slowly shallowing seawater over time. The lower few feet of the Bedford formation contains burrows and brachiopod and clam molds. Siltstone beds, being more resistant to weathering, typically jut out of the formation when exposed along streams. Sandstone becomes a significant stratum in the middle part of this formation. In Cuyahoga County the sandstone was quarried and used as a building stone under the trade name of Euclid bluestone.

Berea Sandstone

Ohio's best-known building stone occurs on top of the Bedford formation. Near South Amherst, the Devonian-age Berea sandstone reaches a thickness of 250 feet in stream and river channels cut into the underlying strata. The thickness of this sandstone is highly variable in north-central Ohio, and the unit becomes discontinuous west of Berea. A series of normal faults are responsible for the irregularity. Normal faults involve the relative movement of blocks of strata down inclined cracks that develop because of stresses within the earth. This causes blocks that were once laterally continuous to be offset. Ripple marks, cross-beds, and other sedimentary structures indicate that this sandstone was deposited on the inner part (closer to the shore) of a large delta lobe originating to the north of present-day Ohio. The weight of sand on the underlying soft mud (the Bedford shale) caused gentle folding in the surrounding sediments. This is called *soft-sediment deformation*; it happens before the sediments become rock. Sometimes mud pushed up through cracks in the sandstone, forming shaley "dikes," much as molten magma forms dikes in sedimentary rocks. This soft-sediment deformation by mud may also account for the tilted beds often seen in quarries of northeast Ohio. Occasionally people find bryozoans, shark teeth, and plant debris, but fossils in this sandstone are generally rare.

In the quarry district around Amherst and Berea, the lower portion of the Berea sandstone is a deformed, silty sandstone. This unit shows ripple marks, occasional burrows, interbeds of shale, and more soft-sediment deformation, including rounded lumps of sandstone extending into underlying shales, and overturned and wrinkly beds. The unit reaches thicknesses over 30 feet, and it represents deposition that occurred on the outer part of the delta, farther out to sea. Similar features characterize the modern Mississippi Delta.

Outside the quarry district, the Berea sandstone is a widespread sandstone-siltstone, varying from 25 to 60 feet thick. A predominance of cross-beds, ripple marks, and other sedimentary features suggests it was deposited in estuaries in northern Ohio and offshore basins in southeastern Ohio.

Sunbury Shale

About 359 million years ago, a pause in delta front deposition in eastern Ohio as the sea grew deeper allowed mud and organic material to accumulate again in quiet water of deeper basins. This marks the transition from Devonian into Mississippian time. Above the Berea sandstone is the 20- to 40-foot-thick Sunbury shale, a black shale very similar to the Ohio shale, but of Mississippian age.

Cuyahoga Formation

About 355 to 350 million years ago, a new episode of delta growth marked the next Mississippian-age unit. The Cuyahoga formation varies widely in thickness and rock type, and time relationships between its members are not clearly understood. In northern Ohio, its type area, the formation is dominantly composed of shale that is locally rich in fossils, and it consists of the Orangeville, Sharpsville, and Meadville members. In central and southern Ohio, it changes to sandstone. The Black Hand sandstone member reaches a thickness of 300 feet in the Hocking Valley; the entire Cuyahoga formation may be 625 feet thick here. The Black Hand sandstone is a prominent cliff-forming rock and often exhibits striking cross-bedding, which is accentuated by weathering that causes

Mississippian-age Black Hand sandstone at the crest of Allen Knob near Lancaster. Honeycomb weathering is typical of certain parts of this widespread formation.

harder layers to stick out beyond those that have eroded away. Farther south, the older Buena Vista sandstone is the dominant member of the Cuyahoga formation. Both the Black Hand and Buena Vista sandstones were once important building stones.

Logan Formation

In central and southern Ohio, a 200-foot-thick sandstone formation overlies the Cuyahoga formation. The Logan formation contains the Berne conglomerate, Byer sandstone, Allensville conglomerate, and Vinton sandstone members and represents a time when the sea was growing shallower and deltas were expanding. Locally the members can contain molds of clams, brachiopods, and crinoids, and various trace fossils.

Conglomerates are sedimentary rocks that contain the coarsest particles and thus represent deposition by very strong currents or waves along a beach or in a stream channel; they usually form close to where the sediments are produced by weathering. The particles are mainly quartz, a very hard and resistant mineral that can stand up to swirling currents and crashing waves. The particles generally have rounded shapes; sharp points and edges tend to be worn down by the turbulence. In stream and river environments this sediment will be restricted to active and abandoned channels; on beaches it will be in the area methodically washed by waves.

Maxville Limestone

The youngest Mississippian-age formation lies in unconformable contact with the Logan formation and is visible from near Zanesville to Logan. The Maxville limestone is discontinuous and averages 50 feet thick where exposed. Unconformities, or erosional surfaces, separate it from underlying and overlying strata and also occur within the formation itself, representing periods when this stratum was being eroded. A few scattered outcrops occur from Hamden to near Wheelersburg in southernmost Ohio. Perry County has many exposures. This limestone usually exhibits thick beds and contains bryozoan, brachiopod, clam, and snail fossils. The Maxville limestone was formed during mid- to late Mississippian time (some 330 to 318 million years ago) when delta growth was apparently stalled and marine deposition took over for a time. The many unconformities suggest that some of the deposition took place in very shallow water and at times perhaps on tidal flats. Due to the limited occurrence of thick limestone in the Allegheny Plateau, this limestone was an important quarry stone.

Pennsylvanian Formations

Beginning around 318 million years ago, another episode of folding, faulting, and volcanic eruptions affected the eastern margin of North America, pushing up the Appalachian Mountains and rejuvenating the streams and rivers flowing to the sea that inundated much of western Ohio. This happened when Africa collided with Laurentia, making the supercontinent Pangea. Deltas extended across Ohio, forming what is the plateau today. Contrary to the uplift, southeastern Ohio and neighboring West Virginia and Pennsylvania were part of the slowly sinking Appalachian Basin, which served as a sediment trap. Thus a small finger of the western interior sea still periodically covered southeastern Ohio until late Pennsylvanian time. The Pennsylvanian rocks in Ohio are a complex of channel sandstones and siltstones; floodplain and back-swamp shales (poorly drained parts of a floodplain); coal seams; underclays (clay seams below coal seams); freshwater limestones; and marine sandstones, shales, and limestones. The rocks occur in repetitive sequences bounded above and below by unconformities; some geologists call the sequences *cyclothems*. Other geologists refer to them as *sequences* of deposition that are bounded by paleosols (ancient buried soil horizons) within clay layers below coal seams. The cyclothems and/or sequences reflect fluctuating sea level during late Paleozoic time.

A typical cycle in eastern Ohio began with a coastal alluvial plain undergoing weathering and soil development. At the same time, mud was being deposited on floodplains and sand in stream channels meandering across sloping areas of land that extended across eastern Ohio. As the streams meandered across this area, the channels gradually shifted in the direction of each meander, eroding the outer riverbanks and making sandbars on the opposite sides. This led to meander cutoffs and oxbow lakes. Abandoned channels filled in with mud as new parts of the floodplain became channels. Calcareous sediments precipitated in the oxbows and ponds on the floodplain, eventually becoming freshwater limestones. Floodplain muds eventually became

Sketch of a swamp forest during Pennsylvanian time depicting the typical plants and animals that are now found as fossils in the rocks of eastern and southeastern Ohio.

nonmarine shale, and channel sands became nonmarine sandstones. Marshy and swampy areas on the floodplains filled with organic-rich sediments, which later became coal. Areas of wet and flooded soils became underclay, often containing plant fossils. Erosion of certain areas and consequent flooding by the sea led to the deposition of marine sands and muds, which became sandstone and shale; sometimes these beds contain fossils of sea creatures.

In other areas, the surface of the land was undergoing weathering and erosion, creating an unconformity that would eventually be buried by future sediments of some sort. Because each cyclothem is unique and sedimentation rates and thicknesses of the constituent rock units would have been dependent on local conditions, it's difficult to estimate the length of time it took for a single cyclothem or repetitive sequence to develop, but certainly we are looking at an interval of a couple hundred thousand years.

Geologists recognize four units of Pennsylvanian strata mainly based on minable coal seams—the Pottsville, Allegheny, Conemaugh, and Monongahela groups. Many of the recognized units of these groups only occur locally and cannot be traced from one exposure or mine to another. These strata form the bedrock in a 40- to 70-mile-band from Geauga County south to Lawrence County.

Ohio lay near the equator during Pennsylvanian time, so the climate was warm and moist. Pennsylvanian plant fossils show few growth rings and had large leaves and thin bark, suggesting tropical to subtropical conditions. Coastal swamp forests were dominated by plants called *lycopods*, or *scale trees*, with fernlike foliage. Some of these plants reached heights of 100 feet. The understory contained tree ferns up to 30 feet high, seed ferns (fernlike plants with distinctive seed pods), ferns, early scouring rushes, and ancestral conifers. The forests teemed with insects; fish inhabited the lakes, ponds, streams, and bays. Amphibians and reptiles were at the top of the food chain. With the exception of plant remains, fossils of the land dwellers are rare. Fragments of cockroaches, a few over 3 inches long, are perhaps most common. In later Conemaugh time, a change in plant fossils, including a reduction in lycopods and expansion of ferns, suggests that the climate became drier. The Pennsylvanian rocks pass imperceptibly into Permian deposition 299 million years ago, leading to questions about the age of the latest Paleozoic rocks. Geologists disagree about what strata represent latest Pennsylvanian deposition and earliest Permian deposition.

Pottsville Group

The Pottsville group is mainly a sequence of shales and sandstones, beginning with the Sharon sandstone and ending with the Homewood sandstone. In between are various coals, clays, limestones, and iron-rich horizons that the iron industry took advantage of. The Sharon, Quakertown, and Lower Mercer coals are the most prominent coal seams, but they are discontinuous and often not of minable thickness. Of the three prominent sandstones, builders used the Massillon sandstone widely as a building stone. Brachiopods, clams, snails, and crinoids are common fossils in several limestones and shales of this group. The Pottsville group varies in thickness from 100 to 350 feet, mainly due to its unconformable contact with the underlying Mississippian Maxville limestone. Where the Maxville limestone is present, the Pottsville group is thinner;

in places where the limestone was eroded away, the Pottsville group is thicker. It stretches along the western edge of the Allegheny Plateau from the Youngstown area to Geauga County, and then south to eastern Scioto County on the Ohio River. The group is thickest in southern Ohio.

Allegheny Group

Mainly limestones, shales, clays, coals, and some iron-rich units form the next group of Pennsylvanian-age rocks. The Allegheny group is a major source of coal and clay where it is present. The Brookville coal forms the base of this group, and the Upper Freeport coal forms the top. Minable coals include the Brookville, Clarion, Lower Kittanning, Middle Kittanning, and Upper Freeport seams. Major clay-bearing units include the Clarion, Lower Kittanning, Oak Hill, Middle Kittanning, Lower Freeport, and Upper Freeport. Builders used the Clarion sandstone to construct many of the early iron furnaces of the Hanging Rock Iron Region, which spans from Hocking County south to Greenup County, Kentucky. The marine limestone beds of this group contain brachiopods and mollusks; the shales overlying coal seams contain plant fossils. The Allegheny group forms a band along the western Allegheny Plateau from Mahoning to Scioto County, varying in thickness from 175 to 280 feet.

Conemaugh Group

Overlying the Allegheny group are similar strata. The lowest unit in the Conemaugh group is the Lower Mahoning sandstone, and the uppermost unit is the Summerfield limestone. Sandstones, shales, and clays are typical, with a few thin coal seams. The lower portion of this group has a few thin limestone beds, marking the waning stages of shallow seas in eastern Ohio. Brachiopods, clams, snails, and crinoids are found in some of the marine units; fossil plants occur with the coals. The Conemaugh group stretches from Columbiana County

Pennsylvanian-age Ames limestone forms a prominent cliff among the Conemaugh strata exposed in this railroad cut near Broadacre.

millions of years ago				
2	**QUATERNARY**			
251	PERMIAN	*Dunkard group*	Greene formation	
			Washington formation	Upper Marietta sandstone Creston Red shale Lower Marietta sandstone Washington coal Mannington sandstone Waynesburg sandstone
299	PENNSYLVANIAN		*Monongahela group*	Waynesburg coal Uniontown coal Benwood limestone Upper Sewickley sandstone Meigs Creek coal Fishpot limestone Pomeroy coal Pittsburgh coal
			Conemaugh group	Summerfield limestone Connellsville limestone Morgantown sandstone Skelley limestone Ames limestone Saltsburg sandstone Cow Run sandstone Portersville shale Cambridge limestone Buffalo sandstone Brush Creek limestone Mahoning coal Mahoning sandstone
			Allegheny group	Upper Freeport coal Upper Freeport sandstone Lower Freeport coal Washingtonville shale Middle Kittanning coal Columbiana shale Lower Kittanning coal Vanport limestone Clarion coal Putnam Hill limestone Brookville coal
318			*Pottsville group*	Homewood sandstone Upper Mercer limestone Lower Mercer limestone Lower Mercer coal Boggs limestone Massillon sandstone Quakertown coal Lowellville limestone Sharon coal Sharon sandstone/conglomerate

Generalized geologic column with some of the more common rocks of Pennsylvanian and Permian age that occur on the Allegheny Plateau in Ohio.

southwest to Lawrence County, ranging in thickness from 350 to 500 feet. The upper half of the Conemaugh group suggests terrestrial conditions and a withdrawal of the Pennsylvanian sea from Ohio.

Monongahela Group

Red and green clays and shales, light-colored freshwater limestones, continuous seams of minable coal, and scattered massive sandstones characterize the youngest Pennsylvanian rocks of the plateau. The Pittsburgh coal forms the base of the Monongahela group and the Waynesburg coal forms the top; in between are the Pomeroy, Meigs Creek, and Uniontown coals. Because it was widespread, relatively close to the surface, and of a consistent thickness, the Pittsburgh seam was and remains the most important coal seam in North America. This group represents deposition that occurred under terrestrial conditions—plant fossils become more indicative of upland flora—and grades into more terrestrial sediments of Permian age. Limestone layers marking deposition in freshwater ponds and lakes are prevalent. The Monongahela group forms a broad band from Jefferson to Lawrence County, just west of the Ohio Valley. It averages 250 feet thick.

Coal Formation

The accumulation of a thick organic mat on the forest floor and in estuaries, swamps, abandoned stream channels, and ponds typified Pennsylvanian time. Chemical processes and the increasing weight of overlying sediment compressed this humus into peat where the water table remained near the surface and oxygen was deficient, retarding the destructiveness of soil bacteria and fungi. As the peat warmed at depth, it slowly transformed into a brownish rock called *lignite*. The lignite, in turn, slowly became subbituminous coal, and finally, bituminous coal. It's estimated that Ohio peat formed at rates of 1 foot every 175 to 200 years, and it took about 10 feet of peat to make a 1-foot coal seam over about 2,000 years. Through time, shifting streams carved deep gashes in the coal and other units, making the correlation of Pennsylvanian and Permian strata a challenge in many areas.

Over fifty seams of coal are known in Ohio, but many are too thin or impure to be successfully mined. In the late 1800s, mining geologists assigned the minable seams numbers. Number 1 coal was the Sharon seam, the oldest Pottsville group seam, and the youngest was Number 12, the Washington coal of Permian age. Although this numbering scheme ended long ago, the coal industry in Ohio still uses it today.

Permian Formations

Around 299 million years ago, Ohio was still part of a vast plain gently sloping to the seashore nearly 1,000 miles to the west. Except for a drier climate, the terrain was similar to that of Pennsylvanian time. Because of this, some geologists suggest that Ohio's Permian rocks were actually deposited in late Pennsylvanian time. The good indices of Permian strata are only found in marine layers, of which Ohio has none, so the age of these rocks remains in question.

Ohio's Permian-age rocks are best known for scattered occurrences of excellent amphibian and reptile fossils, especially sail-back reptiles. Permian limestones, coals, sandstones, and red and buff shales cap the uplands above the Ohio Valley from Belmont County to Meigs County. Collectively, they compose the Dunkard group, which is divided into the Washington and Greene formations.

Washington Formation

Sandstones, red shale, coal, and freshwater limestones are Ohio's oldest Permian-age rocks. Variable thicknesses of sandstone today mark deposition that occurred in stream channels. Sometimes strong currents deeply scoured the channels, leaving thin layers, while shallow, slow-moving water left thicker deposits. The pinching out of sandstone layers in a lateral direction marks what was the edge of a channel. Mud and silt deposited on the banks and floodplains of streams eventually became shale and siltstone. Some shales oxidized during exposure, leading to their reddish color. They are commonly referred to as *redbeds*. Calcium carbonate precipitated in many lakes and ponds, forming discontinuous beds of freshwater limestone. In swamps and other wetlands the accumulation of organic matter led to peat formation; eventually, with burial, thin coal seams formed.

The Washington formation thickens from 270 feet in Belmont County to 380 feet near Marietta. A conspicuous unit is the Waynesburg sandstone, a massive, cross-bedded, gray, cliff-forming rock marking the base of Permian strata. Redbeds dominate the upper part of the formation near the Ohio Valley, leading to landslide hazards throughout the area. These clay-rich beds often

A pelycosaur, or sail-back reptile, was one of the larger inhabitants of the swamp forests of Permian time in ancient Ohio.

take on more water than surrounding strata and become horizons along which landslides occur. Farther north limestones replace the redbeds.

Greene Formation

A mixture of sandstone, coal, and limestone is the youngest bedrock in Ohio. It was deposited in an environment that was similar to the Washington formation's. The Greene formation caps the highest hills from Belmont to Meigs County.

Coal Mining

As early as the 1740s, settlers knew that coal existed in Ohio; however, the first recorded coal production dates back only to 1800. Certainly it found local use long before this date. Early settlers excavated back into hillsides and streambanks by pick and shovel to get at coal they observed at the surface. They started these coal banks to simply supply their homesteads with fuel.

Commercial mines developed in the 1860s as railroads being built across the state demanded coal and, in turn, provided efficient transportation from mines to consumers. At the same time, the fledgling iron and steel industry spread through the Cuyahoga, Hocking, and Mahoning Valleys. Early surface mines led to drift mines that followed seams horizontally into hillsides. Since the Pennsylvanian strata are tilted to the southeast, the result of the sinking of the Appalachian Basin that started during Ordovician time, slope mines, which followed inclined seams deeper into the earth, evolved. Another type of underground mine involved sinking a vertical or inclined shaft to a certain depth where miners then followed a buried coal seam. Most companies used an underground method called *room and pillar*, in which miners created a grid of rooms, leaving 30 to 50 percent of the coal as columns to support the roof of the mine. Underground mines dominated the industry until the 1940s.

Mechanized surface mining started in the 1880s, but it did not become common until the late 1940s. Earthmoving equipment removed overburden and shoveled the underlying coal into waiting railcars and trucks, increasing the

A typical underground mine, Ohio Collieries Mine No. 266 mined the Middle Kittanning coal between 1888 and 1926 near Glouster.

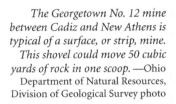

The Georgetown No. 12 mine between Cadiz and New Athens is typical of a surface, or strip, mine. This shovel could move 50 cubic yards of rock in one scoop. —Ohio Department of Natural Resources, Division of Geological Survey photo

efficiency of the mining operation and eliminating the hazards of explosions, toxic gases, and collapse, which were common with underground mining. The cost, however, was the destruction of the original landscape. Through the years, strip-mining equipment became gargantuan; by 1970 draglines and shovels could move hundreds of tons of overburden in one bite.

To remain competitive with strip mines, some underground mines turned to the *longwall method*, in which miners excavated a rectangular mass of coal, as much as 3 miles by ⅛ mile, along a long coal face, leaving little roof support. This allowed them to recover nearly 80 percent of the coal. After an area was mined out, miners removed temporary roof supports they had put in place and allowed the area to slowly collapse. Where coal seams were too thin and overburden too thick for conventional mining, auger mining found use. This method involved drilling into coal seams with a machine-mounted auger bit, just as a carpenter would auger into wood. The coal would spiral out along the bit. Miners could then move the machine laterally and drill another hole.

Ohio coal production remained minor compared to other U.S. regions until the mid-1800s, when it began to rise, peaking during World War I with nearly 48 million tons in 1918. After falling during the postwar years, production rose again to a record 55 million tons in 1970. Environmental constraints placed on the industry and cheaper fuel sources led to an overall decline in production after 1970.

Environmental Problems and Reclamation

Mining is by nature a destructive process and leaves lasting effects on the earth's surface. Underground mining leads to subsidence over abandoned workings, dangerous open shafts and drifts, and acidic mine drainage. Many early mines were not charted, or old maps have long since disappeared, so the below-surface configurations of many Ohio mines are unknown. The decay of wooden support timbers and removal of coal pillars can lead to collapse, causing the overlying ground to sink. Improperly stabilized shafts can collapse suddenly. Exposing long-buried layers of coal to air and water leads to oxidation and the release of sulfurous acids into surface streams. The orange and yellow colors of the water and bank materials marks the precipitation of iron hydroxide (yellow boy) and other compounds, which are toxic to many organisms. Yellow boy comes from the weathering of pyrite (iron sulfide) in waste rock from mining operations or from flooded mine tunnels. Obviously little can live in an environment with these toxins.

Surface mining scars the landscape as well, but modern methods of reclamation can eventually restore the surface to premining conditions. This was certainly not the case in earlier times, when some companies left open pits and rock waste piles, which eroded rapidly to badlands. Ohio's first surface mining legislation, the Strip Coal Mining Act passed in 1947, was somewhat weak but a step in a direction toward responsible mining practices. Today, Ohio's mining laws rank among the strictest in the nation. Included among the current legislation's many regulations is the posting of significant bonds for each acre mined, which ensures the restoration of a mine to approximate premining landscape; and the successful revegetation of a mine area within five years after a mine has closed.

The Clay Industry

Coalfields yielded another important natural resource—clay and shale to make pottery, bricks, tile, and other ceramic materials. Although Devonian- and Mississippian-age shales and Pleistocene glacial clays found throughout much of Ohio were important locally, the vast majority of Ohio's ceramic products came from the Pennsylvanian underclays (clay directly below most coal seams) and shales of the Allegheny Plateau. Ohio was the nation's leader in clay and shale production for most of the twentieth century, reaching a peak of over 650 million tons in the mid-1950s.

Clay and shale come from both surface and underground mines, and they are often mined with coal. Industry can use clay directly from the ground, but shale needs to be crushed first. Clay becomes plastic when it is mixed with water, making it easy to shape when wet, but it hardens with drying and heating. A type of clay called *fireclay* can withstand high temperatures without deforming,

The Continental Pottery, one of a number in East Palestine, circa 1907.

so Ohio industries used it to make bricks for furnaces and stacks. Flint clay is a hard variety of fireclay that lacks the plastic properties of other clays. The fireclays mainly come from underclay and shale layers within the Pennsylvanian strata of Ohio.

Early settlers viewed all clays and shales as potentially useful, but by the late 1800s companies focused on nine Pennsylvanian-age layers: the Sciotoville, Lower and Middle Mercer, Tionesta, Brookville, Clarion, Lower and Middle Kittanning, and Oak Hill clays. The Allegheny group of middle Pennsylvanian time was the greatest clay-bearing unit in Ohio. It included the Lower Kittanning clay, which was Ohio's most important clay because it was the most widely available clay with acceptable characteristics; Ohioans used it in sewer pipe, fireproofing, and firebricks and building bricks. Records indicate that locals started using it as early as 1826. Clay and shale industries are still important to southeastern Ohio's economy. In 2004, companies reported nearly 2.3 million tons of production, and Ohio ranked eighth out of forty-one producing states in fireclay production.

NOTABLE PENNSYLVANIAN CLAYS AND THEIR USES

Clay	Use
POTTSVILLE GROUP	
Lower and Middle Mercer clays	Early stoneware, building bricks and blocks, and sewer tiles.
Sciotoville clay	Refractories and heat-resistant ceramic materials.
Tionesta clay	Most important clay of this group. Many uses; often mixed with other clays to enhance color and other properties.
ALLEGHENY GROUP	
Brookville clay	Stoneware, building bricks and blocks.
Clarion clay	Firebrick, sewer pipe, and various refractories.
Oak Hill clay	Refractory ware.
Middle Kittanning clay	Pavers, face brick, and fireproofing material.

The Pleistocene Ice Age

Ice sheets of Pleistocene time battered the edges of the Allegheny Plateau during the last 2.5 million years. Silt and clay of the pre-Illinoian glacial stage, deposited in ice-dammed lakes over 240,000 years ago, fill many valleys along the Ohio River from Cambridge to Portsmouth. The exact boundaries of the earlier ice sheets are in question, but eroded remnants of ground moraine from Illinoian time, 230,000 to 125,000 years ago, stretch from central Columbiana County to just south of Canton. A wider patch of Illinoian ground moraine extends from the Bellville area south to Junction City and then southwest from Lancaster to Chillicothe.

After a long interval of ice-free conditions when the climate warmed and animals and plants reestablished themselves, Wisconsinan-age ice sheets advanced in three lobes—the Grand River, Cuyahoga, and Killbuck—across the northern Allegheny Plateau. Geologists can chart at least two major advances of Wisconsinan ice with intervening minor advances and retreats. The moraine sequence of western Ohio is better known; geologists are still trying to understand the glacial activity that occurred along the plateau. Ice, like water, will seek the path of least resistance, and therefore it extended farther south in Ohio where the terrain is lower and covered by softer sedimentary rocks.

Glacial stratigraphy is complex, and geologists are still refining their understanding of it. Most people will notice little difference between the various tills and outwashes in Ohio; it takes the experienced eye of a glacial geologist and sometimes radiometric dating to distinguish them. The best record is of units from 24,000 to 16,000 years old, those that haven't been eroded away. Areas that existed between ice lobes filled with a mixture of ice- and water-deposited debris and often contained broken chunks of ice that slowly melted away, leaving pockmarks, or kettles, in the ice-free landscape. Kettles often later became the ponds and lakes that we see on the edges of the plateau today. During the melting of the ice sheets, numerous lakes backed up in valleys of the Allegheny Plateau, especially between moraines and the edge of the ice sheet. These bodies of water deposited lake clays and silts. Of course, after Wisconsinan ice left the Lake Erie Basin, ancestral lake waters covered the present-day Ohio lakeshore from Conneaut to Cleveland. The ice also modified the courses of the Cuyahoga, Chagrin, Grand, and other rivers, many of which cut new paths flowing parallel to the ice edge and moraines during glacial retreat. Many north-flowing rivers flowed south after the ice retreated.

The Teays Drainage

Water-well drilling records across the state record the depth to bedrock and the thickness of overlying glacial deposits, as do geophysical techniques, such as bouncing radar waves off the bedrock surface or penetrating it with seismic waves. By interpreting these records geologists have deduced that the bedrock surface underlying the sediments and soils of Ohio is irregular. Troughs carved in the bedrock mark earlier stream and river courses that developed in preglacial times, millions of years ago. It's difficult to imagine the landscape of Ohio before the glaciers came—before there was a Lake Erie and Ohio River, the receptacles of Ohio's present drainage. Midwestern geologists of the late 1800s, including a number associated with the Ohio Geological Survey and Ohio colleges and universities, tackled the puzzling complex of old and new stream valleys in the Allegheny Plateau, trying to unravel its mystery.

The most notable drainage in the central and southern Allegheny Plateau of Ohio was the Teays River system, which formed during Tertiary time, 65 to 2 million years ago. During this time, the river had a source in what is now northwestern North Carolina, flowed north through Virginia and West Virginia, and entered Ohio at Wheelersburg. From Wheelersburg north to Petersburg, the ancient Teays Valley is prominent at the surface; smaller streams and ditches occupy its flat bottomlands today. At Petersburg the valley heads west to Piketon, where it intersects the more modern Scioto Valley. The Teays and Scioto Valleys are superimposed as far as Chillicothe, but the larger and deeper Teays Valley is filled with glacial debris from Chillicothe northward before disappearing from sight under the glacial sediments of western Ohio.

Across Ohio some 150 to 500 feet of sediment fills the buried valley. Studies of well records and other drillings have led geologists to deduce two possible northern routes of the Teays River from where it is no longer visible: one route would have it traveling northwest across central Ohio to Mercer County and Indiana, where it more or less followed the route of the Wabash River to Lafayette, Indiana, and then across central Illinois to the join the route of the Mississippi River; the other has it traveling north to connect with the northeast-flowing Erigan River, which drained what is now Lake Erie and Lake Ontario. The river's course may have been 800 miles long depending on which one it followed.

The first Pleistocene ice sheet blocked the northward flow of the Teays River, which caused a lake to form, flooding the valleys across southern Ohio and neighboring Kentucky and West Virginia. During 6,500 years or so, fine sediment accumulated in this lake, Glacial Lake Tight, named after Denison

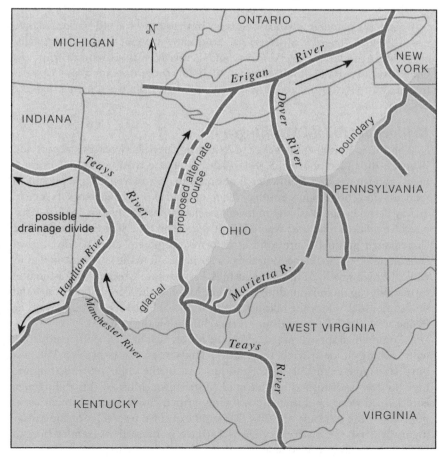

The preglacial drainage of Ohio. —Modified from Ohio Department of Natural Resources, Division of Geological Survey publication

University professor William George Tight, to become what is known as the Minford clay (a clay that would become important to several Ohio industries). Through drilling records, the Minford clay is shown to reach thicknesses of over 250 feet under Madison County and averages around 80 feet in southern counties. Glacial Lake Tight eventually rose to higher levels, covering some 7,000 square miles of southern Ohio and adjacent parts of Kentucky and West Virginia, forming new outlets to the south and cutting new channels through the plateau that were lower in elevation than the Teays system. This new drainage system was called the Deep Stage, the beginning of the Ohio River system. After the Deep Stage, much of Ohio's drainage pattern shifted from northerly to southerly.

INTERSTATE 70 AND U.S. 40
Bridgeport—Buckeye Lake
96 miles

I-70 generally parallels the oldest highway in the state—the old National Road, now U.S. 40. I-70 and U.S. 40 traverse around 80 miles of unglaciated Allegheny Plateau underlain by Permian- and Pennsylvanian-age strata from the Ohio River to east of Gratiot. Illinoian till masks the surface for about 12 miles from east of Gratiot to east of Jacksontown. Wisconsinan sediments underlie Buckeye Lake.

Bridgeport to Old Washington
I-70 and U.S. 40 follow the valley of Wheeling Creek between Bridgeport and Blaine; I-70 is high above U.S. 40, which lies on the twisting floodplain for 5 miles. Sandstones and freshwater limestones of Pennsylvanian age jut out in roadcuts along both routes. South of the routes and at Brookside, workers used to quarry Permian-age Washington formation limestone. Pennsylvanian-age Pittsburgh coal lies at road level where U.S. 40 leaves the Wheeling Creek valley. The Conemaugh group Bellaire sandstone forms prominent cross-bedded brown cliffs along U.S. 40 in Blaine. Wheeling Creek heads off to the northwest at Blaine, while the highways head west into the hills. Permian-age Greene and Washington formations cap the surrounding hills, while Monongahela group limestones and sandstones stick out of the hillsides. Pittsburgh coal lies at road level west of the Blaine area, and Bellaire sandstone forms cliffs below the bridges.

Commercial sprawl spreads along the routes near St. Clairsville, competing with mining at destroying the natural landscapes. Cross-bedded Waynesburg sandstone occurs along U.S. 40 just east of the Ohio 149 junction and tops the few undisturbed hills around Morristown. Most of this rolling area is reclaimed mine land that is now being farmed. Some Pennsylvanian rocks poke above the roadside at I-70's Ohio 149 exit. Five reservoirs of the Muskingum Watershed Conservancy District stretch from north of Hendrysburg to Dellroy. More strip-mined land dominates the landscape near I-70's Ohio 800 exit. The Egypt Valley Wildlife Area utilizes reclaimed land near Hendrysburg.

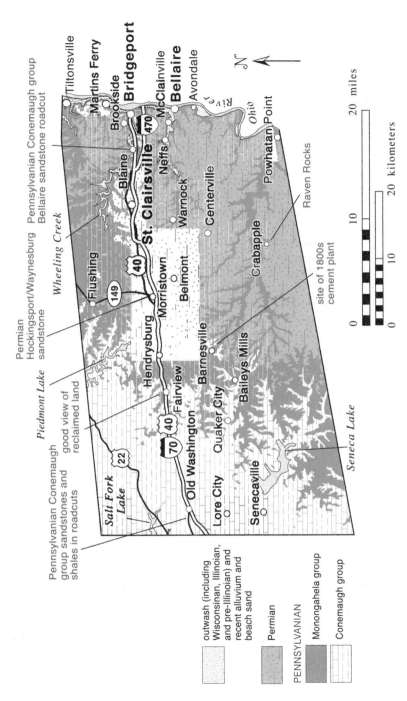

Geology along I-70 and U.S. 40 between Bridgeport and Old Washington.

South of I-70 and U.S. 40 there are shaft and drift mines in Pittsburgh and Upper Freeport coals at Baileys Mills, Kings Mine, Lore City, Quaker City, and Senecaville. Most of them were abandoned by the early 1900s. The largest of the Muskingum Watershed Conservancy District's reservoirs, Seneca Lake, is south of Old Washington.

Belmont County Coal and Other Mineral Resources

Belmont County contains widespread coal. Since 1905 this county has been the leading producer in the state. Settlers mined coal for decades before the first commercial operation opened around 1830 at Bellaire. Of the fifteen coal seams that occur in the area, most are thin and inconsistent; only two have seen significant use—the Pennsylvanian-age Meigs Creek and Pittsburgh coals of the Monongahela group. Early on, miners followed these seams back into the hillsides with slope mines and also tapped them from above with deep shaft mines. Into the 1900s miners worked the coal by pick, shovel, and dynamite, loading it into mule-powered carts, and eventually hoisting it to the surface. Coal was floated down the Ohio River by flatboats and later barges from Bellaire, Bridgeport, Martins Ferry, and Powhatan Point. The opening of the Central Ohio Railroad in 1858 allowed mining companies to ship coal to Zanesville and points west. By the late 1880s other rail lines provided convenient transportation throughout the county, leading to the opening of new coalfields.

The Pittsburgh seam, the only seam in the area mined until 1880, was used as a steam coal, first for industry and steamboats and later for railroad locomotives. By 1888 northeastern Belmont County contained the largest mines in this seam in Ohio. Armstrongs Mills, Badgertown, Barnesville, Belmont, Flushing, and Warnock, all central–Belmont County communities, also had significant operations. Large corporations took over the mine operations in the early 1900s. The county's coal production climaxed between 1916 and 1924.

MINING HIGHLIGHTS OF BELMONT COUNTY

- Hanna Coal once operated some thirty mines in Belmont, Harrison, and Jefferson Counties.

- North of the viaduct at Blaine there was a Lorain Coal & Dock drift mine that extended over 4 miles into the hillside following the Pittsburgh seam.

- The first continuous mining machines in Ohio appeared at the Saginaw mine of Oglebay Norton Company near St. Clairsville in 1949.

- Ohio's second-worst coal mine disaster took place March 16, 1940, at the Hanna Coal Company's Willow Grove No. 10 Mine between St. Clairsville and Neffs. A tremendous explosion, the result of coal dust ignition, killed 72 men and trapped 101 for some five hours. The mine remained open until 1954.

The demand for more coal during World War II brought the rapid spread of strip-mining throughout the area, first between St. Clairsville and Cadiz, and later elsewhere. Although early strip operations led to many environmental problems, mine reclamation was already being practiced by some companies in the 1930s and 1940s. A major problem, though, was the air pollution generated from burning sulfur-rich coal. The 1970 passage of the Clean Air Act limited sulfur dioxide emissions and cut into the marketability of Ohio coal. The industry still survives as research into the clean burning of coal progresses.

Two-thirds of Ohio's coal is still mined using underground techniques. Aside from air pollution from hoists and coal handling and the erosion of land, a major consequence of underground mining is sinking land over the mine workings. This is especially prevalent when an operation has used longwall mining methods. Subsidence, be it sudden or slow, has already led to many problems in southeastern Ohio. Subsidence from proposed mining potentially threatens a number of pristine sites throughout southeastern Ohio, including Raven Rocks, an area of rock shelters near Crabapple, and Dysart Woods, old-growth woods near Centerville.

In 1858, Parker & Sons Cement Works opened just north of Barnesville and quarried Pennsylvanian Fishpot limestone for use in a natural cement product. Around 11,000 barrels went to the Baltimore & Ohio Railroad during the construction of its bridge at Bellaire. Cement was a rare commodity in the coalfields where limestone was not common, so this plant prospered into the 1880s. Quarry operations crushed Fishpot limestone for aggregate at a number of quarries southeast of Barnesville. Builders used Pennsylvanian-age Bellaire sandstone from quarries at Avondale, Bellaire, Martins Ferry, and McClainville in bridges at Bellaire, Bridgeport, and Wheeling. A plant at McClainville also used the Summerfield and Connellsville shales to make common bricks. In the 1930s Bellaire and Steubenville were the main consumers of the forty-five thousand bricks this plant produced per day.

Old Washington to Zanesville

Along U.S. 40 just west of Old Washington, Conemaugh group sandstones and shales are visible in roadcuts. Reclaimed hillsides, devoid of Upper Freeport coal, rise east of the I-77 interchange. The hills around Cambridge exhibit Conemaugh group rocks, but the valleys are cut in Allegheny group rocks.

At Cambridge, Wills Creek winds under the routes, another example of a misfit stream in a former meltwater valley. Pre-Illinoian lake sediments, mainly Minford clay, fill valleys south, east, and west of Cambridge. In 1873 a group of Cambridge businessmen saw the potential profit in establishing a glass-making facility in their community since the necessary coal to fire the furnaces lay under the countryside. The Cambridge Glass Company was opened in 1902; by 1910 it had its own coal mine, which was a necessity since each furnace consumed over 50 tons daily. The company also used natural gas that it tapped on its property. The company became famous for its finely made crystal. Declining sales led to its closure in 1958, and the factory was demolished in 1989. On the west side of Wills Creek valley a roadcut on U.S. 40 exposes Conemaugh-age Mahoning sandstone and shale at road level, overlain by the thin, sporadic Mason coal and Buffalo sandstone. The Mason shale atop the coal is rich in plant fossils.

Strip mines in the Upper Freeport coal are scattered between Cambridge and Zanesville; Conemaugh sandstones occasionally poke through the vegetated hillsides. Allegheny group rocks form the valleys. Drift mines in the local Harlem

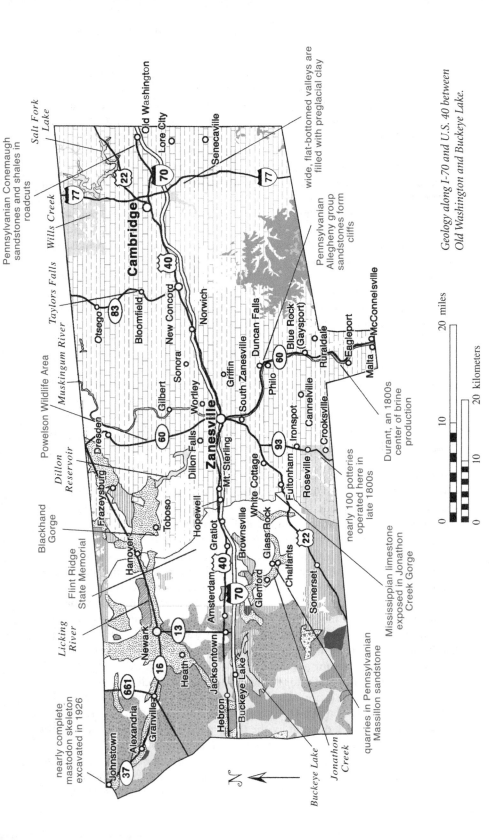

Geology along I-70 and U.S. 40 between Old Washington and Buckeye Lake.

Pennsylvanian Conemaugh sandstones and shales in roadcuts

wide, flat-bottomed valleys are filled with preglacial clay

Pennsylvanian Allegheny group sandstones form cliffs

Durant, an 1800s center of brine production

nearly 100 potteries operated here in late 1800s

Mississippian limestone exposed in Jonathon Creek Gorge

quarries in Pennsylvanian Massillon sandstone

Blackhand Gorge

Flint Ridge State Memorial

Powelson Wildlife Area

nearly complete mastodon skeleton excavated in 1926

Salt Fork Lake

Wills Creek

Taylors Falls

Muskingum River

Dillon Reservoir

Licking River

Buckeye Lake

Jonathon Creek

Old Washington

Lore City

Senecaville

Cambridge

Otsego

Bloomfield

New Concord

Norwich

Sonora

Gilbert

Wortley

Griffin

South Zanesville

Duncan Falls

Blue Rock (Gaysport)

Philo

Ruraldale

Eagleport

Malta

McConnelsville

Zanesville

Dillon Falls

Mt. Sterling

White Cottage

Glass Rock

Fultonham

Ironspot

Cannelville

Crooksville

Roseville

Dresden

Frazeysburg

Toboso

Hanover

Hopewell

Gratiot

Amsterdam

Brownsville

Glenford

Chalfants

Somerset

Newark

Jacksontown

Heath

Hebron

Granville

Alexandria

Johnstown

0   10   20 miles

0   10   20 kilometers

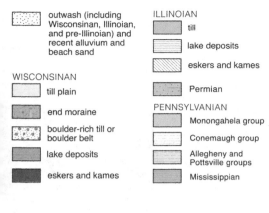

outwash (including Wisconsinan, Illinoian, and pre-Illinoian) and recent alluvium and beach sand	**ILLINOIAN** — till
	lake deposits
	eskers and kames
WISCONSINAN — till plain	Permian
end moraine	**PENNSYLVANIAN** — Monongahela group
boulder-rich till or boulder belt	Conemaugh group
lake deposits	Allegheny and Pottsville groups
eskers and kames	Mississippian

A 2-foot-thick seam of Upper Freeport coal near Cambridge. Mahoning sandstone lies above and underclay below.

coal of the Conemaugh group were once common around Norwich, Sundale, and New Concord. Gray fossiliferous Ames limestone rims the hills and ridges around New Concord. If you follow Ohio 83 north from New Concord, you'll reach Taylors Falls, along White Eyes Creek northwest of Bloomfield, a 40-foot cascade over Pennsylvanian-age sandstone. Farther north, Otsego is the site of an important oil pool discovered in 1903 in Devonian-age Berea sandstone. Nearly 170 wells had been sunk to this layer by 1917; although local production was never more than 75 barrels of oil a day, the field proved to be the most profitable in Muskingum County.

Pennsylvanian-age Conemaugh strata poke from the hillsides near New Concord. In 1986 a landslide in these weak beds destroyed the westbound lanes of I-70. It took crews a month to repair the road. More lake basins filled with Minford clay from pre-Illinoian glaciation form valley bottoms around Zanesville. Between Zanesville and Dresden, east of Ohio 60, there are reclaimed strip mines that are now part of the Powelson Wildlife Area. North of the Muskingum River crossing in downtown Zanesville is the earthen Dillon Dam. The dam, part of the Muskingum Watershed Conservancy District, impounds the Licking River, creating Dillon Reservoir. North of the routes near Frazeysburg, drift mines in Quakertown coal were common in the late 1800s.

New Concord's Meteorite

On March 1, 1860, at least thirty chunks of a large meteorite bombarded a 30-square-mile area extending from New Concord southeastward. Witnesses reported that they heard a series of cannonlike noises and saw a fiery streak.

Pennsylvanian-age Conemaugh group strata form the sides of I-70 near New Concord. The layers contain great thicknesses of clay and shale and weather readily.

People as far away as Bellaire, 50 miles, and Newark, 38 miles, heard the explosions. Some people thought it was an earthquake, and fire alarms sounded in Barnesville, 28 miles away, and Old Washington, 17 miles away. Nine pieces of the meteorite weighed between 15 and 103 pounds each; locals reported seeing many smaller particles. The meteorites were black, angular masses generally with gray interiors of nickel-iron composition—what geologists deduce the earth's core is composed of. Many of the specimens disappeared from the site as souvenirs, and entrepreneurs exhibited pieces and charged people admission to see them. Several of the pieces ended up in museum and university collections. Other pieces are probably still buried in the countryside.

Zanesville Area

Pioneers often sailed up the Muskingum River from Marietta to enter the lands west of the Ohio River since there were few trails or roads through the Allegheny Plateau. Where the Licking River flows into the Muskingum, a settlement developed. The settlement became Zanesville in 1797, and it served as the capital of Ohio from 1810 to 1812. Like many Ohio residents and entrepreneurs, Zanesville locals made use of the region's mineral resources. Using mainly Pennsylvanian-age rocks, they built their communities and an economy.

Brick making began around 1802, followed shortly by the fashioning of crude crockery items. The first Zanesville pottery opened in 1820, leading to the nickname Clay City. Tionesta clay from drift mines in the Drake-Ellis area north of Zanesville became the raw material for potteries in Cambridge, Roseville, and Zanesville. Artisans used Pleistocene-age Minford clay and Homewood shale of the Pottsville group to manufacture flowerpots at South Zanesville in the 1930s.

Workers of the Townsend and Harris brick plants, both on the south side of Zanesville, dug Lower Freeport shale of the Allegheny group for pressed brick. The brick plants also used Bolivar, Oak Hill, and Upper Freeport clay of the same group. The Townsend Company ran a mechanized coal mine to supply its kilns with Middle Kittanning coal and sold its excess. The Harris Mine near Griffin provided the Harris brick plant with Upper Freeport coal between 1918 and 1928. The Harris plant had closed by 1928, but the Townsend plant continued making bricks as Zanesville Clay Products, eventually using clays that lie below the Middle Kittanning coal. A plant in South Zanesville used the Putnam Hill and Oak Hill shales of the Allegheny group to make pavers and common bricks.

Lower and Upper Mercer and Putnam Hill ores fed the charcoal-fired Dresden and Dillon Furnaces near Zanesville from 1847 to 1850. Lower and Upper Mercer ores near Hopewell stocked the furnaces of Zanesville's Ohio Iron Company, starting in 1871. The plant, located along the Muskingum River just north of downtown, was a landmark with its 62-foot stack. A rolling mill, where iron is formed into sheets and bars, was also on the site. Coal that fed the furnace came from Straitsville, and coke from Connellsville, Pennsylvania, since good coke was not known to exist in the area and railroads provided an almost direct route. Workers used local limestone as a flux, and they produced some 40 tons of iron each day in the early 1870s.

Because of the size of the Middle Kittanning seam, it played the greatest role in the industrial growth of Zanesville and certainly was the most important coal for domestic use in Muskingum County. It had few impurities and burned more cleanly than other coals in the area. The Clarion sandstone is a prominent cliff-forming rock around Zanesville. A 30- to 60-foot escarpment of this sandstone stretches from Zanesville north to Gilbert along the Muskingum River. Twenty to 40 feet of this rock occurs in a cliff along Brush Creek south of town. Look for this sandstone in sills and capstones of the older buildings of Zanesville. Builders also used it in some local canal locks. The most notable use of this sandstone was in the 1886–88 construction of New Philadelphia's courthouse; the courthouse was the most important building in most communities and had to be an imposing structure—one that would draw attention. The 1874 Zanesville courthouse has walls composed of Mississippian-age Maxville limestone and is trimmed with Pennsylvanian-age Clarion sandstone.

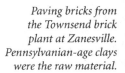

Paving bricks from the Townsend brick plant at Zanesville. Pennsylvanian-age clays were the raw material.

The Putnam Hill limestone is one of the more-easily traced Pennsylvanian beds in the Zanesville area. This fossiliferous gray marine limestone surrounds Putnam Hill, its namesake, south of I-70 and U.S. 40 on the west side of the Muskingum River. People made lime from the Putnam Hill limestone, and in the early days of Ohio's iron boom it was used as flux in charcoal-fired furnaces.

The hill that separates South Zanesville from the Muskingum River contains sand deposits on the north slope that were mined for molding sand in the early 1900s. The deposit consists of wind-transported sand. Other Pleistocene sand deposits suitable for molding sand were found at Ellis and Dresden, up the Muskingum Valley.

OTHER QUARRIES AND MINES OF THE ZANESVILLE AREA

Town	Rock Removed
Dresden	Middle Kittanning coal.
Dillon Falls, Frazeysburg	Drift mines in Bedford coal were developed in the late 1800s.
Jackson (now a ghost town)	Upper Freeport coal.
New Concord	Fossiliferous Cambridge limestone and Portersville shale from a quarry northeast of town.
Roseville	Middle Kittanning coal.
Sonoma (now a ghost town)	Upper Freeport coal.
South Zanesville	Middle Kittanning coal.
Wortley	Several mines in Lower Kittanning coal; quarries in Clarion sandstone.
Zanesville	Tionesta coal from drift mines in Salt Gum Hollow on the south side of town. Middle Kittanning coal came from roughly one hundred small mines east of town. Upper Freeport coal came from many small mines east of town. The Homewood sandstone was quarried west of town for building stone.

A Side Trip to McConnelsville on Ohio 60

From Zanesville to McConnelsville, Ohio 60 closely parallels the eastern bank of the Muskingum River. Cliffs of Pennsylvanian-age Lower Freeport sandstone tower above the river near Duncan Falls and Philo. Blue Rock (Gaysport) was once the site of Oil City and was named for a bluish shale that occurs above the Anderson coal. In the southeastern corner of Muskingum County, far east of Blue Rock, there is a field of Meigs Creek coal. The coal averages 4 feet thick. South of Stone on the west side of the river there are long-abandoned quarries in the Cow Run sandstone along Cedar Run. People used the stone in bridge piers, foundations, and canal locks, and as flagstone, in the late 1800s. The sandstone forms erosion-resistant cliffs and steep bluffs that contain slump blocks that broke loose along joints in this part of southern Muskingum County. The Cow Run sandstone varies from 20 to 30 feet in thickness and is coarser grained than most Pennsylvanian sandstones. Anderson coal, the most consistent Conemaugh group coal in the area, occurs here below the sandstone. At the top of the coal, the Portersville member contains pyritized marine fossils in shale and minute marine fossils in scattered limestone masses. The limestone marks the return of a shallow sea that flooded the former coal swamp, but apparently conditions were not favorable enough to allow the organisms to grow to normal size. Their remains form what paleontologists call a *dwarf*, or *stunted*, assemblage. The shale beds of the Portersville member probably formed in local areas where water circulation was poor and oxygen deficient, so nothing much was living. Shells of dead animals were replaced with pyrite (iron sulfide).

The Blue Rock Mines occur where Blue Rock Creek meets the Muskingum River. They were the site of a remarkable rescue of four coal miners in 1856. Twenty men and boys were working in one of the mines when the ceiling gave way; sixteen of them escaped from the collapsing mine. After fourteen days of digging, rescuers found the trapped miners some 700 feet back from the mine's entrance. Soon after they were rescued, a second collapse occurred. All of the miners recovered from the harrowing experience.

The Muskingum River of pioneer Ohio was a wild and free waterway, cascading across many resistant sandstone beds as it flowed to the Ohio River. The town of Duncan Falls formed along the river's most notable waterfall because people were forced to portage around it. Navigation between Marietta and Dresden was limited and difficult. During the early 1800s the common solution to this problem was to build a canal. The closeness of bedrock to the surface in the area made this an expensive proposition, so early settlers devised a dam and lock system. Ten dams between Marietta on the Ohio River and Dresden created lengthy pools where rapids once thrashed in the river. Ten locks improved navigation on the river by 1841. The 112-mile Muskingum River Parkway—as the system of locks and dams is known today—is a National Historic Civil Engineering Landmark that now serves the needs of recreational boaters.

Oil fever reached the middle Muskingum Valley in 1864 when locals observed gas bubbling from the headwaters of Mans Fork near Ruraldale. Drillers struck oil 80 feet down in the Pennsylvanian-age Cow Run sandstone. A hundred wells went in within a year, and the boomtown Oil City sprang up where Blue Rock is today. Unfortunately, the wells were short-lived and prospectors moved elsewhere. By the early 1900s most oil discoveries popped up farther west near Cannelville, Fultonham, and Roseville in the deeper Devonian-age Berea sandstone and Silurian-age Clinton sands. Oil City washed away in the 1913 Flood, taking with it the surface remnants of the oil boom. Workers struck natural gas in the Muskingum Valley in 1830 near Malta while searching for brine from which to distill salt. A salt plant operated near Malta until 1878. Flaring gas wells lit the way for early steamboats traveling up the Muskingum River.

Upriver of Malta, about 2 miles south of the Muskingum and Morgan County line, lies the small community of Durant, largely bypassed by travelers and forgotten by most Ohioans. Durant, however, was a stronghold of brine

WHERE OIL CITY ONCE STOOD
BEVERLY O, 1913

The raging Muskingum River put an end to what remained of Oil City in March 1913.

production in the 1800s. Thirty-seven kettle furnaces operated around Eagleport and Durant from 1825 to 1905. The industry went into rapid decline after the Civil War. As local forests were logged, wood for the furnaces disappeared and operators had to turn to coal, which was more expensive because it had to be shipped in by flatboat from mines upriver at Philo.

Durant was the home of Big Bloom Salt Works until 1903. The company changed little over time. When the doors closed, large iron kettles were still being used to evaporate the brine. A more modern plant replaced it in 1904, but it is long gone as well. Workers drew the brines for these operations from the Mississippian-age Black Hand sandstone. The salt of the brines came from seawater that was trapped in the pores of sediment as it was deposited in ancient Paleozoic seas about 340 to 335 million years ago. These natural brines undergo great chemical alteration over geologic time, so today's brines differ greatly from the original composition of the Paleozoic seawater.

A few miles upriver from McConnelsville was where Devil's Tea Table once stood, long a geologic attraction of the Muskingum Valley. It was a 30-by-20-foot slab of Pennsylvanian sandstone on a pedestal of shale standing about 25 feet tall at the summit of a high hill. Continued weathering of the shale left too little to hold the sandstone up, and the structure collapsed into a pile of rubble in 1906.

Zanesville to Hebron

U.S. 40 follows Timber Run, which cuts into Pottsville group shales from Zanesville to Mt. Sterling. Workers relocated the stream when they realigned U.S. 40. Hopewell, formerly Coaldale, was once a mining center in a small field of Clarion coal. The coal averaged 6 feet thick and was in demand by farms and communities along the National Road for many years. The field was mined out by the early 1900s, and it's now farmland. Between Hopewell and Gratiot the routes pass onto a 12-mile stretch of Illinoian till, the farthest east glaciers made it in this area. North of here the Licking River cuts into Mississippian-age sandstone at Blackhand Gorge on the western edge of the Allegheny Plateau. A number of prominent sandstone knobs mark the transition from the plateau onto the western till plains, both north and south of the highways near Brownsville. South of I-70 there are other quarrying sites at Glass Rock and near Glenford where the Pennsylvanian Massillon sandstone was mined for glass sand and building stone. Flint Ridge State Memorial, where Native Americans once quarried flint, Ohio's gemstone, is north of Brownsville. The Brownsville area is also a small oil field, which lies in the underlying Silurian-age Clinton sands. Hills like Buzzard Glory Knob, west of Brownsville, have caps of Vanport flint, part of the Pennsylvanian Vanport limestone. Between Brownsville and Linnville, Lower Mercer limestone yields nice fossils, including large crinoid stems, brachiopods, bryozoans, and snails that lived in and around the fluctuating Pennsylvanian sea.

Just west of Amsterdam, on U.S. 40, a teardrop-shaped hill behind a small brick church is a mound of the Hopewell culture. Ohio 13 crosses the routes along the Wisconsinan-age terminal moraine that forms the eastern edge of Buckeye Lake, a canal reservoir that lies south of the highways. This reservoir was necessary for the Ohio & Erie Canal because the canal crosses the divide between the Muskingum and Scioto River basins near Etna on U.S. 40. The site of the lake was once part of what locals called the Great Swamp. About 350 feet of sand and gravel lie hidden below Wisconsinan till on the east side of Hebron, marking the buried valley of the pre-Illinoian Newark River, which

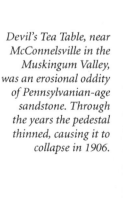

Devil's Tea Table, near McConnelsville in the Muskingum Valley, was an erosional oddity of Pennsylvanian-age sandstone. Through the years the pedestal thinned, causing it to collapse in 1906.

once flowed southwest to Chillicothe. At Newark impressive earthworks of the Hopewell culture offer insight into the mound-building activities and culture of this early people. Quarriers pulled Mississippian-age building stone from quarries at Granville and Newark, north of the roads.

Fultonham and White Cottage

The Fultonham Stone Company quarried Mississippian-age Maxville limestone at White Cottage, which is south of Zanesville off U.S. 22, as building stone in the late 1800s; other quarries opened along Jonathan Creek and Kent Run between here and Fultonham. As much as 35 feet was once exposed in the quarries along the creeks. The 1874 courthouse in Zanesville contains this stone, although people used it more commonly in bridge abutments, foundations, and retaining walls. The company also burned it for lime and crushed it for railroad ballast and road gravel, and by the 1900s, the company used the limestone to manufacture portland cement. A company affiliated with Pittsburgh Plate Glass operated a quarry and underground mine in East Fultonham in the 1940s, using the Maxville limestone in cement making and as additives in the chemical industry. During the early 1900s stoneware and brick plants used the clay and shale under the Flint Ridge coal of the Allegheny group at Fultonham and White Cottage. A Fultonham plant used Lower Mercer shale to Manufacture brick facing.

Strip-mined land lies south of Fultonham. Lower and Middle Kittanning coals outcrop near the hilltops south of U.S. 22 as far south as New Lexington. Because the two seams average only 25 feet apart, miners stripped them collectively. West of Fultonham, near the Muskingum and Perry County line, Jonathon Creek lies in a 200-foot-deep gorge north of U.S. 22 where the creek cuts through a pre-Illinoian drainage divide. Mississippian-age Maxville limestone outcrops at stream level, and Allegheny group strata cap the highest hills. The Lower Mercer limestone of the Pottsville group is particularly fossil rich here.

A Side Trip to Crooksville and Roseville

In the late 1880s, some eighty-five to ninety small potteries dotted the countryside around Roseville, which is south of Zanesville off Ohio 93. Most of them employed only one or two workers. Although various Pennsylvanian-age clays were used to manufacture stoneware, the Tionesta and Lower Kittanning clays of the Allegheny group were most popular. In the late 1940s, only eight companies in Crooksville and four in Roseville remained active.

The Roseville Pottery opened in 1890, initially producing standard stoneware. In 1910 the entire plant moved to Zanesville, where workers crafted the well-known "art pottery" until 1954. In 1910 the Nelson McCoy Sanitary Stoneware Company was established in Roseville. The company mined and sold clay while also making lines of utilitarian and decorative stoneware. Struggling in its later years, the pottery closed its doors in 1990.

The Ransbottom Pottery began in 1900 when the Ransbottom brothers purchased the Oval Ware and Brick Company at Beem City, now Ironspot. Production expanded rapidly, and by 1916 workers were making a boxcar load of stoneware jars every hour. Since then it has held the position as the number one manufacturer of stoneware jars in the world. In 1920 it merged with Robinson Clay Products to become Robinson-Ransbottom Pottery. The company continues to offer an extensive line of products. I highly recommend taking a factory tour.

A Side Trip on Ohio 668
Ohio's Gemstone

Just north of I-70 and U.S. 40 near Brownsville, there is a long, resistant ridge of Pennsylvanian-age rocks called Flint Ridge, named after the Vanport limestone and its colorful flint. The flint is unique in that it forms an 8-by-3-mile sheet that blankets the area; most flint occurs in nodules or discontinuous streaks. The sheet ranges from 1 to 12 feet thick. It grades from milky white to light yellow, brown to reddish brown, to multicolored with veins and cavities of clear quartz crystals.

The stone is noted today for its beautiful colors and patterns. Paleo-Indians knew about it some 10,000 years ago. They began using it on a large scale around 3,000 years ago, and ancient pits that the Indians dug to mine the flint dot Flint Ridge. Archaeologists find Vanport flint implements among the remains of many cultures across the eastern United States, showing how widely Indians traded this useful stone. European settlers found the porous stone made good millstones. The archaeologic significance of the Flint Ridge area led the Ohio Historical Society to purchase land here in 1933; eventually it became Flint Ridge State Memorial. Trails lead to old pits, some of them filled with water, and a museum built over one of the excavations is open during warmer months. Flint became the official state gemstone in 1965, and specimens of it are in museums around the world.

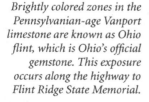

Brightly colored zones in the Pennsylvanian-age Vanport limestone are known as Ohio flint, which is Ohio's official gemstone. This exposure occurs along the highway to Flint Ridge State Memorial.

Collecting is prohibited at the memorial, but specimens are for sale in the gift shop and at privately owned collecting sites in the area. The flint outcrops along the road just north of the memorial along Ohio 668. Look closely; some of the flint contains abundant fusulinids, fossilized microorganisms about the size and shape of wheat grains that lived in the Pennsylvanian sea some 318 to 300 million years ago. East along Flint Ridge—the ridge runs east to west—into Muskingum County, an older flint occurs at the horizon of the Upper Mercer limestone. This flint is black with an occasional cavity filled with quartz crystals.

Blackhand Gorge

Continuing north on Ohio 668 from Flint Ridge State Memorial leads to what remains of Claylick, underlain by fine lake sediments that were deposited in a glacial lake. Just east of here is the west end of Licking Gorge, also called Blackhand Gorge. The best scenery, however, is at the east end of the gorge near Toboso, off Ohio 16 and 146. A portion of the gorge became Blackhand Gorge State Nature Preserve in 1975.

Before the Ice Age, streams drained westward across Licking County. The broad valley that stretches from Newark to Coshocton, now the route of Ohio 16, marks the path of a major preglacial river, the Cambridge River. The valley is commonly referred to as Hanover Valley. Illinoian-age ice and then Wisconsinan ice blocked this valley at Newark and sent meltwater eastward down the valley. A pileup of outwash sands and gravels in Hanover Valley ponded meltwater in Brushy Fork Valley to the south at the western end of the present-day gorge. A significant lake developed. Torrents of water drained from this glacial lake into lower valleys south of the outwash-filled Hanover Valley. Continued downcutting formed the rocky gorge between Hanover and Toboso. It is likely that the gorge was eroded on and off during glacial and interglacial time with its final, significant downcutting related to late Wisconsinan events. Eastward drainage continues today as the Licking River.

Over 80 feet of Mississippian-age strata of the Cuyahoga and Logan formations are exposed in the gorge. The discovery of a large pictograph, or painting, of a black hand on the wall of the gorge led to the naming of the rock at river level—the Black Hand sandstone. In the gorge the Black Hand contains a lot of pebbles, is strikingly cross-bedded, and contains a few trace fossils—evidence that much of this sandstone was deposited in a delta. Resting unconformably on top of the Black Hand is 5 feet of the Berne conglomerate member of the Logan formation. Look closely at the boundary for evidence of this unconformity. The boundary rises and falls, suggesting that channels were carved into the top of the Black Hand sandstone when the delta sands were above sea level, and cross-beds of the sandstone end abruptly at the contact with the Berne member. Above the Berne member the rock layers are composed of finer sand, marking the Byer sandstone. Marine fossils occur in both the Berne and Byer members, indicating that the area was again covered by the Pennsylvanian sea.

Besides the geologic story, the gorge has a long and interesting human history. Native Americans knew the gorge. Pictographs indicated its significance as a central Ohio trail marker for them. Workers constructing the Ohio & Erie Canal through the gorge in the late 1820s were the first to modify it. They blasted a towpath in the wall at Black Hand Narrows for mules pulling the canal

Black Hand Rock, an erosional remnant of Mississippian-age Black Hand sandstone east of Newark. The Licking River curves around the rock; the ledge near the river's level was the towpath of the Ohio & Erie Canal. An interurban rail line passed around the north side of this rock; the only interurban tunnel in Ohio is hidden from view to the left. —Wilber Stout photo, Ohio Department of Natural Resources, Division of Geological Survey

packets, which accidentally destroyed the famous black hand pictograph, and built two locks and a dam. Since there was no easy way to dig a canal through the rock-bottomed gorge, the river became part of the canal for this stretch. Locks at the east and west ends of the gorge, where the canal connected with the river, permitted canal packets and barges to be lowered or raised to the level of the of the river or canal. A dam upstream of Toboso slowed the current of the river and created an area of slack water so canal boats could navigate this short stretch of river. Remnants of the east lock remain along Canal Lock Trail; the west lock lies in a restricted area of the preserve. The stone for these locks, as well as the wall below the towpath, is Black Hand sandstone that was quarried nearby.

To see the towpath that workers carved into the base of Black Hand sandstone, follow Blackhand Trail. A hand-built stone wall is visible below the towpath. The water level would cover the lower part of this wall had the dam not washed away in an 1898 flood.

The Central Ohio (later Baltimore & Ohio) Railroad forged a path through the gorge in 1850. At the east end of the gorge, opposite Black Hand Rock, the railroad blasted a 65-foot-deep cut through the Mississippian-age sandstone. The rails were relocated in the 1950s, during the construction of Dillon Reservoir, and now a bike trail passes through the cut. The Columbus, Newark & Zanesville Electric Railway began running interurban cars through the gorge in 1904. They followed the north bank of the river, skirting Black Hand Rock, but then tunneled 327 feet through the next rocky promontory. The interurban

Strongly cross-bedded Black Hand sandstone in Blackhand Gorge. Weathering has accentuated the intersecting layers.

turned the gorge into a tourist attraction. A picnic area was built above the tunnel, and cottages and a dance hall sprang up nearby. The electric railway was abandoned in 1927, and a road replaced the line. The road was closed when the Dillon Dam project was started in the late 1940s.

The Blackhand or Quarry Rim Trails of the Blackhand Gorge State Nature Preserve lead to abandoned quarries in the Black Hand sandstone just west of the former railroad cut. The E. H. Everett Company of Newark, a successful bottle manufacturer, quarried and crushed the Black Hand sandstone for glass sand here starting around 1890.

South of Ohio 16 on the hillside between Toboso and Hanover is the shale quarry of the former Hanover Brick Company. Workers quarried the Vinton member of the Logan formation here and mixed it with Illinoian-age till to manufacture brick and tile. In later years the operators trucked in Pennsylvanian-age shales and clays from a pit near Frazeysburg.

Glass Rock and Other Sand Plants

The Pennsylvanian-age Massillon sandstone is the reason behind Glass Rock's name. Since the 1880s, quarriers, here and in nearby Glenford and Chalfants south of I-70 and U.S. 40, have removed this sandstone from the earth. In later years Glass Rock housed the plant of the Central Silica Company, which produced crushed sand for making glass and other industrial purposes. An abandoned quarry near Glenford exposed over 120 feet of Pottsville group strata and underlying Mississippian-age Logan formation. Nearly 24 feet of Massillon sandstone rests unconformably on top of the Logan formation in the bottom of this pit. Some 320 to 315 million years ago, erosional forces stripped away the layers that separate these strata at other Ohio locales. A quarry operation still quarries the Massillon sandstone at Chalfants and crushes it at Glass Rock for use in glass plants, foundries, and other industries. Another company, Ohio Flint and Glass Sand, operated a quarry south of Somerset and trucked the rock to a crushing plant at Rushville, outside of Columbus, in the 1920s.

A Side Trip to the Granville-Newark Area

Head north on Ohio 13 from Jacksontown. A company at Heath produces salt from natural brines in Silurian-age Lockport dolomites. Road crews use the precipitated salt for highway snow and ice control. One mile south of Newark, east of Ohio 13, is Rock Hill, an area of stone quarrying in the mid-1800s. Older stone buildings downtown contain Byer sandstone from these quarries.

Downtown Newark sits in a wide, flat valley underlain by glacial deposits of Illinoian and Wisconsinan age. The valley is wide because it is the confluence of the North and South Forks of the Licking River and Raccoon Creek, which form the Licking River. Tremendous volumes of meltwater flowed eastward during times of glacial melting.

The gravel terraces of the valley margins were important sites to Native Americans 2,000 years ago; a complex of circles, octagons, and other earthworks covered a 4-square-mile area. As Newark grew, the Hopewell culture earthworks disappeared, but luckily some survive at the Great Circle Earthworks (formerly Moundbuilders State Memorial), Octagon Earthworks, and Wright Earthworks, all part of the Newark Earthworks State Memorial. Unfortunately, the solitude that Native Americans may have experienced at these places has been lost because urban sprawl surrounds them. Head west on Ohio 16.

Iron smelting, using limonite from the Mississippian-Pennsylvanian unconformity, was one of the earliest mineral industries in Licking County. Wagons hauled the stripped Pennsylvanian-age Harrison ore to the Mary Ann Furnace northeast of Newark and the Granville Furnace on Parnassus Hill, a bedrock hill on the southeast edge of Granville. Pig iron was the main product of these short-lived enterprises. Both furnaces were in ruins by the mid-1800s.

In the early years of Granville, settlers bisected a rocky promontory that juts out into the valley between Granville and Newark with a road; this rock exposure became known as the Dugway. In 1962 the new Ohio 16 was built through it. In the roadcut, Logan formation Byer and Berne members are underlain by about 20 feet of Black Hand sandstone. Look in the talus for fossil brachiopods, crinoids, and clams from the Berne and Byer members.

Granville, located on the terraces and hills above Raccoon Creek, is the home of Denison University, certainly a cradle of geologic studies in Ohio. Geology and mineralogy courses appeared in the school's catalog as early as 1854. Geology field trips were part of classroom activities as early as the Civil War. An important early journal of the school, the *Bulletin of the Scientific Laboratories of Denison University*, began in 1885 and was full of geologic articles about Ohio.

Mississippian-age rocks compose the bedrock of this region: the Cuyahoga formation lies in the Raccoon Creek valley, and the Black Hand sandstone occurs in a roadcut near the Ohio 661 and Ohio 16 intersection. Over these rocks the valley is composed of 200 or more feet of gravel and sand deposited by meltwater and present-day Raccoon Creek. Lake Hudson, southeast of Granville, is an old gravel pit that filled with water. Gravel terraces carved by meltwater flank the valley. Prominent kames occur upstream toward Alexandria. The Logan formation caps the hills above Raccoon Creek at Granville. A number of quarries worked the Allensville and Byer members of this formation for building stone before 1900. The Granville Inn on West Broadway is a good example of a structure built with Byer sandstone.

In 1926, while burying farm animals on a Johnstown-area farm northwest of Granville, a tenant farmer uncovered the huge skull of an unknown beast less than 2 feet down in glacial till. It was quickly identified as a mastodon (*Mammut americanum*); word spread and soon people were driving from across the state to see the bones. Within ten days over ten thousand people had gazed at the remains. A local entrepeneur sold more than ten thousand photos, and the farmers made over $1,000 dollars in parking fees in three days. Digging continued for about two months. The skeleton was complete except for a few tail vertebrae and foot bones—the most complete mastodon skeleton found in the state at that time.

Initially, a Newark businessman purchased it a few days after its discovery. He, in turn, sold it to the Cleveland Museum of Natural History before the year ended. Johnstonians may disagree, but it's a blessing that the remains reached a museum where professional care was available. The skeleton went first to the

The skull of the Johnstown mastodon before it was completely excavated.

American Museum of Natural History in New York City to be prepared for display, but it returned to Cleveland in 1927 and went on permanent display in Kirtland Hall of the museum in 1928. It continues to be an important part of the Cenozoic mammal exhibit. The Johnstown Historical Museum also has a nice display about the mastodon's excavation.

A second significant mastodon find happened in December 1989 at Burning Tree Golf Course near Heath. An excavator struck a skull while digging a pond. Within three days a team of archaeologists and volunteers extracted most of the skeleton. The preservation of a portion of the intestinal contents of this mastodon allowed scientists a rare opportunity to investigate ancient intestinal bacteria, DNA, and this mastodon's diet. Radiocarbon dating of pollen within the intestine, twigs with the skeleton, and bone collagen showed that the mastodon died around 11,500 years ago. Its diet was wetland plants, including various sedges and water lilies. Meticulous laboratory work in the 1990s, using state-of-the-art microbiological techniques, eventually led to the isolation of living bacteria from samples of intestinal matter, bone-associated matter, and sediments not in close contact with the fossil remains. A scientist from Ohio Wesleyan University partially reconstructed a mastodon gene from DNA from intestinal matter in 2000. Although there was interest in displaying the Burning Tree mastodon locally, it was sold to a Tokyo museum.

INTERSTATE 71
Cleveland— Columbus
145 miles

Pleistocene-age lake sediments and lake-planed till underlie downtown Cleveland south to Berea. The occasional excavation at a construction site might offer a glimpse of these sediments, but most of it is covered by asphalt and concrete. From Berea to Columbus I-71 crosses glacial till and outwash of the Wisconsinan-age Killbuck and Scioto lobes of ice.

Cleveland Area
Downtown Cleveland rises from the narrow, flat plain that forms the shore of Lake Erie onto the dissected edge of the Allegheny Plateau. The plateau is capped by hard Pennsylvanian-age Sharon conglomerate. The lake-facing escarpment of the plateau is not a sheer cliff, but rather descends to the lake plain in two steplike rock terraces. The Devonian-age Berea sandstone underlies the upper terrace, and the Devonian Euclid bluestone (part of the Bedford shale) underlies the less obvious lower terrace. Sandwiched between these layers is about 50 feet of the easily eroded Bedford shale. From 150 to 350 feet of Mississippian-age Cuyahoga formation separates the Sharon and Berea layers, and a thick sequence of Devonian shales underlie the Euclid member. The terraces are more apparent on the east side of downtown and east of I-71, but they disappear as I-71 approaches the bridge over the Rocky River. The Rocky River, Big Creek, Cuyahoga River, Doan Brook, and Euclid Creek all flow out of deep notches cut in the face of the plateau near Cleveland.

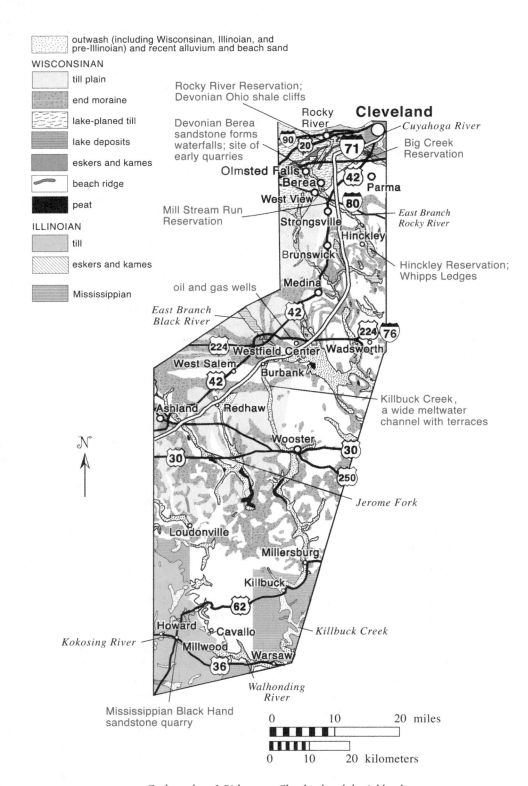

WISCONSINAN

- outwash (including Wisconsinan, Illinoian, and pre-Illinoian) and recent alluvium and beach sand
- till plain
- end moraine
- lake-planed till
- lake deposits
- eskers and kames
- beach ridge
- peat

ILLINOIAN

- till
- eskers and kames

Mississippian

Rocky River Reservation; Devonian Ohio shale cliffs

Rocky River

Cleveland

Cuyahoga River

Devonian Berea sandstone forms waterfalls; site of early quarries

90 20 71

Big Creek Reservation

Olmsted Falls

42

Berea **Parma**

Mill Stream Run Reservation

West View

80

East Branch Rocky River

Strongsville

Hinckley

Brunswick

Hinckley Reservation; Whipps Ledges

oil and gas wells

Medina

East Branch Black River

42

224 76

224 **Westfield Center** **Wadsworth**

West Salem

42 **Burbank**

Ashland **Redhaw**

Killbuck Creek, a wide meltwater channel with terraces

Wooster

30

30

250

Jerome Fork

Loudonville

Millersburg

Killbuck

62

Howard **Cavallo**

Killbuck Creek

Kokosing River **Millwood**

Warsaw

36

Walhonding River

Mississippian Black Hand sandstone quarry

0   10   20 miles

0   10   20 kilometers

Geology along I-71 between Cleveland and the Ashland area.

Geologists speculate that most rivers and streams, including Cleveland-area rivers and streams, had longer courses before glaciation. They started farther south of Cleveland, on the plateau, and connected with the pre-glacial Erigan River drainage, which ran to the St. Lawrence Valley through where Lake Erie is now. Evidence for this comes from drilling records and geo-physical measurements that indicate that much of Ohio is underlain by buried valleys—large valleys carved deep into bedrock and filled with sediment. The advance of ice sheets into the state caused drainage to reverse, from northerly to southerly, and the lower parts of preglacial valleys filled with sediment and ice. The complex movements of ice to and fro across northern Ohio during the Ice Age, down preglacial drainage networks and onto the edges of the Allegheny Plateau, resulted in alternate cutting and filling of the preglacial valleys. Later, as ice sheets receded into Canada, rivers and streams continued flowing in a southerly direction and north-flowing rivers and streams once again developed. Ancestral lakes that eventually became Lake Erie further affected valley and channel development of the rivers and streams near Cleveland. The rise and fall of lake levels extended or shortened these drainage channels that flowed north.

Some streams reoccupied their former valleys, while others cut new courses. The Rocky River mostly cut a new channel in relatively soft Devo-nian-age shales. Where a channel is new, it is narrow and steep walled; where it follows a preglacial valley, it's broader with gentler slopes. The Cuyahoga River, in an old channel, cut a twisting, broad channel with steep bluffs mainly in Pleistocene lake sediments through the downtown area. The valley bottoms, or flats, are 100 to 180 feet below the plateau and up to a mile wide. Tributaries of the Cuyahoga River, like Mill and Tinkers Creeks, often cascade over bedrock at the edge of the Cuyahoga Valley, having cut new channels after glaciation.

Around 1910, twenty-three clay plants served the Cleveland area, making construction brick, pavers, drain tile, and other products from glacial clays and the Devonian-age Chagrin, Cleveland, and Bedford shales. Two early promi-nent geologists called Cleveland home—John Strong Newberry and Colonel Charles Whittlesey. Newberry headed the second Geological Survey of Ohio and served on the faculty of the Columbia School of Mines (now Columbia University). Newberry was the first person to find dinosaur bones in Utah and wrote the earliest detailed geologic description of the Grand Canyon. He described many Ohio fossils, especially of fish and plants, and laid the ground-work of Ohio paleontology. Whittlesey was known for his beach ridge studies, archaeological surveys, contributions to Henry Howe's *Historical Collections of Ohio* (a two-volume history of Ohio), and an early history work on Cleve-land. He also was an assistant on the first Ohio Geological Survey (1837–38). They're both buried in Cleveland's Lake View Cemetery. Whittlesey's tomb-stone is a jasper conglomerate, a glacial erratic from Ontario that once rested in his front yard.

Jared Potter Kirtland (1793–1877), a famous Cleveland naturalist, is also buried at Lake View Cemetery. He was an assistant in charge of zoology for the first Ohio Geological Survey and hosted such famous geologists as Louis Agassiz, known for his study of glaciers, and Charles Lyell, known for his concept of uniformitarianism. Kirtland founded the former Cleveland Academy of

Science, which eventually led to the establishment of the Cleveland Museum of Natural History. The museum incorporated in 1920 and had a geology department in place by 1922. Later the department was organized into separate departments of invertebrate paleontology, mineralogy, paleobotany, and vertebrate paleontology.

From its juncture with I-77 in downtown Cleveland, I-71 parallels the west edge of the Cuyahoga Valley and then turns west up the valley of Big Creek to Linndale. Wisconsinan-age ground moraine lies to either side of the East Branch Rocky River and I-71, sandwiched between the complex of Wisconsinan end moraines along the Allegheny Plateau of Cleveland's southern suburbs and the lake-planed till to the north. I-71 cuts through the end moraines from just south of Strongsville to the Ohio 18 interchange near Medina.

Cleveland Metroparks—the Emerald Necklace

The vision of chief engineer of Cleveland city parks, William Stinchcomb, in 1905, was to create parks in the Rocky River and Chagrin River valleys to the west and east of the rapidly growing city. Stinchcomb's vision, and the concern of homeowners along the Rocky River that their tranquil valley might be spoiled by commercial and industrial development, led to the formation of a county park board in 1912. This board envisioned a series of parks surrounding the city, including in the Rocky River, Euclid Creek, and Tinkers Creek areas. The board paid for a survey of natural lands in the county worthy of preservation. In 1917 the Cleveland Metropolitan Park District was established, the first such system in Ohio. In 1920 the district owned 109 acres in the Rocky River and Big Creek valleys, but by 1930 it owned 9,000 acres forming nine large, unconnected reservations: Bedford, Big Creek, Brecksville, Euclid Creek, Hinckley, Huntington, North Chagrin, Rocky River, and South Chagrin.

During the Depression, workers from the Works Progress Administration, Civilian Conservation Corps, and other relief organizations fashioned the land into parks. In the 1940s the park board set out to complete the Emerald Necklace—a chain of parks encircling Cleveland—by acquiring land connecting the reservations and by building parkways. Cleveland's metroparks consist of twelve reservations and the Cleveland Metroparks Zoo, and the system is connected to Summit County's metroparks system to the south by Cuyahoga Valley National Park—the largest strip of parkland in the state at 33,000 acres. These parks are great places to see some of Ohio's geological wonders up close, like fossil remains of giant predatory fish of the Devonian sea, scenic waterfalls, a maze of intersecting joints and rock shelters in Pennsylvanian sandstone called *ledges*, and sites of old building stone quarries.

Rocky River Reservation—Devonian
Fish and the Cleveland Shale

This narrow, winding park borders the Rocky River from I-90 south to Bagley Road, just west of I-71. Sixty-foot cliffs of Cleveland shale, the uppermost member of the Devonian-age Ohio shale, dominate the downstream portion of the park, hiding the remains of ancient fish and other creatures that frequented a sea that covered the Cleveland area 365 million years ago. Fish fossils and

The Devonian-age Cleveland shale member of the Ohio shale forms high banks along the Rocky River at Rocky River Reservation. Many fossil fish have been found in such outcrops. —Joe Hannibal photo, Cleveland Museum of Natural History

imprints can be found throughout the entire Ohio shale, but nowhere are they common or easy to find. Here along Rocky River and at other sites in Cuyahoga and Lorain Counties, pioneer collectors combed the slopes looking for fossils as early as 1870. Some surprisingly complete fossil fish skeletons came from the cliffs in the reservation, and many of them are now at the American Museum of Natural History in New York City and the British Museum in London. Later fossil finds were acquired by the Cleveland Museum of Natural History. When workers laid I-71 through the Cleveland shale of Cuyahoga County in the mid-1960s, the museum didn't miss the opportunity to add more fossils to their holdings.

Just outside the Rocky River Nature Center in North Olmsted there is a bronzed, outdated replica of *Dunkleosteus*, perhaps the best-known inhabitant of this Devonian sea. This fish was an arthrodire—a type of bony-plated fish—that reached a length of 18 feet and was certainly at the top of the food chain. It was a good swimmer and was speedy enough to catch sharks. Arthrodire skeletons are the most common vertebrate fossils found along the Rocky River: at least twenty-two species are known. Shark teeth and fin spines are a close second. Near the nature center there are steps that lead to the summit of the forested Fort Hill, where three built-up ridges, possibly ancient fortifications, continue to mystify archaeologists. At the southern end of the reservation, north of Bagley Road and off Barrett Road, is Berea Falls, where the East Branch Rocky River tumbles over the lip of the same Devonian Berea sandstone that was once quarried just to the south.

This fossil of Dunkleosteus terrelli, *a large Devonian-age arthrodire, came from the Rocky River valley. It's on display in Kirtland Hall of the Cleveland Museum of Natural History.*
—Cleveland Museum of Natural History photo

Big Creek—More Devonian-Age Fish

Early fossil collectors also favored a site east of I-71. Big Creek and its tributaries yielded carbonized compressions of Devonian sharks. Compressions are fossils that have been squeezed by overlying sediments, causing a flattening and slight distortion in a lateral direction of what used to be a three-dimensional animal. Much of the original organic matter remains, but typically it has been changed to coal. These fossils captured not only the hard parts but the animal's actual body outline as well. Fossil hunters uncovered complete sharks over 5 feet long. The collecting along Big Creek differed from that along the Rocky River in that the fossils occurred within flat concretions in low shale banks. The concretions formed around decaying fish bodies as calcite crystallized in the enclosing mud. The flattened elongate shapes of the fish bodies led to the flatter concretions. In some places nearly every concretion contained a shark body or arthrodire bone. The Cleveland Museum of Natural History resorted to a power shovel to uncover concretions from one of the streambeds before the area was built over in the mid-1920s. This area is now part of Linndale.

Upstream and south of I-480, Big Creek is part of Big Creek Reservation. The narrow park developed mainly in Wisconsinan ground moraine and extends from Brookpark Road south to Valley Parkway. A discarded block of glacial ice formed the depression that is now Lake Isaac, a kettle lake along Big Creek Parkway.

Berea—Grindstones and Berea Sandstone

Settlers of western Cuyahoga County made immediate use of the Berea sandstone they found along the Rocky River and its tributaries for foundations, hearths, and grindstones. John Baldwin, who settled in the Berea area in 1828 and later established Baldwin College (now Baldwin-Wallace University), was reportedly the first local to manufacture grindstones, in the early 1840s. Grindstones are rocks that have a gritty texture of hard quartz particles, a uniform texture that does not become smooth upon abrasion. Baldwin shaped the original stones with an axe, but he quickly developed a way of turning the stones on a water-powered lathe. The demand for stone to make grindstones led to the opening of quarries along the East Branch Rocky River in the 1840s. James Wallace is credited with opening the first sandstone quarry in Berea in 1844. He later founded Wallace College (now part of Baldwin-Wallace University). The nearby Clark and Brown Quarries followed in 1845 and 1846. Quarries near Olmsted Falls also produced grindstones by the 1860s. Workers crafted some of the stone into whetstones; smaller blocks of stone were also used to sharpen tools.

Between 1850 and 1870, people began using the Berea sandstone as building stone. The construction of railroads through the area led to the stone's wide distribution. In 1858 James McDermott & Company began operating and proceeded to develop into one of the largest building stone producers in Berea. The Berea Stone Company incorporated in 1871, taking over three companies and five more Berea-area quarries in the next ten years. By 1884 the largest sandstone quarry in Cuyahoga County was at Berea; small grindstones and building stone were its main products, and over one hundred men worked there. Another quarrying center was south of West View, where the sandstone reached a thickness of 70 feet.

Mill Stream Run Reservation

From Berea to Strongsville another metropark parallels the interstate—Mill Stream Run Reservation. It begins at Bagley Road, the southern boundary of the Rocky River Reservation. Mill Stream Run is a typical tributary of the East Branch Rocky River: it's a deep gash through glacial till and Mississippian-age shales and siltstones of the Cuyahoga formation. The Meadville shale and underlying Sharpsville siltstone members form the wall of the run on the east side of Strongsville. Following the East Branch Rocky River downstream into Berea,

Cleveland Stone Company No. 9 Quarry at Berea. —Ohio Department of Natural Resources, Division of Geological Survey photo

Pennsylvanian-age Sharon conglomerate forms Whipps Ledges in Hinckley Reservation.
—Kevin Diego photo

the exposed rocks get older. The Orangeville shale forms the river bottom in Berea. Berea Falls lies just north of downtown where East Branch Rocky River crosses over the resistant Devonian-age Berea sandstone, which is also exposed along the river. Baldwin and Wallace Lakes are former sandstone quarries that the Works Progress Administration converted to lakes around 1935.

Hinckley Reservation and Other Ledges

Heading east or west on Ohio 303 out of Brunswick leads to sandstone cliffs on either side of the Cuyahoga Valley. Although Hinckley's most famous for its buzzards, it's the geology of the area that attracts these birds. The buzzards nest among the jointed sandstone/conglomerate cliffs and shelves, known as Whipps Ledges, along the East Branch Rocky River south of the dam that creates Hinckley Lake. The tops of these resistant outcrops of Pennsylvanian-age Sharon conglomerate tower 350 feet above the lake. Look for cross-beds and honeycomb weathering. Below the conglomerate is the Meadville shale member of the Mississippian-age Cuyahoga formation. The boundary between the formations is an unconformity marked by natural springs. A considerable thickness of late Mississippian rocks—limestone, sandstone, and shales deposited between 340 and 320 million years ago—is missing even though it occurs in other places in Ohio.

Worden's Ledges is on the south edge of the park. A landowner carved the rocks into various images between 1945 and 1955, before this scenic area was preserved. The Medina County Park District is developing Princess Ledges Nature Preserve west of Brunswick; it offers more ledge scenery.

Medina to Westerville

The interstate traces the southern flanks of Wisconsinan end moraines of the Killbuck lobe from Medina to West Salem. At the I-76 interchange the route crosses a small oil and gas field that surrounds Westfield Center west of I-71. Sand and gravel mining scars are scattered throughout the Little Killbuck Creek valley. Lakes abound in this hilly region. Many of them are impoundments,

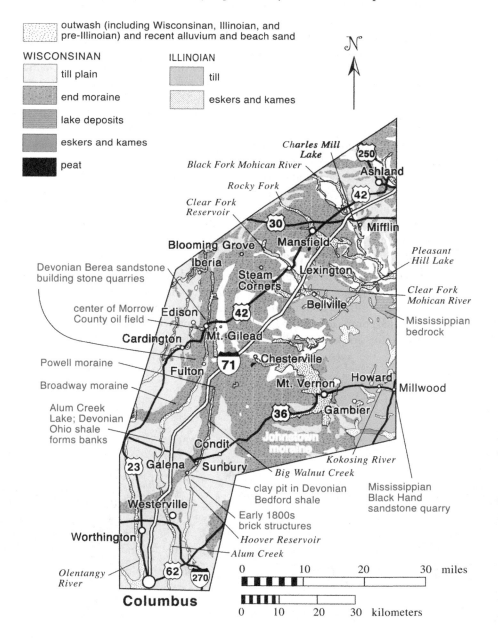

Geology along I-71 between the Ashland area and Columbus.

but others are kettles that formed as the stagnant ice blocks, left behind by the receding ice sheet, melted.

I-71 parallels the Killbuck Valley for a short stretch around the morainic hills of Burbank. The channelized Killbuck Creek flows south from Burbank to Wooster in a wide, terraced meltwater valley cut into Mississippian-age shales and sandstones. Numerous oxbow lakes and meander scars mark its former course. South of the interstate, near Redhaw, there is another oil and gas field. The field was discovered in 1915 and continues to produce from the Silurian Clinton sands. I-71 then crosses 36 miles of outwash-filled valleys and Mississippian sandstone uplands from Ashland to Chesterville—all sandwiched between where the Scioto and Killbuck lobes once rested. Kames and kame terraces are common near the valley walls of several meltwater channels. Just northeast of the interchange with U.S. 30, I-71 bridges the Black Fork Mohican River, which south of the interstate becomes Charles Mill Lake, a reservoir of the Muskingum Watershed Conservancy District. Rocky Fork meanders under the route on the southeast edge of Mansfield. Southeast of Lexington the valley of Clear Fork Mohican River, another meltwater channel, exhibits nice terraces near the interstate; sand and gravel pits tap this outwash. From northeast of the Lexington exit to southwest of it, the route skirts the edge of Illinoian till for about 8 miles. This till that rested between the Killbuck and Scioto lobes wasn't mixed by Wisconsinan ice. This rugged landscape lies along the western edge of the Allegheny Plateau from Lexington south to Newark and beyond.

Bellville—Gold in Them "Thar" Hills

After seeking a fortune in the California gold rush, a Bellville doctor discovered gold in his own backyard in 1853. He discovered the yellow metal in Deadmans Valley in the hills north of town. He and others panned the tributaries of Clear Fork, which commonly yielded small flakes; some people discovered a few nuggets. One fortune seeker sank at least one 40-foot shaft, but water problems

View to the north from a prominent Wisconsinan kame south of Lucas showing the typical glaciated Allegheny Plateau. Just to the south is the glacial boundary and the unglaciated plateau.

forced its early closure. The Bellville gold, like other Ohio gold, was of glacial origin. Ice movement transported the gold in quartz particles to the edge of the ice sheets, where it was concentrated in streambeds.

Building Stone Quarries of the Mt. Gilead and Mt. Vernon Areas

In Morrow and Delaware Counties, I-71 crosses and closely parallels Devonian-age Berea sandstone that is close to the surface. The Whetstone Creek valley, on the northeast side of Mt. Gilead, was the site of early building stone quarries. Twenty feet of stone was exposed along this creek, but it turned out to be of poor quality. Most of it went to bridge and building foundations and the quarrying operations suspended activity by 1900. Between Edison and Iberia, quarries worked the stone in the 1870s, but most of them closed by 1900. There were other quarries west of Steam Corners, south of Blooming Grove, and along Alum Creek west of Fulton.

Less than a mile north of Fulton, along an old railroad grade, there is the water-filled quarry of the Fulton Stone Company. Nearly 40 feet of Berea sandstone forms the walls of this pit, while 8 to 25 feet of till lie at the surface. The lower 27 feet of the sandstone is massive, lacking vertical fractures and prominent bedding planes. The Fulton Stone Company began quarrying around the Civil War. At first, workers quarried the stone by hand. The company introduced steam power in 1890 and saws in 1897. In the early 1900s, forty men worked the pit, producing 75,000 cubic feet of flagging, foundation stone, curbing, and building stone annually. Bucyrus, Columbus, Marysville, and other towns along the former Toledo & Ohio Central Railroad used it.

Quarries worked the Mississippian-age Black Hand sandstone in Millwood on U.S. 36, an industry dating back to 1906 in this town. The pureness,

The Fulton Stone Company quarried Devonian-age Berea sandstone at Fulton. The main product was building stone.

A pit dug into the Devonian-age Bedford shale. It belonged to the Galena Shale Tile and Brick Company in Galena.

intermediate grain size of the rock, and lack of a strong cement makes it an excellent glass sand. Not far from Millwood the unit becomes unsuitable because it lacks these qualities. The Black Hand sandstone is 40 to 60 feet thick along the Kokosing River in this area, and Berne sandstone overlies it. Much of the sand went to glass plants in Columbus, Coshocton, and Mt. Vernon; some went to make water filters at Delaware, Lima, and Struthers in the early 1900s. Other quarries operated in Cavallo, Gambier, Howard, Mt. Vernon, and Warsaw.

Morrow County's Oil Boom

I-71 follows the Powell moraine from Chesterville to north of Worthington. During Ohio's oil boom of the 1800s, Morrow County was not known for its natural gas or petroleum. Drilling records show that the county's wells were mainly dry holes after 1850. After the Trenton oil boom of 1890–1910, a major discovery of oil in the subsurface Ordovician Trenton limestone in northwestern Ohio, not much hope for economic prosperity existed for the area.

In 1958 a well on a small farm in Bennington Township hinted at deep-seated oil below the Ordovician-age Trenton limestone. A well drilled on the flat farmland west of Mt. Gilead in 1961 hit the jackpot. At 3,174 feet down drillers struck an oil pocket; oil and flaming natural gas gushed from the well, surprising everyone. The well produced 400 barrels of oil a day, and by 1963 it produced nearly a 250,000 barrels. The oil and gas came from the Cambrian-Ordovician Knox dolomite, which was deposited in Paleozoic seas. The irregular thickness of the formation is the reason why other wells sunk in 1961 and 1962 yielded little.

In 1963, oilmen drilled productive wells at Cardington and near Edison, and the boom was on. Drillers, geologists, investors, and others flocked to the area. Soon every speck of available land was under lease. Mt. Gilead's population more than doubled, and wells were put down with little thought to spacing and environmental concerns. They flared off the natural gas, which occurred in considerable quantities. Morrow County's oil rush was much like a Texas-style oil rush—a free-for-all with holes drilled everywhere and little thought of the future—since Ohio had virtually no oil and gas regulations in non-coal-bearing counties. Of two thousand wells drilled in the county between 1959 and 1964, some four hundred wells yielded around 40,000 barrels of oil in summer 1964. Emergency legislation in 1964 and 1965 sought to control the rampant development of the field, but by then the boom was ending. Some people still pump oil from deep beneath the county, but most of the scars of the oil boom are gone, reclaimed by nature and the hand of humans. Morrow County has produced some 40 million barrels of oil since that first well was sunk in 1959.

Sunbury and Galena—Sandstone and Bricks

The Devonian-age Berea sandstone also occurs close to the surface in the Sunbury area. Quarries opened along the valleys of Big Walnut and Rattlesnake Creeks and spread east of town. By 1855, Sunbury was the home of two major sandstone producers: the Sunbury Stone Company and Westwater Quarries, both of which were located on the west bank of Big Walnut Creek directly east of downtown. Flechner's Quarry, a midsize quarry, was located just south of downtown Sunbury, and a number of other small quarries stretched along Rattlesnake Creek and Big Walnut Creek toward Condit. The arrival of the Cleveland, Akron & Columbus Railroad in 1873 increased business for these quarries, and a stonecutting mill was erected. The stone for bridges along this rail line came mostly from the Sunbury Stone Company's quarries. Most quarrying in the Sunbury area ceased by 1903, and it's hard to find evidence of this once-flourishing industry.

It's not lead ore, as the town's name implies, but bricks that brought fame to Galena. Brick structures on the village square date as far back as 1826. Early residents knew of the suitability of local surface clays of glacial origin for brick

Galena's village center exhibits what was once a local product. The inn to the left was built between 1826 and 1828; the building to the right now serves as a post office.

manufacturing long before a commercial brick plant opened. Itinerant masons plowed up the clay, mixed it with water, placed the wet clay in wooden molds, and dried them in the sun. They then stacked the bricks loosely in layers separated by straw and kept a fire going under the stack for a week or so, at which point the bricks were ready for use.

The Galena Tile Company opened in 1893, producing both tile and brick. The plant initially operated only during the summer, but the 1924 discovery of Devonian-age Bedford shale just east of town sparked the industry to grow. The Galena Shale Tile and Brick Company built a short narrow-gauge railroad to connect the shale pit east of town to its kilns. Eighty people worked for the plant in the 1930s, making one hundred thousand bricks each day. In later years the plant made various bricks, including some with a unique glaze. In the 1980s, Galena Brick Products began buying crushed shale from a number of locations, and the local shale pit closed. The plant closed in the 1990s, but a few of the buildings still remain on Holmes Street.

INTERSTATE 77
Cleveland—Marietta
167 miles

I-77 originates in the heart of Cleveland on the shores of Lake Erie and passes over glacial sediments that fill a buried valley underlain by Devonian-age bedrock. Wisconsinan and Illinoian ice sheets barely made it past Canton, so I-77 crosses a band of Pennsylvanian-age bedrock from North Industry to Marietta, and rock of Permian age caps the uplands in Washington County.

Cleveland to Canton

From Lake Erie's shore to Royalton Road, I-77 passes over glacial lake deposits and alluvium of the Cuyahoga River, though much of this is covered by pavement and buildings. Hilly terrain at Royalton Road marks the Defiance end moraine, which parallels the Cuyahoga River and I-77 south to Peninsula. The moraine is narrow, and the interstate passes over it quickly and is on till plain again. Wisconsinan end moraines, including the terminal moraine—the end moraine marking the farthest advance of the ice sheet across Ohio—abut the higher Allegheny Plateau at Ghent Road. Many kames and kettle lakes are just west of the interstate between Ghent Road and Ohio 18. Between Ohio 18 and Ohio 241 south of Akron, the highway crosses a mixture of Illinoian- and Wisconsinan-age tills and outwash.

Natural gas comes from 3,000 to 4,000 feet below the interstate around the Ohio 241 exit. The gas-bearing formations are Silurian- and Devonian-age dolomites and sandstones. The field dates back to the 1930s. Wisconsinan outwash dominates the area near the Akron-Canton Regional Airport, which rests on artificial fill material. Abandoned mines in Pottsville group coal and clay seams are under the valley of the West Branch Nimishillen Creek east of the airport and I-77. Natural gas wells were prevalent from North Canton to Canton from the 1930s to 1960s. The natural gas came from Silurian-age rocks 4,000 to 5,000 feet down. Wisconsinan kames dominate the area north and west of North Canton and Canton, but at the North Canton exit the hills have been leveled for a shopping area.

Salt under the City and Lake

In 1886 a well drilled in Mill Creek valley near Newburgh Heights tapped a vast resource underlying Cleveland and adjacent Lake Erie—rock salt. The Cleveland Rolling Mill Company sunk the well hoping to find natural gas. Nominal gas yields appeared at 3,000 feet, but salt appeared before it about 2,000 feet below the surface. Drillers tapped four salt beds, ranging from 5 to 164 feet thick, down as far as 2,500 feet. This began the production of artificial brine, a process in which workers pumped water down a well to the buried rock salt, dissolving it, and pumped the brine back to the surface. The rock salt precipitated out of the Silurian sea that covered Ohio. Before this discovery, locals obtained salt from natural salt seeps and subsurface natural brine. The Newburg Salt Company operated from 1889 to 1902, pumping brine from this first well plus two others the company drilled in the Mill Creek valley. Other companies soon appeared, including the Cleveland and Union Salt Companies, drilling wells a few miles north of the original well.

The great thickness of rock salt led companies to develop underground mines in Cleveland. The International Salt Company opened a mine on a peninsula known as Whiskey Island in the Cleveland Flats, just west of the Cuyahoga River, in 1962. Shafts extend 1,700 feet down in this mine to tunnels that run $2\frac{1}{2}$ miles out under the lake and then parallel the shoreline. Miners remove about half of the rock salt, leaving the remaining salt as pillars to support the mine's roof. The current mine is one of the largest underground salt mines in North America and is a major producer of road salt.

Oil and Gas around Cleveland

The discovery of oil and gas in Silurian-age Clinton strata of central Ohio between 1900 and 1907 renewed interest in drilling in the Cleveland area. Prospectors deepened the old Newburgh Heights salt wells but had little success. In 1911 the Newburg Brick and Clay Company struck oil and gas 2,520 feet under its property near the intersection of Canal and Warner Roads in the Newburg sand (a drillers' term for part of the Silurian Lockport formation), and productive wells popped up in Lakewood in 1913 and 1914 in the same horizon. The strong initial play of the Stadler well—drilled to the Newburg horizon—in South Brooklyn in January 1914 started a minor gas boom in Cleveland; by April of that year, gas flowed from fifty-five wells. By January 1916 the western suburbs had around nine hundred wells. The average life of a well in Cleveland was about eight months. Some wells initially produced over 10 million cubic feet of gas per day. Drilling continued into 1916, but the boom ended as the success rate of the wells dropped below 50 percent. Two wells on the east side of town penetrated the Ordovician-age Trenton horizon—the formation that fed western Ohio's economy through the early 1900s—but both failed to yield oil.

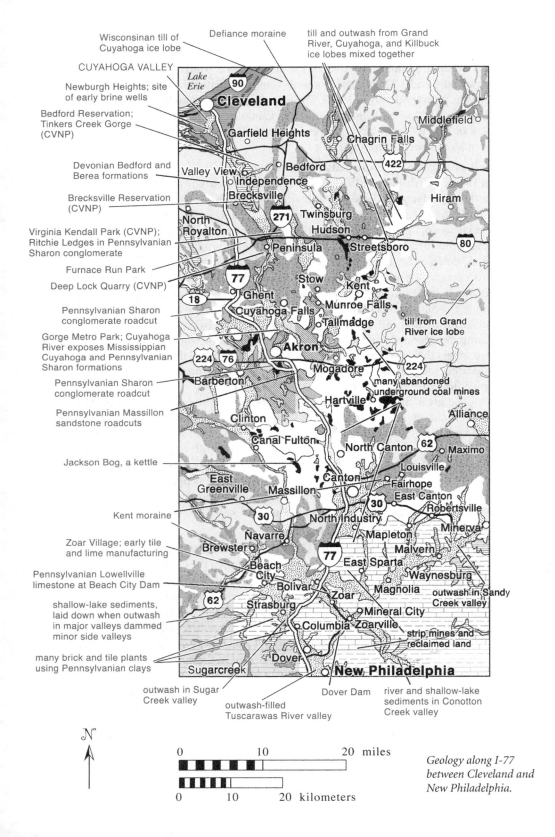

Wisconsinan till of
Cuyahoga ice lobe

Defiance moraine

till and outwash from Grand
River, Cuyahoga, and Killbuck
ice lobes mixed together

CUYAHOGA VALLEY

Lake
Erie

Cleveland

Newburgh Heights; site
of early brine wells

Bedford Reservation;
Tinkers Creek Gorge
(CVNP)

Garfield Heights

Middlefield

Chagrin Falls

Devonian Bedford and
Berea formations

Valley View
Independence

Bedford

422

Brecksville Reservation
(CVNP)

Brecksville

Hiram

North
Royalton

271

Twinsburg
Hudson

80

Virginia Kendall Park (CVNP);
Ritchie Ledges in Pennsylvanian
Sharon conglomerate

Peninsula

Streetsboro

Furnace Run Park

77

Stow

Kent

Deep Lock Quarry (CVNP)

18

Ghent

Munroe Falls

till from Grand
River ice lobe

Pennsylvanian Sharon
conglomerate roadcut

Cuyahoga Falls

Tallmadge

Gorge Metro Park; Cuyahoga
River exposes Mississippian
Cuyahoga and Pennsylvanian
Sharon formations

Akron

224   76

Mogadore

224

Pennsylvanian Sharon
conglomerate roadcut

Barberton

many abandoned
underground coal mines

Pennsylvanian Massillon
sandstone roadcuts

Clinton

Hartville

Alliance

Canal Fulton

North Canton

62

Maximo

Jackson Bog, a kettle

Louisville

East
Greenville

Massillon

Canton

Fairhope

East Canton

Kent moraine

30

North Industry

30

Robertsville

Minerva

Zoar Village; early tile
and lime manufacturing

Navarre
Brewster

Mapleton

Malvern

Pennsylvanian Lowellville
limestone at Beach City Dam

Beach
City

East Sparta

Waynesburg

shallow-lake sediments,
laid down when outwash
in major valleys dammed
minor side valleys

62

Bolivar

Strasburg

Zoar

Magnolia

Mineral City

outwash in Sandy
Creek valley

many brick and tile plants
using Pennsylvanian clays

Columbia

Zoarville

strip mines and
reclaimed land

77

Dover

Sugarcreek

New Philadelphia

outwash in Sugar
Creek valley

outwash-filled
Tuscarawas River valley

Dover Dam

river and shallow-lake
sediments in Conotton
Creek valley

N

0   10   20 miles

0   10   20 kilometers

*Geology along I-77
between Cleveland and
New Philadelphia.*

Newburgh Heights and Vicinity

Mill Creek Falls, a 45-foot waterfall, marks the site of many early mills that led to the establishment of Newburgh over two hundred years ago. The falls pours over a hard sandstone unit within the Devonian-age Bedford formation called the Euclid bluestone. This siltstone was quarried at two sites along Mill Creek near Warner Road around 1900. The Caine Stone Company operated a stone-cutting mill in Newburgh in 1913. Nearby a brick and tile plant worked the Bedford shale. As Cleveland industry surrounded the area, Newburgh—now Newburgh Heights—became part of the city, and sewage, garbage, and construction waste polluted Mill Creek as Cleveland grew.

Upstream of the falls in Garfield Heights, Mill Creek has carved banks in glacial sediments that date back to Illinoian time. Above the Illinoian sediments a Wisconsinan-age loess, windblown silt, yields fossil insects, spiders, mites, centipedes, pond and land snails, mastodon bone fragments, and plants.

A Side Trip to Cuyahoga Valley National Park

In 1974 President Ford signed legislation that created Ohio's largest parkland—Cuyahoga Valley National Recreation Area. It became a national park in 2000. It covers 33,000 acres and stretches 22 miles from the Cleveland suburbs to those of Akron, tying together a number of metroparks and state lands with newly acquired properties in the Cuyahoga River valley. The area offers scenic waterfalls and gorges, ledges, an old furnace site, and abandoned sandstone quarries. The Towpath Trail, which follows the historic Ohio & Erie Canal, bisects the park.

If visiting the park from the north, you should start your journey at the Canal Visitor Center in Valley View. Devonian-age Berea sandstone forms the lower part of this structure and a nearby lock. Quarries in the Cuyahoga Valley at Independence were active in the late 1800s. The upper and lower portions of 30 to 40 feet of Berea sandstone in these quarries made good grindstones and building stone. The quarries shipped the stone by wagon and canal to Cleveland as early as 1840. In the early 1900s the Euclid bluestone of the underlying Bedford formation was used for flagging. A brick plant in Independence took advantage of the shales of the Bedford formation.

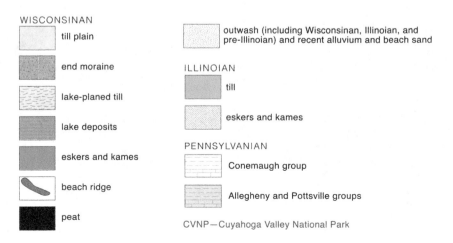

WISCONSINAN
- till plain
- end moraine
- lake-planed till
- lake deposits
- eskers and kames
- beach ridge
- peat

outwash (including Wisconsinan, Illinoian, and pre-Illinoian) and recent alluvium and beach sand

ILLINOIAN
- till
- eskers and kames

PENNSYLVANIAN
- Conemaugh group
- Allegheny and Pottsville groups

CVNP—Cuyahoga Valley National Park

From the visitor center follow Canal Road south and take Tinkers Creek Road east to visit the Bedford Reservation, a Cleveland metropark that surrounds Tinkers Creek Gorge. Tinkers Creek begins its twisting course near Hudson and Streetsboro, on the plateau, and ends cutting a 150- to 200-foot-deep, tree-lined valley to the Cuyahoga River west of Bedford. The fast-flowing water attracted mills to the upper gorge at Bedford. This became a favorite picnic site for Cleveland residents, known as Bedford Glens, during the early 1900s. The lower gorge consists of a series of waterfalls and rapids through Devonian sandstones and shales from the Berea sandstone down to the Chagrin shale. For a close look at the rocks, follow the trail that leads into the gorge from the Willis Street parking lot. The overlook on Gorge Parkway provides an overall view of this national natural landmark.

You can reach Bridal Veil Falls, the highest cascade in the park at 45 feet, by following a short boardwalk trail off Gorge Parkway. The lip of the falls displays Berea sandstone with ripple marks made by seas of Devonian time. Below this the water cascades over steplike layers of siltstone and shale. A plan to dam Tinkers Creek and flood this area was proposed in the early 1960s; fortunately, it was not approved.

The Brecksville Reservation, Cleveland's largest metropark, is on the west side of the valley off Ohio 82 near Brecksville. An overlook on the north edge of the park offers a view of Chippewa Creek Gorge. The waterfall at the head of this gorge tumbles over Berea sandstone. Below the waterfall, the creek cuts through Bedford, Cleveland, and Chagrin shales and siltstones, also of Devonian age. As the creek undercuts the softer shale, large slump blocks of Berea sandstone slide onto the creek bottom. A rock shelter, eroded into the Bedford shale at the Berea-Bedford contact, occurs along Deer Lick Cave Trail about 1.2 miles south of the Brecksville Nature Center. Skilled stonemasons of the Works Progress

A restored Ohio & Erie Canal lock in Cuyahoga Valley National Park near Valley View. Although most of this lock is concrete, some of the earlier building stone is still evident. The building to the left is Canal Visitor Center.

Administration and Civilian Conservation Corps built the center in 1939 with Berea sandstone they quarried on the reservation. Here and there in the park there are boulders of granite and other igneous and metamorphic compositions, glacial erratics dropped by the ice sheets that once covered the area.

Southeast of the Brecksville Reservation, Riverview Road follows the west side of the meandering Cuyahoga River. At Boston turn west on Boston Mills Road to reach the parking area for Blue Hen Falls. From the parking area a short trail leads to a 15-foot drop over Berea sandstone onto Bedford shale. Continue down the trail along Spring Creek to Buttermilk Falls, a 20-foot cascade over Bedford shale and siltstone. Furnace Run Park, a Summit County metropark, is farther west on Boston Mills Road. Rock Creek Trail from the Brushwood Area of Furnace Run Park leads past Bog Iron Pond, a marshy area probably excavated by iron miners in the 1800s. Sometimes a low-grade iron ore, from a combination of organic and inorganic processes, accumulates in wetlands. The name Furnace Run implies that there was an old charcoal-fired furnace in the area, but information on it is sketchy.

Brandywine Falls, the highest waterfall in the national park at 65 feet, is on the east side of Cuyahoga Valley National Park where I-271 crosses Brandywine Road. Boardwalks offer unobstructed, safe views of this waterfall in Indian Creek. Berea sandstone forms the lip of the falls, with Bedford and Cleveland shales below. Hike the 1½-mile trail into the inner gorge to get a closer view of the geology.

Lock 29, just north of Ohio 303 in Peninsula, is perhaps the best preserved of the many Ohio & Erie Canal locks in Cuyahoga Valley National Park. The lock dates to 1882 when stonemasons reconstructed it using Berea sandstone that was quarried by the local Peninsula Stone Company. The skill of the stonemasons is evident at this lock; the blocks have hand-tooled drafted margins and

Ripple marks and small potholes in Devonian-age Berea sandstone at the lip of Bridal Veil Falls in the Bedford Reservation.

dressed faces. Stonemasons used special hammers and chisels to create bevels, texture, and decorative patterns on the surfaces of quarried blocks once they were set in place. A dressed face was one that had had its irregularities removed through tedious chiseling along bedding planes.

Peninsula Stone was one of several companies that operated sandstone quarries around Peninsula. Records indicate that it had a quarry as early as 1837 and distributed stone by canal to Cleveland and Akron. The Cleveland Stone Company had acquired properties in Peninsula by the 1900s. Quarries Number 15 and 16 operated just north and west of town. Deep Lock Quarry, another Summit County metropark, is south of Peninsula and east of River-view Road. The park has an entrance sign made of a grindstone, one of the products of this quarry. Opened in 1829, this quarry has the typical benched walls of a building stone quarry. Channeling machines left vertical grooves in the walls, and natural joints are evident on the quarry floor. Initially, stone-masons used the quarry's Berea sandstone along the canal; later, the quarry's owners focused production on grindstones for stripping the hulls from grain at the mills in Akron. The devastating 1913 Flood on the Cuyahoga River ended production at this site.

Peninsula was also the site of a glacial lake, which formed during late Wisconsinan time when meltwater ponded between the present-day Grand River and Killbuck lobes. Under the ski slopes north of town, lake clays and silts lie sandwiched between two Wisconsinan tills that rest on top of Chagrin shale. Similar lake deposits occur at several other sites from here to Akron.

Deep Lock Quarry near Peninsula provided building stone and, later, grindstones of Berea sandstone.

Virginia Kendall Park, the first unit opened in the national park, is southeast of Peninsula. Another national park visitor center is on Ohio 303, and it is a good starting point if you are visiting the park from the south. The Ritchie Ledges of Pennsylvanian-age Sharon conglomerate rise above the Mississippian-Devonian rocks of the valley at this park. Joints in the rock allowed deep crevices, shelters, and slump blocks to develop, as is typical of Pennsylvanian rocks across northeastern Ohio. Follow the Ledges Trail for many views of heavily jointed rock with cross-beds, ripple marks, and honeycomb weathering. From an overlook in the park you can see similar ledges on the opposite side of the Cuyahoga Valley, including Whipps Ledges and Worden's Ledges at Hinckley Reservation.

Ice Box Cave is not a cave but a crevice that reaches 50 feet back into the Sharon conglomerate. This crevice formed when blocks of Sharon conglomerate broke loose from the wall of this valley along a joint, or plain of weakness, and slowly slid downslope on top of the weaker underlying Mississippian shale. Along the wall of Ice Box Cave there is what appears to be a channel filled with pebbly sediment. Geologists interpret this rock as having been a pebble-filled pool at the end of a sandbar. Currents swirling around the bar in whirlpool-like fashion formed it, allowing only the coarsest sediment to settle out. Along the wall there is also honeycomb weathering and some impressive iron banding. This kind of weathering is related to the migration of iron ions to the surface, where they precipitate as hematite (an iron oxide) and arm the rock with a hard crust. Areas that lack the iron crust crumble more readily and erode back into the wall. The swirling patterns of reddish brown and yellowish bands are related to the movement of iron ions through the sandstone and precipitation that occurred under changing chemical conditions, which either oxidized or reduced the iron.

Akron Area

Akron sits within the portion of the Allegheny Plateau that was glaciated. During Wisconsinan time, a small lobe of ice—the Cuyahoga lobe—situated between the Grand River and Killbuck lobes, stopped just north of Akron, depositing a stony till. Wisconsinan outwash followed and fills a preglacial valley, at places over 150 feet deep, between Cuyahoga Falls and Stow. A number of kettle lakes, namely Crystal, Silver, and Wyoga Lakes, formed from leftover ice blocks, occur on either side of Ohio 8. The Portage Lakes west of I-77 are also of ice block origin. Pennsylvanian-age Massillon sandstone appears in I-77 roadcuts at the Ohio 21 junction, just south of the I-76 junction, along the I-277 bypass of Akron, and at the Ohio 764 interchange. The Sharon conglomerate also appears along the Cuyahoga River where it flows through town.

The Cuyahoga River originates in Geauga County, flows southwest to Kent, then west to Cuyahoga Falls, south through downtown Cuyahoga Falls, west across northern Akron, and then turns north again to flow through Cuyahoga Valley National Park, inscribing a roughly U-shaped course. The river was an early source of power where its channel narrowed and followed a rock-walled gorge through the northern edge of Akron. The town of Cuyahoga Falls developed around this gorge in 1812, and by 1840 factories and mills blanketed the river margins. The "falls" are actually rapids that extend 2 miles across resistant

Pennsylvanian Sharon conglomerate. Munroe Falls, upstream, was also named for the rapidly flowing water. The portion of the Cuyahoga River valley below Prospect Street in Cuyahoga Falls became High Bridge Glens in 1882, when a dance hall and various amusements were added to supplement natural attractions like Fern Cave, a small, damp rock shelter in the Sharon conglomerate cliff noted for its many small ferns; Observation Rock, offering scenic views of the Cuyahoga Valley; and Weeping Cliffs, where groundwater seeped down the face of Sharon conglomerate along a line of springs. The 1913 Flood greatly damaged this popular park, leading to its closure in the 1920s. Today the Glen Trail of Gorge Metro Park leads through portions of the old Glens.

Gorge Metro Park, separating Cuyahoga Falls and Akron, stretches from State Road east to Ohio 8 along a dammed portion of the Cuyahoga River. Before the dam, the gorge had swirling rapids, the Big Falls (where the dam is now), and islands formed from large slump blocks. Two hundred feet of Mississippian and Pennsylvanian rocks form the walls of the 1½-mile-long gorge. Hike downstream of the dam to see the oldest rocks of the park—the Orangeville shale of the Mississippian Cuyahoga formation. The Sharpsville and Meadville members occur above the Orangeville member. The rock forming the upper 35 feet of the gorge walls is Pennsylvanian Sharon conglomerate. As elsewhere in northeastern Ohio, an unconformity occurs at the base of this conglomerate where it meets the more impermeable Mississippian Meadville shale and is marked by a line of springs. The conglomerate occasionally

Mary Campbell Cave is a rock shelter capped with Sharon conglomerate in Gorge Metro Park in Akron. —Annabelle Foos photo

contains fossilized logs of an ancient swamp plant called a *scale tree*, or *lycopod*. Its bark had a distinctive diamond-shaped pattern. Cross-bedding is also quite evident in this formation. The best-known geologic feature in Gorge Park is Mary Campbell Cave, originally called Old Maid's Kitchen, which is located along the Gorge Trail. It has been a popular site since the mid-1800s, when visitors scrambled down from the plateau above. It's a rock shelter that formed when blocks of the conglomerate collapsed and fell from the jointed ceiling. The floor of the shelter is the Meadville shale member. If you look closely, you might see that the shale is fossiliferous; small brachiopods are common. The Little Cuyahoga River also passes through Akron in a deep cut it incised in the plateau. East of the Ohio 8 bridge over the Little Cuyahoga River is the site of Old Forge, one of the early charcoal-fired furnaces in Akron.

As elsewhere, locals used the Sharon conglomerate as sand aggregate for cement and as an ingredient of firebrick in the Akron area. A pond on the Stan Hywet Hall and Gardens grounds east of Fairlawn is a sandstone quarry, which was operated by the Akron White Sand and Stone Company in the 1890s. Wolf's Quarry near Akron, which operated in the mid-1800s, marketed a unique purplish sandstone. Much of it went to Cleveland. Other quarries operated along the gorge through Cuyahoga Falls. Quarries also produced building stone in the center of downtown Akron where office buildings and the Municipal Building now rise. East Bowery Street, which cuts through the heart of Akron, was known as Quarry Street for many years, attesting to the number of quarries that existed in town.

Pennsylvanian-age clays are another resource of the Akron area. In 1823, workers fired up kilns at Mogadore, east of I-77. Shallow pits in this town exposed a 6- to 10-foot-clay layer at the Quakertown coal horizon. The clay was ideal for stoneware, and potteries opened in Akron, Cuyahoga Falls, Middlebury (later East Akron), and Tallmadge. So many potteries produced wares that the local market became saturated, so entrepreneurs peddled stoneware to increasingly distant communities by wagon and, later, by canal and railroad. In the 1870s, Summit County potteries led the state in stoneware production. The success of the potteries eventually led to the exhaustion of the local clay, and area companies stopped producing kitchenware in 1915.

Factories also fashioned Pennsylvanian-age clays and shales, including the Anthony and Bear Run shales, into roofing tile and sewer pipes beginning in the 1860s. Sprawling plants and clay pits dotted East Akron. Firms included the Akron Sewer Pipe, Buckeye Sewer Pipe, Crouse Clay Products, Robinson Clay Product, and Windsor Brick Companies. Other plants were in Barberton, Mogadore, and Tallmadge. One by one these plants closed, because by 1910 most of the easily acquired clay was exhausted. The Camp Brothers, Robinson Clay Product, and U.S. Stoneware Companies were the last to operate in the Akron area.

Some of the earliest coal mining in Ohio took place near Tallmadge around 1810. The Sharon coal appeared around the margins of Coal Hill, an isolated knob between Tallmadge and Cuyahoga Falls, and companies had numerous drift mines in operation by the 1820s. They sold most of the coal to local blacksmiths. Early miners of this seam, Asaph Whittlesey and Henry Newberry of Cuyahoga Falls, both sired sons who became famous geologists—Charles Whittlesey and John Strong Newberry of Cleveland. Attempts to convince steamship owners to use this Sharon coal as fuel failed until the 1840s; wood was too cheap and abundant. After the 1840s, however, wood was no longer so abundant. The opening of the Ohio & Erie Canal provided Tallmadge coal companies access to a ready market—Cleveland. The Talmadge [*sic*] Coal Company had control of most of these mines by 1838 and supplied most of Cleveland's coal needs until 1845 when coal from the famous Brier Hill mines of Youngstown entered the Cleveland market. By the 1880s, the Tallmadge-area coals were depleted.

Coal Hill became the site of Akron's first enclosed shopping mall—Chapel Hill Mall. In East Akron, the Middlebury shaft tapped a 5-foot seam and was one of the largest in the district, but it and other mines were nearing exhaustion by the early 1880s. The Brewster Mines lie to either side of the I-76 interchange. The mines tapped the Sharon coal and were closed by 1875. In 2001 subsidence over an old mine near the Akron-Canton Regional Airport forced I-77 lane closures while the mine was filled with a fly-ash slurry, a waste product of power plants.

Jackson Bog

There are high sandy ridges off Ohio 687 west of I-77 at North Canton, part of the Kent moraine complex; they formed between the Killbuck and Grand River lobes of the Wisconsinan ice sheet. It's an area of mixed sediments that were deposited by ice and meltwater. Blocks of ice melted away to form kettles; sand and gravel that filled holes and crevasses in the ice slumped to the ground as kames as the lobes retreated. Boreal plants colonized areas in front of the ice sheet as it retreated northward, and in a few areas like Jackson Bog State Nature Preserve they continue to flourish today.

North of the bog there is a long kame. Springs at its base discharge cool, calcium-rich groundwater to the bog. Jackson Bog is a misnomer; it's actually a fen—a wetland underlain by calcareous sediments, or marl. Beyond the marly soil, vegetation is more typical of northeastern Ohio. Fens, like Jackson Bog, are rare in Ohio because communities have drained most of them. Large segments of similar wetlands disappeared to make way for commercial strips in the late 1990s; sand and gravel operations like those east of Ohio 77 also took their share.

Canton—Paving Brick Capital of the World

As with Summit County to the north, Stark County environs are blessed with near-surface Pennsylvanian-age clays and shales; glacial cover is thin. Early settlers found that local clays made good bricks and stoneware. Potteries sprang up in the early 1800s and brick masons flourished. The first known brickyard in Canton appeared around 1820 somewhere east of present-day McKinley Avenue and south of 12th Street NW. Slowly the potteries and small brick firms gave way to larger brick and stoneware plants. At least fifteen new companies began to manufacture clay products in Canton between 1886 and 1895. The town's production peaked between 1891 and 1900. Around 1900 the clay industry started to mechanize and mergers and closures resulted. Metropolitan Paving

Brick Company, formed in 1902, and Belden Brick Company, formed in 1912, emerged as the leaders in Canton. They mainly used the Lower and Middle Kittanning clays and Clarion shale. Metropolitan focused on pavers, while Belden offered a diversity of construction brick products. In 1923 Metropolitan alone produced 92 million pavers, shipping them throughout the United States.

Little trace of the brick industry in Canton remains. The I-77–Fulton Road interchange is on the site of the old Williams Brick plant; Fawcett Stadium, a football stadium built in 1939, resides in the plant's shale pit. Water-filled pits remain on the southeast edge of town and are easily seen from the interstate. Metropolitan Block pavers are common in streets across Ohio and elsewhere, but the company was out of the brick business by the early 1980s. Belden Brick continues to operate plants near the towns of Sugarcreek and Strasburg.

Canton to New Philadelphia

I-77 and the city of Canton lie on Wisconsinan outwash that was dumped between the Grand River and Killbuck lobes. A narrow, thin band of Illinoian till extends from Canton to North Industry, where the highway passes onto unglaciated Allegheny Plateau. I-77 follows the Tuscarawas Valley from Canton to Bolivar, and then it crosses over to the valley of Sugar Creek and follows its valley from Strasburg to Dover and New Philadelphia. The Tuscarawas River begins east of the interstate near Hartville and flows west and south nearly 130 miles before entering the Muskingum River at Coshocton. Tributary valleys below Navarre contain lake sediments from the time when thick valley train deposits—deposited by glacial meltwater—blocked their creeks from entering the Tuscarawas River and Sugar Creek.

Beneath the Fohl Street interchange, south of downtown Canton, a mine tapped the Middle Kittanning coal until 1932. Just north of Strasburg and at Columbia there are the remnants of firebrick plants, which used Lower Kittanning clay. Many other towns also had clay plants at one time (*see* table on p. 215). South of Bolivar, reclaimed land and surface mines in the Strasburg and Lower Kittanning coals are evident.

A Side Trip in the Tuscarawas Valley

At the Bolivar exit, Ohio 212 leads south to Zoar, in the Tuscarawas Valley, where Zoarists manufactured roofing tile and lime as early as 1820. The Zoarists were separatists who came to the U.S. to practice their religion. They came from Germany, where roofing tile had been used for centuries. Upon settling in Ohio, they immediately prospected for suitable building materials. They dug clay from fields and hauled it to tile yards where it soaked for twenty-four hours in pits. Workers then kneaded it with their bare feet. Once covered with straw, the clay stayed ready for use. The Zoarists used wooden molds to shape the tiles, which were then air-dried. In the final step, they burned the tiles in updraft kilns. Although the Zoar settlement disbanded in 1898, their workmanship survives in the Ohio Historical Society's Zoar Village; many privately owned Zoarist structures have been restored.

The Zoarists found a good source of lime in the local Pennsylvanian-age Upper Mercer limestone. Lime Kiln Lake, a water-filled pit, is a legacy of bygone days. Zoarists also discovered Blackband ore lying just above Upper Freeport coal in the 1830s. The ore is black shale impregnated with 25 to 40 percent iron. Mines dotted the hilltops between Mineral City and Dover, west of I-77 and southwest of the town of Stone Creek, along Oldtown Creek, and along Postboy Creek east of Postboy. The ore went to furnaces at Dover, Glasgow, and Zoarville. The Dover Furnace, erected in 1855, consumed 13,000 tons of Pennsylvanian iron ore, 7,000 tons of limestone for flux, and 20,000 tons of coal yearly to make 5,000 to 6,000 tons of foundry iron. Tuscarawas Coal and Iron Company quarried a sandstone, probably the Homewood sandstone, for firestone in the early 1870s near Zoarville. Fairfield Brick Company operated from 1925 to 1985 at Zoarville, producing various face and construction brick.

Strip mines gird many of the hillsides between Zoar and Zoarville. Ohio 212 travels through the Muskingum Watershed Conservancy District, bottomlands that were set aside in 1933 for flood control and conservation. The devastating 1913 Flood initiated the planning for the conservancy's extensive system of dams, reservoirs, and recreation areas. It's the largest in the state, covering eighteen counties. If major flooding occurred, the impoundment of water behind the Dover Dam south of Zoarville would submerge Ohio 212 between Zoar and Zoarville.

Egypt, Johnstown, Mineral City (formerly Mineral Point), New England, and Somerdale are other early coal and iron mining towns. Egypt and New England are bona fide ghost towns. The Fairfield Furnace, circa 1853, was 3 miles southwest of Mineral City. It produced pig iron. Near town there were firebrick plants and coal and iron mines in Lower and Middle Kittanning coal and clay. Eagle Hill, which rises above Ohio 800 near Dover Dam, was the site of an old coal mine and sandstone quarry. The mine and quarry disappeared during strip-mining that came later. The foundations of many local buildings came from stone from this quarry.

Tuscarawas Valley Drainage

At Strasburg, Sugar Creek looks incapable of carving its wide valley. That is because it's a misfit stream that appeared after drainage changes during Wisconsinan time. The mature valley belongs to the ancient north-flowing Dover River, a tributary of the preglacial Erigan River.

Before the Ice Age, some 2 million years ago, the north-flowing Dover River drained the central part of Tuscarawas County. The preglacial Cambridge River system drained the southwestern corner of the county, and Zoar Creek, which flowed into the Dover River north of Tuscarawas County, drained the northeastern corner. A pre-Illinoian ice sheet in northeastern Ohio had blocked the Dover River by at least 240,000 years ago, causing a reversal of flow. This became the Newark River, which flowed through a low spot in the former drainage divide between Uhrichsville and Newcomerstown and eventually connected with the Teays River drainage network in central and western Ohio. The headward erosion (erosion of a stream in its source, or headwaters, region) of the preglacial Newcomerstown Creek and an unnamed tributary of the Dover River had probably carved this low notch through the ridge before the glacial advance. The tributaries of the Dover River became barbed tributaries of the Newark River. Barbed tributaries project off the main channel in a downstream direction instead of upstream, which is the normal direction.

Zoar Creek's flow was also reversed at this time; it eventually broke through a drainage divide near Zoarville and cut a new channel southwest to Dover, where it joined the Newark River. By Wisconsinan time the path of the modern Tuscarawas River was defined, utilizing several preglacial channels. As the Killbuck lobe receded northward, headward erosion extended the head-waters of the river northward into Stark County.

New Philadelphia to Cambridge

At New Philadelphia, I-77 rejoins the Tuscarawas Valley. Dover and New Philadelphia were major ceramic centers because of the Pennsylvanian-age clays underneath and around them. Dover was also the site of the first Ohio salt furnace, which was erected in 1865. The brines that this furnace boiled came from about 900 feet down, in the Berea sandstone. Three furnaces operated in Dover in the late 1880s until a drop in salt prices caused their owners to abandon them. At New Philadelphia the Tuscarawas River heads to the southeast while I-77 follows the narrower Stone Creek valley south through strip-mined land. High banks of Pennsylvanian-age rocks are overgrown with grass in many

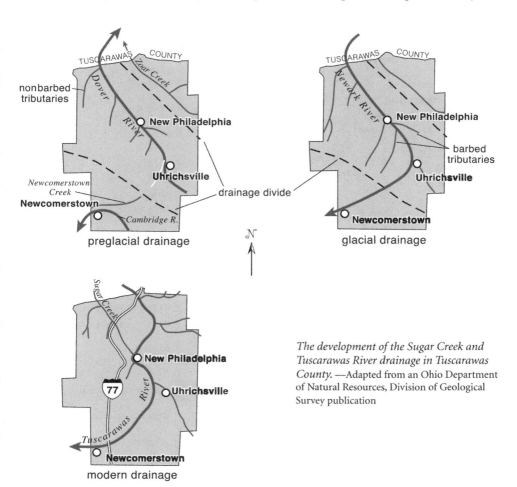

The development of the Sugar Creek and Tuscarawas River drainage in Tuscarawas County. —Adapted from an Ohio Department of Natural Resources, Division of Geological Survey publication

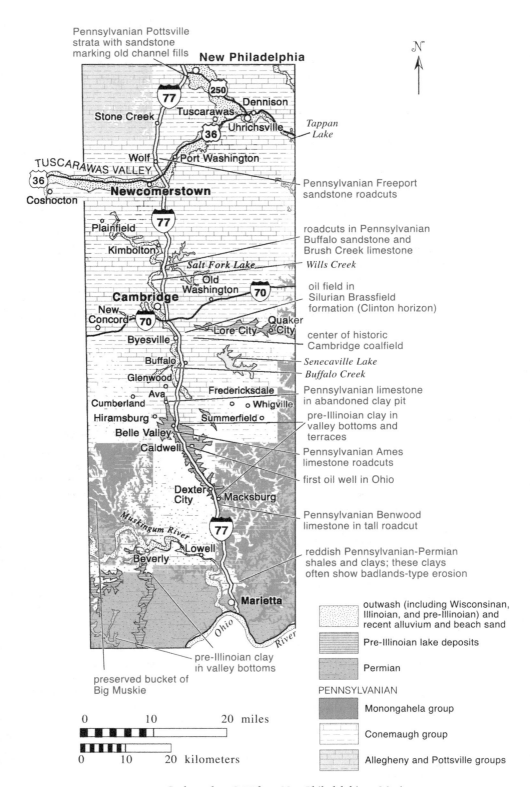

Geology along I-77 from New Philadelphia to Marietta.

places. Between New Philadelphia and the town of Stone Creek lies the ghost town of Blackband, named after a local Pennsylvanian-age iron ore. Pioneers in the hills of Tuscarawas County were aware that the landscape was blessed with mineral riches, including iron ore, as early as the 1840s. Blackband came into existence as a station on the Cleveland & Marietta Railroad in the 1870s. About 5 miles west of town there was a high ridge underlain by a 10-foot layer of Blackband iron ore, an unspecified thickness of Mountain iron ore, and an underlying 4 feet of Upper Freeport coal of the Allegheny group. The iron occurred in black, organic-rich shale, so it didn't look like iron ore. Upon weathering the shale crumbled, leaving tiny rusty flakes of ore. The Mountain ore is a calcareous layer with nodules of siderite, an iron carbonate. Settlers exposed the ore by stripping away the hillside. The coal was used locally for heating homes and businesses; the ores were hauled by wagons downhill to the railroad depot at Blackband for shipment, mainly to Dover and Massillon, where it was turned into pig iron or mixed with other components to make casting iron. Iron mining at Blackband, as well as other Tuscarawas County locations, ended as higher-quality iron ore began to come from the newly opened Lake Superior iron ranges in the later 1800s.

A clay pit operates at Stone Creek. Pennsylvanian Upper Freeport sandstone outcrops between the Stone Creek and Newcomerstown exits; it fills ancient channels that streams and rivers cut into an ancient delta some 320 million years ago. Near Wolf it forms Table Rock, one of a number of erosional remnants that formed as joints were eroded into the sandstone and eventually reached the softer shale underneath. I-77 swings east around Newcomerstown, once again crossing the wide Tuscarawas Valley. A brick plant operated just west of I-77 at Mizers (now part of Newcomerstown).

Farther south, roadcuts near Kimbolton expose thick Pennsylvanian Buffalo sandstone that exhibits nice cross-beds. The courthouse in Cambridge is made of this sandstone, which came from quarries north of town and at Cumberland. Cambridge became the site of many glass plants and is well-known for Cambridge glass. The Cambridge Glass Museum, featuring Cambridge's glass-making history, is on Jefferson Avenue (*see also* the I-70 and U.S. 40 road guide). Just northeast of Cambridge is Salt Fork State Park, Ohio's largest state park. It lies on the shore of Salt Fork Lake, created in 1967 with the completion of the earthen Salt Fork Dam. Pennsylvanian sandstone abounds along the Pine Crest Loop Trail. The Kennedy Stone House, built with local sandstone in 1837, is the highlight of the Stone House Loop Trail.

A Side Trip on U.S. 36

From Newcomerstown, U.S. 36 heads west down the Tuscarawas Valley to the headwaters of the Muskingum River at Coshocton. At Coshocton, U.S. 36 follows the Walhonding River valley, which is carved in Mississippian-age Cuyahoga formation, northwest to Nellie. North of Randle, where Killbuck Creek joins the Walhonding River, there is a broad bottomland. Tremendous volumes of meltwater gushed through here as the ice sheets receded in late Wisconsinan time.

Hilly strip-mined and reclaimed land dominates the landscape from Newcomerstown to Coshocton. Numerous small coal mines peppered the

Coshocton hills in the late 1800s (*see* table on p. 215). The Middle Kittanning coal was the most extensively mined seam in this area.

Between 1858 and 1860, miners dug into the Bedford coal of the Nellie and Warsaw area to produce coal or illuminating oil. The Bedford coal, a cannel coal made up of the microscopic remains of algae, formed a 6-foot seam along Simmons Run, south of Warsaw. Soon a dozen distillation plants operated at various local mine sites, making illuminating oil. The promising industry received a deathblow when Colonel Edwin L. Drake discovered petroleum in western Pennsylvania in 1859, and oil became more readily available and cheaper; most of the plants came down during the Civil War, and some of their workings were recycled and cast as ammunition.

Cambridge Coalfield

The Cambridge area has a long coal mining history. Miners dug Pennsylvanian-age coal from many locations. Between I-70 and the Buffalo interchange, I-77 crosses over one of the more-important early mining districts in Ohio. The Cambridge coalfield, once underlain by 4 feet or more of Upper Freeport coal, was centered around Byesville. Miners concentrated on the seam along Leatherwood Creek east of Cambridge and along Wills Creek south of town with drift mines. Important mines in the 1880s were the Akron and Cambridge, Buffalo, Cambridge, Central, Guernsey, Manufacturers, Nicholson, and Scotts (associated with a pioneer saltworks).

South of Belle Valley on Ohio 821 a coal pavilion commemorates the local industry. Roadcuts on both sides of the interstate south of Belle Valley display the fossiliferous Ames limestone, which represents one of the last advances—recorded in the rock record—of a Paleozoic sea into Ohio. Shales between the Ames and Cambridge limestones became pavers at Glenwood beginning around 1906, and more bricks composed of the same shale were produced at an Ava plant.

Big Muskie

In 1952 the Ohio Power Company began large-scale stripping of the Meigs Creek coal at the Cumberland Mine near Cumberland. A cross-country conveyor transported the coal about 15 miles to a power plant at Relief in the Muskingum Valley. Ten years later the Central Ohio Coal Company, a subsidiary of American Electric Power, mined Meigs Creek coal at its Muskingum Mine southwest of Cumberland. Bucyrus-Erie, a Wisconsin-based manufacturer of mining shovels and draglines, built the world's largest walking dragline—a large excavating machine—at the mine between 1967 to 1969. Known as Big Muskie, the dragline was as wide as an eight-lane highway and its cab was six stories up. Each scoop of its shovel removed 325 tons of dirt and rock. It could move over 19,000 tons of material in an hour. Seven men composed the crew of this electric-powered behemoth. Everyone who saw the machine left the mine in awe; the machine could literally move mountains. The company also constructed the first electric automated railroad in North America to move coal from the mine to a coal preparation plant 15 miles away

The dragline remined areas that miners had already worked with smaller shovels. Big Muskie easily removed overburden so smaller shovels and dozers could take up the coal, and then the dragline replaced the rock and soil. Big Muskie operated from 1969 to 1991, when surface mining ended at the Muskingum Mine due to environmental restrictions and changes in the power industry. The machine sat idle as reclamation of the Muskingum Mine continued. Then in 1999 the machine was dismantled on site, as is common with huge mining machines since it is not feasible to move them whole. The only part the company preserved was the huge bucket, which they trucked, very slowly, to reclaimed mine land along Ohio 78, 9 miles east of McConnelsville. American Electric Power made it the centerpiece of a park commemorating employees of the Central Ohio Coal Company in 2001. Some 10,000 acres of land mined by Big Muskie and other earthmovers became The Wilds, a wildlife preserve for endangered mammals, in 1986.

Ohio's First Oil Well

West of the Alleghenies salt was a precious resource in the early 1800s. People needed it as a preservative, and it commanded a high price when it was shipped in from the East Coast. In 1814 two Noble County men embarked on a venture to provide salt locally. They knew of a salt lick along a tributary of West Fork Duck Creek in what is now Caldwell. They drilled a well, striking not only brine but also gas and oil at about 200 feet; it became Ohio's first oil well.

Outbursts of gas occasionally blew water 30 feet in the air, creating quite a spectacle. The initial well produced around 1 barrel of oil each week. The oil they recovered was foul smelling, and they sold it under the name Seneca Oil, a medical cure-all. They boiled the brine in kettles to produce salt, which sold for $2 dollars a barrel. In 1816 the thriving saltworks drilled a second, deeper well near the first one. The saltworks burned to the ground in 1831 and was never rebuilt.

Drilling returned to the Duck Creek valley in 1860, prompted by Colonel Edwin L. Drake's 1859 discovery of oil in western Pennsylvania. One prospector found a good yield of oil and gas within 60 to 300 feet of the surface. Soon der-

---

OTHER MINERAL INDUSTRY TOWNS ALONG I-77

Towns with a Clay Plant

Alliance (Lower Kittanning clay)	Malvern (Lower Kittanning clay)
Carrollton (Lower Kittanning clay)	Mapleton (Lower Kittanning clay)
Coshocton	Maximo (Lower Kittanning clay)
East Canton (Lower Kittanning clay)	Minerva (Lower Kittanning clay)
East Greenville (Lower Kittanning clay)	Newcomerstown
East Sparta (Lower Kittanning clay)	North Industry (Lower Kittanning clay)
Fairhope (Lower Kittanning clay)	Port Washington
Louisville (Lower Kittanning clay)	Robertsville (Lower Kittanning clay)
Magnolia (Lower Kittanning clay)	Waynesburg (Lower Kittanning clay)

Towns with a Coal Mine

Bakersville	Mohawk Village	Tunnel Hill
Conesville	New Moscow	Warsaw
Linton Mills	Plainfield	Wills Creek

ricks sprouted like weeds between Caldwell and Macksburg. A particularly good yield of oil came from a well along Cow Run near Macksburg in Washington County, from a shallow Pennsylvanian-age sandstone that has since been named Cow Run sandstone. Wells were drilled deeper and deeper between Macksburg and Caldwell, striking oil at several depths in various Pennsylvanian-age sandstones. In its prime the Macksburg field, which this oil field came to be known as, produced 3,500 barrels daily. It continued to produce through the 1880s and led future prospectors to explore more Pennsylvanian-age sandstones of southeastern Ohio.

Most of the drilling today is to the Silurian-age Clinton horizon some 5,500 feet down. Wells around Byesville are in this horizon. The Byesville Museum on East Main Street has interesting mining displays. You can still see the 1816 well at a park east of Caldwell at the Ohio 78 and 564 junction; road crews destroyed the older well while constructing Ohio 78.

Cambridge to Marietta

The area around Caldwell was an important coalfield from the 1870s to 1914. Miners excavated Upper Freeport, Pittsburgh, and Meigs Creek seams at Belle Valley, Caldwell, Fredericksdale, Hiramsburg, Summerfield, and Whigville. East of Caldwell along Ohio 78, there are outcrops of the fossiliferous Ames limestone. From Caldwell to Marietta the roadcuts along I-77 have a reddish color. Many of

Oil wells tap Pennsylvanian-age sandstones north of Marietta in 1907.

the roadcuts show badlands-type erosion where the clay-rich strata have been riddled with many small incisions that channel rainwater down the hillsides.

The Monongahela group and Permian strata, a mixture of coal, sandstones, shales, limestones, and mudstones (fine-grained rocks without any evident stratification), along this stretch of interstate are thought to represent deposition that occurred in a series of northeast-trending deltas that entered a slightly salty bay and lake in southeastern Ohio. The last marine incursion of this region occurred earlier, in Conemaugh time, when a small embayment led to the deposition of the Ames limestone. The deltas were continually shifting, slightly up and down the bay coast and bayward and landward through Monongahela time. By early Permian time the deltas had merged into a wide fluvial plain dotted with small freshwater lakes. The higher parts of these deltas and the later fluvial plain were well drained, and oxidation of sediments was common. Iron in the sediments reacted with oxygen to form ferric oxides, which led to the red coloring of certain mudstones and shales. The Creston Red shale of the Permian Washington formation are a good example.

Geologists have always debated the origins of the so-called redbeds. Some believe the color developed at the time of deposition, while others believe it happened afterward. The green Permian mudstones that usually occur with the red mudstones, on the other hand, formed when the reddish sediments were chemically reduced, which was common where lakes and swamps were prevalent.

The sandstones seen in the roadcuts along this stretch of I-77 have evidence that indicates they were deposited in river and stream channels. They have ripple marks and cross-bedding and can be traced long distances as lens-shaped masses surrounded by shales and other rocks. The larger streams forming these deltaic deposits were probably the size of the present-day Scioto River. The shales and mudstones represent deposition on floodplains. Some contain mud cracks and plant fossils. The thin coal seams were formed in back-swamps on the floodplains, and the thin limestone beds mark the sites of lakes or ponds. An occasional fossil of a freshwater snail, clam, or ostracod may occur. Landslides are prevalent along this stretch. As I-77 descends into the Ohio Valley, Marietta stretches across the floodplain of the Ohio River, protected by massive floodwalls.

INTERSTATE 80 AND U.S. 422
Pennsylvania Line—Cleveland
73 and 79 miles

I-80 and U.S. 422 traverse nearly 80 miles of glaciated Allegheny Plateau from the Pennsylvania line to Cleveland. Farther north, U.S. 6 and U.S. 322 follow similar paths. Tills of the Wisconsinan-age Grand River lobe and Pennsylvanian-age coal sequences underlie the routes between the Pennsylvania border and the Cuyahoga Valley in Cleveland. Tills of the Cuyahoga lobe and Mississippian-age rocks characterize the routes as they cross the Cuyahoga Valley.

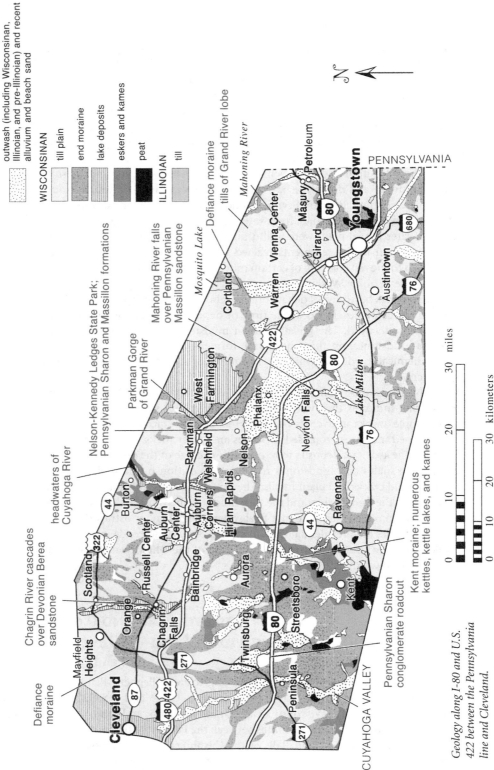

N

PENNSYLVANIA

Defiance moraine
tills of Grand River lobe

Mahoning River

Masury
Petroleum

80

Youngstown

680

Vienna Center

Girard

Austintown

76

Mosquito Lake

Warren

Cortland

80

Mahoning River falls
over Pennsylvanian
Massillon sandstone

Defiance moraine

Parkman Gorge
of Grand River

Nelson-Kennedy Ledges State Park;
Pennsylvanian Sharon and Massillon formations

West
Farmington

422

Lake Milton

Phalanx

Newton Falls

76

headwaters of
Cuyahoga River

Parkman

Nelson

Welshfield

44

Burton

Auburn
Center

Auburn
Corners

Hiram Rapids

Ravenna

44

Chagrin River cascades
over Devonian Berea
sandstone

Scotland

322

Russell Center

Bainbridge

Aurora

80

Kent

Streetsboro

Kent moraine: numerous
kettles, kettle lakes, and kames

Defiance
moraine

Mayfield
Heights

Orange

Chagrin
Falls

271

Twinsburg

87

Cleveland

480
422

Peninsula

271

CUYAHOGA VALLEY

Pennsylvanian Sharon
conglomerate roadcut

0   10   20   30 miles

0   10   20   30 kilometers

*Geology along I-80 and U.S.
422 between the Pennsylvania
line and Cleveland.*

I-80 follows an upland area of hilly moraine material from the state line to the Austintown area. A long viaduct carries the interstate over the outwash-filled Mahoning Valley at Girard. U.S. 422 stretches through similar hilly moraine. Just southeast of Youngstown this terrain gives way to kames and outwash terraces in the Mahoning Valley, all of which are of Wisconsinan age.

Just south of the Newton Falls and Warren interchange, I-80 crosses the Mahoning River. Numerous oxbow lakes occur along its floodplain north of here between Newton Falls and Warren. The Mahoning River cascades over Pennsylvanian Massillon sandstone at Newton Falls, an early milling and canal town south of I-80. South of Newton Falls the Mahoning River returns to a meandering course, and oxbow lakes and meander scars indicate that it has changed its course many times. A concrete dam built at the falls now covers most of the outcrop. Grand River lobe sediments dominate the landscape of both routes from just northwest of Youngstown to the upper Cuyahoga Valley. Kames, kettles, end moraines, and ground moraine characterize the interlobate terrain that developed between the Grand River and Cuyahoga lobes. Just west of the Ohio 8 exit on I-80, high cliffs of Pennsylvanian Sharon conglomerate appear under an arch bridge over the turnpike. The Defiance moraine occurs on either side of the Cuyahoga Valley between Ohio 8 and 21 on I-80, and between Chagrin Falls and Shaker Heights on U.S. 422.

The glacial history is complex in this part of the state. Wisconsinan ice formed the Defiance moraine around 32,000 years ago. About 28,000 years ago the ice sheet retreated an unknown distance to the north of Ohio. During this recessional period, tremendous volumes of meltwater carrying vast volumes

Pennsylvanian-age Massillon sandstone pops from the bank at Newton Falls. Note the impressive cross-bedding.

of outwash spread down north-south trending valleys of pre-Wisconsinan streams like the Cuyahoga. After the meltwater floods dissipated, the region continued to erode, and winds blowing southward across sheets of outwash deposited loess across northeastern Ohio. Four more advances and recessions of the Wisconsinan ice sheet occurred from the Pennsylvania border to about Cleveland: about 23,000, 19,000, 17,000, and 15,000 years ago. Each one deposited another till layer. About 14,500 years ago the last ice covered the area, and after it receded proglacial lakes began developing in the Erie Basin.

Building Stones of the Youngstown-Warren Area

The Austin flagstone came from the Mississippian-age Cuyahoga formation. This sandstone broke into thin slabs that could be used for walks, chimneys, and facing on buildings. A quarry between Warren and present-day Mosquito Lake supplied this building stone in the late 1800s, which builders used widely in the Warren area. The Pennsylvanian-age Sharon conglomerate was the stone of choice 10 miles west of Warren where it occurs along a ridge that rises some 100 feet above the floodplain of the Mahoning River, stretching from Newton Falls north to Phalanx. People quarried stone from the ridge wherever it was accessible and used it for bridges and foundations.

Between the 1860s and 1900, quarries at Vienna Center provided Pennsylvanian Massillon sandstone for bridge and building construction during the local coal boom. Quarries in the Mill Creek valley, south of Youngstown, and in Masury also produced Massillon stone; the quarry in Masury supplied building stone and flagstone for the Sharon, Pennsylvania, area in the early 1900s. Lack of railroad transportation, water problems, impure stone, and competition from the Berea-Amherst quarrying district farther west led to the early demise of the building stone industry in this area. The eastern industrialized area of Masury called Petroleum hints at the attempt to open an oil field along Little Yankee Run during the 1920s.

Sandstone of the Grand River Valley

The Pennsylvanian-age Sharon conglomerate occurs along the western bluffs of the Grand River valley from Phalanx north to Thompson. Although it's called a conglomerate, the Sharon rock is often a sandstone, sometimes pebbly, in northeastern Ohio. The formation is often 96 to 99 percent silica and is primarily made up of grains of quartz and quartzite. The sediments forming the Sharon strata were derived from erosion that was occurring during the last major plate collision, which formed the Appalachian Mountains. During this time Africa collided with the southeastern coast of North America, causing gently inclined sheets of rock called *thrust sheets* to slide toward the Appalachian Basin, in which sediment was accumulating. The movements of these thrust sheets and other plate adjustments along eastern North America led to sea level changes, which in turn affected the development of streams that were draining this region. The Sharon conglomerate is thought to have accumulated in fast-flowing streams that carried a lot of eroded sediment. Gravel bars and sandbars were common. The channels were continually changing, with some areas being scoured by strong currents while others accumulated sand. Cross-bedding, ripple marks, and cut-and-fill structures are typical.

As was the case with any hard rock, early settlers of the region used the Sharon conglomerate as building material. Look for it in the foundations of older buildings and in bridge piers throughout northeastern Ohio. The purity of this material and the close proximity of the Cleveland and Youngstown steel industry led to its use as an industrial sand. Quarries in this stone operated near Bainbridge, Burton, Geauga Lake, Nelson, and Phalanx.

Nelson-Kennedy Ledges State Park

Nelson-Kennedy Ledges State Park is south of U.S. 422 and Parkman. In this park, downcutting of the Cuyahoga River to the west and the Grand River to the north has created ledges that are isolated from the rest of the plateau. The highest hills of this area are capped by Pennsylvanian-age Homewood sandstone and exhibit two distinct ledges below this; the higher ledge is Massillon sandstone and the lower, more prominent ledge is Sharon conglomerate. Sugarloaf, at 1,380 feet, is the highest point in the area. The ledges are underlain by softer shale layers that erode back, leaving the ledge rock unsupported. The sandstone and conglomerate cracks along joints and drops or slides down along the edges of the ledges. Deep, narrow passages develop between the blocks, some as much

The Pennsylvanian-age sandstone forming the lower part of Nelson Ledges at Devil's Den alternates between pebbles and sand-sized grains.

as 60 feet deep. Trees and other plants hasten the erosion process by widening the joints; you can see roots covering some of the walls of the cracks. A jumble of sandstone and conglomerate blocks lie at the bottom of the rocky cliffs. Rock shelters and waterfalls are also present. In the park, Cascade Falls tumbles 60 feet, while Minnehaha Falls consists of two falls: one that drops 25 and the other 20 feet.

Nelson Ledges has been a tourist attraction since the 1850s. A short-lived gold rush hit the area in 1870, but the shiny metal turned out to be pyrite, or fool's gold. When the Ohio State Park system formed in 1949, the ledges region became part of it. Nelson-Kennedy Ledges State Park preserves both Nelson Ledges and Kennedy Ledges, which only exhibit the Sharon conglomerate. South of the park is a campground that surrounds an abandoned quarry that opened in the early 1900s. The Sharon sandstone had a higher content of quartz here than in the park, and it was used as an industrial sand.

Ohio's Northernmost Coal Mine

Burton, north of U.S. 422, has the distinction of being the location of Ohio's northernmost coal mine. It opened in 1884 on a farm 2.7 miles northwest of the town square. Coal was scattered at various locations in this northern Ohio region, but apparently locals figured out that the Sharon coal was thicker around Burton. Welsh miners who had relocated to Burton from exhausted coalfields at Mineral Ridge west of Youngstown worked the Burton-area coal. Eventually six shafts and four slopes, or inclined tunnels, were opened. The mine never produced a great amount of coal, and it was out of business after about twenty years. Not a trace of it remains today.

The Upper Cuyahoga River

The Cuyahoga River's source is just north of East Claridon on U.S. 322 amongst wetlands and small kettle lakes in a wide outwash-filled valley. Downstream at the town of Hiram Rapids the river flows over Pennsylvanian-age bedrock and in a narrow V-shaped valley. Most upland streams in Ohio that have headwaters on the plateau begin in V-shaped valleys and progress into wider valleys downstream. Glaciers are the reason for the peculiar nature of the Cuyahoga River. During the Ice Age, ice blocked most northerly flowing rivers in Ohio, causing water to pool along the ice's margins. In many places, the water cut gorges through the plateau as it forced new drainage routes to the south. After the glaciers receded, many rivers and creeks flowed north again, joining younger V-shaped valleys with the more established, wider ones. Standing Rock, a large cross-bedded remnant of Pennsylvanian Sharon conglomerate that Indians once used as a trail marker, juts out from the river near Standing Rock Cemetery in Kent south of I-80. Sharon strata are also exposed on the west bank of the river across from Standing Rock and below the Ohio 59 bridge in downtown Kent.

Kent Moraine

West of the Ohio 44 overpass, I-80 passes over 12 miles of the Kent moraine, a complex of kames, kettles, and knobs of till. The Cuyahoga River, flowing southwest at this point, carves through the moraine's center. U.S. 422 crosses 6 miles of Kent moraine between Welshfield and Auburn Center. Just north of Auburn Corners is Punderson State Park, which contains Ohio's largest and deepest kettle lake. The lake and numerous smaller versions around it are about 12,000 years old and formed when large blocks of ice broke loose from the melting glaciers and were partially buried in outwash. Once they melted, depressions remained in the landscape. Some of them filled with water, forming lakes, while others remained dry. Large piles of sand and gravel that rested on these discarded ice chunks and filled crevasses on the larger ice sheet eventually slumped to the surface, forming hills of outwash, or kames.

Chesterland Caves and Twinsburg Quarrying

Flat-topped ridges of rock layers that jut out of the landscape contrast sharply with the wet bottomlands in northeastern Ohio. Nowhere is this contrast better displayed than in Geauga County. Vertical and horizontal cracks, or joints, in Pennsylvanian-age Sharon conglomerate and Devonian-age Berea sandstone have created many striking rock formations, or ledges. Around 1900 these rocks became destinations for weekend visitors. Chardon had Rocky Cellar, until it was marred by quarrying in 1890, and Griswold Hollow, with a 45-foot waterfall; Russell Center had Ansel's Cave and 30-foot Ansel's Ledge Falls (now part of West Woods Reservation, a Geauga County park); Bass Lake, south of Chardon, has a natural bridge over which Bass Lake Road passes; and Chesterland and Scotland had Chesterland Caves.

Chesterland Caves is another area of striking ledges in the Pennsylvanian Sharon conglomerate. The caves are enlarged joints along the edge of a bluff and not underground features like those found in karst landscapes that formed because of the dissolution of bedrock. The flat-topped rocks provided people with plenty of space to spread picnic blankets in the late 1800s

Standing Rock in Kent is a waterworn erosional remnant of Pennsylvanian-age Sharon conglomerate.

and early 1900s. It was a cool, shady spot on a hot summer's day. Shale beds within the Sharon strata caused springs to form at the base of the ledges. Entrepreneurs marketed bottled water from the Chesterland Caves to Cleveland residents. A resort opened in 1900 in the Chesterland-Scotland area, spurred by the opening of a local interurban rail line. Public access to the Chesterland Caves was lost in 1926 when the resort was sold. Today the area is privately owned, and it's difficult to tell that the area around Caves Road and Sherman Road was once so popular.

East and west of Twinsburg, north of I-80 and south of Chesterland and U.S. 422, quarries along ledges provided Pennsylvanian-age Sharon sandstone and conglomerate in the late 1800s. Before the arrival of the railroad in 1880, the stone was hauled to Macedonia for shipment. Look for this conglomerate in walls and foundations of area buildings and in railroad bridges. The stone at Twinsburg consisted of about 15 feet of sandstone underlain by 6 feet of basal conglomerate. A portion of the ledges is preserved in Bennett-McDonald Ledges, east of Liberty Road in Twinsburg.

Chagrin Falls

At one time an ancient river, the preglacial Chagrin River, passed through the Chagrin Falls area in a major valley west of town. This valley now underlies the downstream portion of the present-day Chagrin River from Miles Road to Lake Erie. It was filled with sand and gravel in Pleistocene time and is one of several that extend southward into the Allegheny Plateau in northeastern Ohio, including ones that underlie the Cuyahoga and Grand Rivers.

The Chagrin River makes a U-shaped bend near the Miles Road bridge southwest of downtown Chagrin Falls. It bends because glacial meltwater ponded along the ice sheet's edge, at about the present-day Lake Erie shoreline, and breached the Defiance end moraine. As the ice retreated northward, a stream began to flow north in the old buried valley of the preglacial Chagrin River, eventually capturing the flow of a southeastward-flowing stream (the present-day upstream portion of the Chagrin River) that passed along the southeast margin of the Defiance end moraine. So today the Chagrin River flows southeast along the moraine from its source near Orange, and then it makes a sharp bend to flow north above the buried valley to the lake, much like the nearby Cuyahoga River, which flows southward from its source near Lake Erie and then bends at Cuyahoga Falls to flow north to Cleveland and Lake Erie. The upstream southeast-flowing portion of the present-day Chagrin River carves into hard sandstone through Chagrin Falls, while the downstream portion has an easier time occupying a wider valley that initially was carved into softer Devonian-age shales by the preglacial Chagrin River.

Just south of the Main Street bridge, Chagrin Falls, the town's namesake, consists of ledges of Devonian Berea sandstone underlain by Bedford and Chagrin shale. Steps lead to the base of the falls. Locally, people used Berea sandstone as flagging until about 1900. In the early 1900s a quarry north of town was a well-visited fossil locality. The Mississippian-age Sunbury shale exposed above the Berea sandstone yielded nicely preserved inarticulate brachiopods.

*The Chagrin River cascades over Devonian-age
Berea sandstone in downtown Chagrin Falls.*

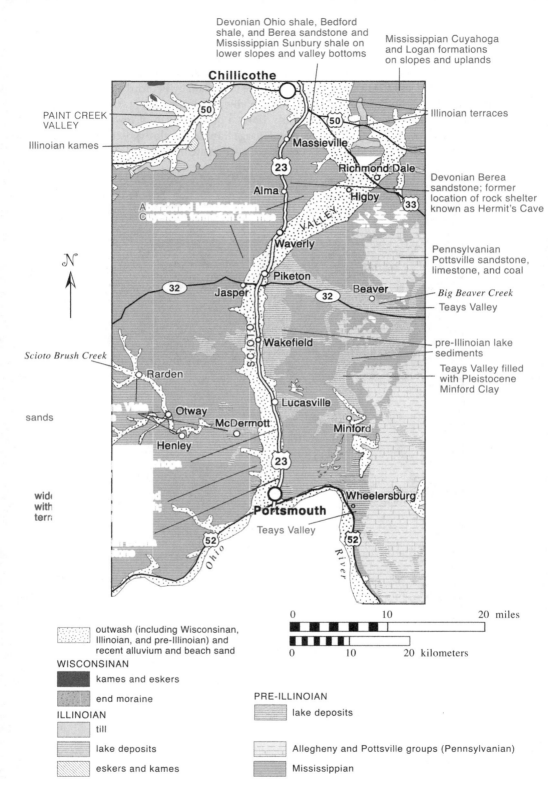

Devonian Ohio shale, Bedford shale, and Berea sandstone and Mississippian Sunbury shale on lower slopes and valley bottoms

Mississippian Cuyahoga and Logan formations on slopes and uplands

Chillicothe

PAINT CREEK VALLEY

Illinoian kames

Massieville

Richmond Dale

Alma

Higby

Illinoian terraces

Devonian Berea sandstone; former location of rock shelter known as Hermit's Cave

Abandoned Mississippian Cuyahoga formation quarries

Waverly

Piketon

Jasper

Beaver

Pennsylvanian Pottsville sandstone, limestone, and coal

Big Beaver Creek

Teays Valley

Scioto Brush Creek

Wakefield

pre-Illinoian lake sediments

Rarden

Teays Valley filled with Pleistocene Minford Clay

sands

Otway

Lucasville

McDermott

Minford

Henley

wide
with
terr

Wheelersburg

Portsmouth

Teays Valley

Geology along U.S. 23 between Chillicothe and Portsmouth.

outwash (including Wisconsinan, Illinoian, and pre-Illinoian) and recent alluvium and beach sand

WISCONSINAN

kames and eskers

end moraine

ILLINOIAN

till

lake deposits

eskers and kames

PRE-ILLINOIAN

lake deposits

Allegheny and Pottsville groups (Pennsylvanian)

Mississippian

0   10   20 miles

0   10   20 kilometers

U.S. 23
Chillicothe—Portsmouth
45 miles

About 8 miles of Wisconsinan outwash, glacial lake sediments, and river alluvium underlie U.S. 23 from Chillicothe to Massieville. The highway then cuts across 10 miles of unglaciated Allegheny Plateau from Massieville to Waverly. Between Waverly and Portsmouth the route follows the Scioto Valley, which is underlain by a thick layer of river sediments and outwash. Tributaries of this river have wide, flat areas at their mouths where lakes existed behind the high piles of outwash in the main valley in Pleistocene time. These flat areas are underlain by pre-Illinoian Minford clay.

Chillicothe to Waverly

The rolling landscape east of U.S. 23 from the Pickaway and Ross County line to the Ohio 159 intersection north of Chillicothe is part of the Wisconsinan terminal moraine of the Scioto lobe, the Cuba moraine. The Wisconsinan moraine abuts a similar but older terminal moraine of the last Illinoian-age ice sheet. The Illinoian ice moved into the hills around Chillicothe, traveling as far as 2 miles south of Massieville. The Illinoian glacial deposits are less prominent due to their greater age and longer exposure to erosion. Chillicothe also marks the boundary between the central lowlands of western Ohio and the Allegheny Plateau of southeastern Ohio. Since glaciers touched only the edges of the plateau, it is a hilly upland heavily carved by streams.

South of Chillicothe as far as Richmond Dale, the Scioto Valley widens to over 4 miles. At the valley's eastern margin there are three higher levels of ground, or terraces, composed of Illinoian and Wisconsinan outwash. The higher terrace is 200 to 250 feet above river level and dates to Illinoian time; the two lower terraces range from 20 to 140 feet above river level and formed

The flat upland marks the highest Illinoian terrace along this tributary of the Scioto River near Higby. Below that is another Illinoian terrace; it is incomplete because of erosion caused by Wisconsinan and recent streams.

during Wisconsinan time. Compare the stream valleys that are cut into the higher terrace to those that are cut in the lower terrace. The higher terrace is decidedly older for it is well dissected, whereas the lower terraces are only partially eroded. The terraces slope toward the south, away from the source of the outwash—the edge of the glacier that once rested at Chillicothe. Because they are younger, Wisconsinan terraces throughout Ohio remain relatively unscathed—that is, there are no torrents of meltwater tearing away at them—and exist at lower elevations. Native Americans built earthworks on the higher terraces, above damaging floodwaters.

U.S. 23 cuts across Wisconsinan terrace sediments and bedrock hills of the Allegheny Plateau just south of Chillicothe instead of following a broad meander of the Scioto River. Illinoian till and glacial erratics cap the hilltops west of the highway between the south edge of Chillicothe and Massieville. Devonian-age Berea sandstone is occasionally exposed along this stretch. U.S. 23 rejoins the river valley just north of Waverly.

Waverly Sandstone Quarries

Locals of the Waverly-Piketon area favored the so-called Waverly stone, a local occurrence of Mississippian-age Cuyahoga formation, as a building stone from the 1830s until around 1912. Three companies quarried about 30 feet of the sandstone along Pee Pee and Crooked Creeks northwest of Waverly in the late 1800s. Wagons hauled stone into Waverly where workers sawed it before shipping it on the Ohio & Erie Canal. Some of the stone went to Columbus for the construction of the state penitentiary. A number of quarries were in the bluffs

Wisconsinan outwash in a sand pit near Higby. Since this was laid down by meltwater, it shows nice layering, or stratification.

bordering the Scioto Valley at Higby and between Waverly and Jasper. Another quarry, overlooking Piketon, furnished bridge stone to the Norfolk & Western Railway until the 1890s.

Waverly to Portsmouth

Up until the Ice Age, the ancient Teays River flowed into Ohio near present-day Wheelersburg from the West Virginia hills, carving a large valley. The Pleistocene ice sheets blocked its northerly flow and permanently changed the region's drainage patterns, and indeed all of Ohio's drainage. Rivers that flowed north flowed south; some valleys were abandoned altogether. Much of the northern portion of this old valley—the Ohio portion—lacks a major river or stream today.

Big Beaver Creek, a Scioto River tributary, occupies the Teays Valley for about 8 miles from Beaver to Piketon, and the Scioto River flows through the old valley between Chillicothe and Piketon, albeit in the opposite direction that the Teays flowed. North of Chillicothe, the Teays channel is filled with glacial sediments and is not recognizable; between Beaver and Wheelersburg, no major river occupies the old valley. Geologists deduce that the Teays flowed northwestward from Chillicothe across central Ohio to Indiana and Illinois, but they are not quite certain. The Minford clay, which fills the Teays Valley between Piketon and Wheelersburg, served as raw material for brick and tile manufacturing at Beaver. The clay was deposited in lakes that formed in the Teays Valley and its tributary valleys in southern Ohio when it was dammed.

Looking north up the Scioto Valley just north of Portsmouth. The broad lowland is the current floodplain. Rail lines are elevated above the flood stage.

A meander scar of the north-flowing preglacial Kinniconick River, an ancestor of the Scioto River, sweeps around the east side of Lucasville and is filled with Minford clay. Nearly 300 feet of the Mississippian-age Cuyahoga and Logan formations border U.S. 23 in a roadcut about 1½ miles south of Lucasville. The lower sandstone is the Buena Vista sandstone, which was a major building stone. Quarries once operated in Henley, McDermott, Otway, and Rarden. The dark shale above the Buena Vista member is the Portsmouth shale, a part of the Cuyahoga formation, which early residents used to manufacture bricks. Between Lucasville and Portsmouth, U.S. 23 hugs the east side of the Scioto Valley while the river sweeps from one side of the floodplain to the other.

More Sandstone Quarries—Scioto County

People began quarrying the Buena Vista sandstone member of the Mississippian Cuyahoga formation in the hills between Rarden and the Scioto River in the early 1880s. Smith & Sons opened a quarry in 1886 at Otway and built a stone cutting mill. Since only a small portion of the stone was of sawable quality, the operation closed in 1898. Quarries operated at Henley from 1890 to 1907. Since better-quality stone could be quarried in Rarden, it became a quarrying center by 1895. Stewart Brothers quarried and sawed stone at their mill in Rarden until 1904 when the new quarry of the Taylor Stone Company opened. Three years later Stewart Brothers closed their last quarry as the stone in it was no longer suitable, but the mill remained open, sawing stone from McDermott-area quarries to the southeast.

The Mississippian-age Cuyahoga sandstone was quarried in the Waller Quarry at McDermott in 1909. —Jesse E. Hyde photo, Ohio Department of Natural Resources, Division of Geological Survey

Smith & Sons also opened a quarry just northwest of McDermott in 1894. In 1905 Waller Brothers Stone Company opened a competing quarry on the other side of town. McDermott became the most important sandstone quarrying site in southern Ohio because it opened a little later. New technology and new markets perhaps led to its success. Quality stone was also harder to come by in Rarden. Quarry operators shipped McDermott-area sandstone throughout eastern North America; some shipments went as far as Alberta, Canada. Quarries in McDermott continue to supply building stone that comes from the hills northwest of town.

U.S. 30 AND U.S. 224/I-76
East Liverpool and the Pennsylvania Line— Crestline and Willard
128 and 125 miles

U.S. 30 enters Ohio at East Liverpool, stretching across unglaciated hills of the Allegheny Plateau as far as West Point. Illinoian-age tills form the roadbed for 4 miles between West Point to just south of Lisbon where a short patch of Wisconsinan-age tills and outwash occur. Illinoian till marks the route between Lisbon and Canton. Rocks peaking through the thin glacial cover are sandstones, shales, and coals of Pennsylvanian age. On the west side of Canton, U.S. 30 crosses onto Wisconsinan sediments of the Killbuck lobe, which it follows to Mansfield. Wisconsinan kames and kame terraces form the pockmarked terrain just west of downtown Canton and in the Tuscarawas Valley at Massillon.

U.S. 224 follows a nearly straight east-west path across eastern Ohio, almost exclusively passing through Wisconsinan Grand River lobe and Killbuck lobe sediments. In the Akron area the route crosses Illinoian-age deposits. Pennsylvanian-age coal sequences make up the bedrock, but exposures of them are few.

East Liverpool to Lisbon and the Pennsylvania Line to Poland

U.S. 30 follows California Hollow, the deep valley of Carpenter Run, between the Ohio River and the Calcutta exit. East Liverpool, once the pottery capital of Ohio, is still a center of commerce on the banks of the Ohio River. Strip mines line the highway from Rock Camp to West Point. Between Rock Camp and West Point, the hills bordering U.S. 30 are a patchwork of strip mines and reclaimed mine lands. In the 1980s geologists found well-preserved insect, spider, and millipede fossils in a shale overlying the Mahoning coal. The most notable fossils included a 4-inch-long cockroach and several ½-inch, nearly complete spiders. Plant fossils, including branches of an early conifer, and teeth of a freshwater shark were also found. In the same area, the overlying Brush Creek shale, a marine unit associated with the Brush Creek limestone, yielded

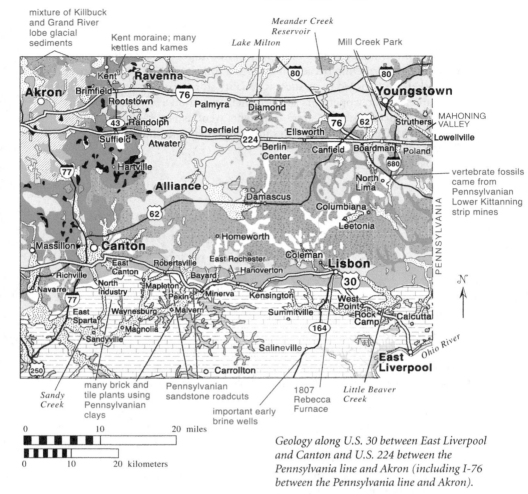

mixture of Killbuck and Grand River lobe glacial sediments

Kent moraine; many kettles and kames

Meander Creek Reservoir

Lake Milton

Mill Creek Park

MAHONING VALLEY

vertebrate fossils came from Pennsylvanian Lower Kittanning strip mines

PENNSYLVANIA

Sandy Creek

many brick and tile plants using Pennsylvanian clays

Pennsylvanian sandstone roadcuts

important early brine wells

1807 Rebecca Furnace

Little Beaver Creek

0   10   20 miles

0   10   20 kilometers

Geology along U.S. 30 between East Liverpool and Canton and U.S. 224 between the Pennsylvania line and Akron (including I-76 between the Pennsylvania line and Akron).

well-preserved corals, brachiopods, clams, snails, and a 3-inch layer of coiled nautiloid cephalopods. Just imagine what's still hidden in the buried strata of southeastern Ohio!

Older underground mines in Upper Freeport coal dot the hills around West Point where Little Beaver Creek meanders by in a wide, deep valley. A preglacial stream cut down to a resistant buried sandstone, where it began to meander and widen this valley. Little Beaver Creek reoccupied the sediment-filled valley after Wisconsinan glaciation ended. North of West Point, strip mines parallel the highway to the southwest along Patterson Creek. U.S. 30 crosses Middle Fork Little Beaver Creek on both sides of Lisbon.

From the Pennsylvania border to Quarry Road, just south of Lowellville, U.S. 224 parallels the deep Mahoning Valley; good views of the valley occasionally appear through the trees. In preglacial times two tributaries that joined near Lowellville flowed northwestward and southeastward in the valley the present-day Mahoning River flows through from the Ohio-Pennsylvania line to Youngstown. These preglacial tributaries fed the ancestral Monongahela

outwash (including Wisconsinan, Illinoian, and pre-Illinoian) and recent alluvium and beach sand

WISCONSINAN

till plain

end moraine

lake deposits

eskers and kames

peat

ILLINOIAN

till

lake deposits

eskers and kames

PENNSYLVANIAN

Conemaugh group

Allegheny and Pottsville groups

River, which flowed out of Pennsylvania and through present-day Hubbard, Youngstown, Niles, and Warren on its way to the Erigan River, where Lake Erie is today. When a preglacial ice sheet blocked the northward flow of the ancestral Monongahela, the valley between Hubbard and Youngstown was abandoned and eventually filled. The Mahoning River eventually occupied the valley of the ancestral Monongahela from Beaver, Pennsylvania, to Warren. The Mahoning River has cut a valley 300 feet down into Pennsylvanian strata north of U.S. 224, southeast of Youngstown. U.S. 224 and I-76 to the south pass through the hilly, dissected Kent end moraine of Wisconsinan age. On either side of the Pennsylvania line and between the highways there is strip-mined land.

Although a Pennsylvanian-age rock member bears the name Lowellville, it was the Vanport limestone that was the major mineral product of this town. The Vanport limestone averages 20 feet thick around Lowellville—the greatest thickness of any of the older Pennsylvanian limestones. The Carbon Limestone Company quarried this limestone in the hills south of Lowellville and marketed it as flux stone, lime, and road gravel. The water-filled quarry is now within a rod and gun club on Quarry Road off U.S. 224. The company also operated quarries at Kansas Corners and east of Lowellville near the Pennsylvania state line.

At Poland, U.S. 224 enters the urban sprawl of Youngstown, which continues uninterrupted for several miles except for the valleys of Yellow Creek in Poland and Mill Creek in Boardman. By 1806 workers had completed the Hopewell Furnace—the first charcoal-fired iron furnace in the state—near the mouth of Yellow Creek in present-day Struthers. The surrounding forests provided the charcoal fuel, and the underlying Pennsylvanian-age strata provided the iron ore. Local limestone served as flux. The furnace produced all the iron necessities of the settlers like pots, kettles, stoves, and plows. A second furnace, the Montgomery Furnace, was built in 1806 between the Hopewell Furnace and the Mahoning River. Both furnaces lay abandoned after 1812, and only a remnant of the Hopewell Furnace remains today. Although short-lived, they were harbingers of the coming iron boom of the Hanging Rock Iron Region.

Lisbon Area

Columbiana County entrepreneurs reaped economic benefits from their mineral resources throughout the 1800s and 1900s. They mainly utilized Pennsylvanian coals and clays of the Allegheny group. Speculators seeking oil switched their interest to salt after a prospect well east of Lisbon, Columbiana County's county seat, gushed natural gas and saltwater in 1866. The well flowed for two years, funneling enough saltwater to the surface to produce 15 barrels of salt per day. The salt came from Silurian strata. Later pumping by the New

Ruins of the Rebecca Furnace near Lisbon are composed of blocks of Pennsylvanian-age Lower Freeport sandstone.

Lisbon Salt Company increased the yield to 65 barrels, which was quite a bit. The salt furnace of this company was the last boiler to operate in Columbiana County, and the plant closed in 1891.

The Rebecca Furnace, the second to operate in the state after the Hopewell Furnace, sprung up along East Fork Little Beaver Creek northwest of Lisbon in 1807. Miners dug Pennsylvanian-age iron ore from weathered soils of the surrounding hills. The furnace prospered until 1830. It was abandoned long before 1900; however, the crumbly foundation remains just south of the junction of Logtown and Furnace Roads. Nearby on Logtown Road, high above a rail trail, are the imposing walls of a quarry where Pennsylvanian-age Lower Freeport sandstone was quarried for building stone, and later for pulp stone. Another quarry lies hidden along Furnace Road. Much of this sandstone went into the locks of the adjacent Sandy & Beaver Canal, which ran between Beaver, Pennsylvania, and Bolivar, Ohio. The Hughes-McKinley house, a private residence that was built in 1807 with this stone, is across the road from the remains of the Rebecca Furnace. There are more stone structures in Lisbon, including the 1805 Old Stone House, a museum, on Washington Street and the Columbiana County Courthouse.

In 1836 the Ohio Cement Company began manufacturing the first hydraulic cement—a cement that sets underwater—in Columbiana County from the 2-foot-thick Lower Freeport limestone. Masons immediately began using the cement in the stonework of the Sandy & Beaver Canal just northwest of Lisbon; a lock set from this cement remains along the abandoned canal on the south side of Furnace Road. To make the cement, workers heated the stone in a kiln and then ground it into a powder. The drift mines of the Ohio Cement Company, closed in the early 1900s, were in a hollow northwest of Logtown. In the early 1900s miners stripped Upper Freeport limestone, as much as 8 feet thick, from hills on the southwest side of Lisbon, where it was converted to agricultural lime. In the 1870s the Eagle Firebrick Company of Lisbon produced firebrick and hydraulic cement using Lower Kittanning clay and Lower Freeport limestone. Lower Kittanning underclay was the raw material of three other Lisbon firebrick plants in the early 1920s. Lower Kittanning coal was also mined from a 3- to 4-foot-thick seam at Lisbon and nearby Logtown and Coleman. Nineteen coal mines operated in the area, the last closing in the early 1940s.

A Side Trip to Salineville

West of the Alleghenies, salt was a precious and necessary commodity in Ohio's early days. Barrels of it carted over the mountains often commanded high prices. Residents of southern Columbiana County discovered salt springs in the early 1800s. Desiring a local source of salt, they attempted to tap the salt by drilling from above but were not successful until 1818. At the salt industry's zenith in Salineville, south of Lisbon on Ohio 164, twenty wells produced brine from the subsurface Silurian group along Little Yellow Creek. Salt production continued until 1865 when locals turned their attention to coal. The rock salt of Michigan and New York had already begun to supplant the old brine plants.

Small coal banks that locals used for fuel existed along Little Yellow Creek for many years prior to the arrival of the railroad in 1852. Here, as elsewhere, it was the railroad that opened up the coalfield to other markets and itself became a primary customer. The early mines around Salineville were drift mines,

The Sterling Mine, located 2 miles west of Salineville, worked the Mahoning coal circa 1910.
—D. Dale Condit photo, Ohio Department of Natural Resources, Division of Geological Survey

Clay works at Salineville in 1910. Note the stacks of hollow blocks and tiles. —D. Dale Condit photo, Ohio Department of Natural Resources, Division of Geological Survey

tunnels into slopes following the 2- to 3-foot-thick Upper Freeport coal or the 3-foot Mahoning seam. The Ohio & Pennsylvania Coal Company consolidated several mines in 1867 and became the first large-scale mining firm in Salineville. It employed two hundred miners by 1875 and the company was producing 800 tons of coal per day. The company also developed the first shaft mines in town. Some twenty-seven shaft mines operated around Salineville. Most of them closed by the early 1900s, and the last one closed in the early 1940s. Surface mines followed; many are along Salineville Road east of town and Ohio 39 west of town.

Ohio Clay Products mined Conemaugh group clays in the early 1900s and produced building tile and hollow blocks. Miners dug the clay from drift mines in the hill behind the plant. Upper Freeport coal was mined as a source of fuel. The company began mining clay some 100 feet above Riley Run, and as they exhausted the clay layers at that height, they moved to the next clay bed underneath. The clays were part of the Pennsylvanian-age cyclothems. A tall concrete stack in town marks the ruins of this plant. Summitville Tiles, in business since 1912, continues to operate a tile factory in nearby Summitville using local Summitville shale of the Allegheny group.

Lisbon to Canton and Poland to Akron

About 1 mile east of Hanoverton, just south of U.S. 30, workers tunneled toward Dungannon through a rocky ridge of the Pennsylvanian-age Conemaugh group while building the Sandy & Beaver Canal. A historical plaque is posted on the south side of the highway. The flooded tunnels are visible along Haessly and Laughlin Mill Roads. A gravel pit near the two tunnels taps outwash in an Illinoian kame.

More reclaimed strip-mined terrain is evident on both sides of U.S. 30 between Kensington and Lynchburg where miners removed 2 to 5 feet of Upper Freeport coal. Illinoian-age kames rise above the highway to the south between East Rochester and Bayard. The removal of sand and gravel has extensively lowered these hills south of Bayard.

North of Bayard there is an exhausted oil field under Homeworth. Prospectors drilling between 1858 and 1860 found evidence that there was oil under Homeworth. It wasn't until 1899, however, when a company struck oil in the Devonian-age Berea sandstone north of town, that people became really interested in Homeworth's oil. Prospectors drilled more oil and gas wells into the Berea sandstone and another sandstone in the underlying Bedford formation east of town. By 1905 the field yielded more than 7,500 barrels per month and a good quantity of gas as well. Nearly 600 wells were sunk in this area into the 1920s, with 272 of them producing; production tapered off during the 1920s.

Cross-bedded Pennsylvanian-age sandstone forms a roadcut just west of Minerva. U.S. 30 enters the drainage of Hugle Run about 1½ miles west of Minerva and follows its flat bottomlands for 3 miles. Just east of Robertsville the highway surmounts a sandstone ridge of exposed Pennsylvanian sandstone of the Allegheny group underlain by shale containing carbonized plant fossil fragments. The road then drops down again and crosses the flat-bottomed valley of Black Run, which is similar to Hugle Run. South of the highway down the valley is the site of an abandoned brick plant on the east side of Robertsville. Another brick plant is on the east edge of East Canton. These plants mined both Lower and Middle Kittanning clay associated with the Lower and Middle Kittanning

Pennsylvanian-age Lower Freeport sandstone and underlying shale just east of Robertsville on U.S. 30.

coals. East Canton is also the home of Stark Ceramics, on the west side of town. U.S. 30 then descends into the complex of kames, kettles, and peat deposits marking the juncture, or interlobate region, of the Grand River and Killbuck lobes of Wisconsinan-age ice at Canton. Meyers and Sippo Lakes, north of U.S. 30, formed as stranded ice blocks slowly melted. Scattered wetlands throughout the Canton area were once smaller kettle lakes that later filled in with vegetation. Other small lakes mark former gravel pits.

South of Youngstown and U.S. 224 lie reclaimed strip mines in the Lower Kittanning coal that contain fish, amphibian, and reptile fossils. The fossil-bearing bed is a cannel coal, meaning it's algae rich, much like the more famous vertebrate fossil site at the ghost town of Linton on the Ohio River. Northeast of North Lima and south of I-76 is a place called Five Points, which is not much more than a crossroads. Paleontologists found some interesting fossils in the cannel coal here in the 1990s. They found two shark species, acanthodians (small, primitive jawed fish), nine species of bony fish, at least thirteen species of amphibians, and one undetermined reptile. Many of the fossils are now part of the Carnegie Museum collections in Pittsburgh.

From Poland to just east of Berlin Center, U.S. 224 crosses over dissected end moraine sediments of the Grand River lobe. From there it passes onto Wisconsinan-age till plain. I-76 leaves the hilly end moraine of the Grand River lobe about 5 miles north of U.S. 224, hitting till plain. The Mahoning River appears again just west of Berlin Center on U.S. 224 and the Newton Falls exit on I-76. In both places the river is disguised as a lake: Berlin Lake on U.S. 224 and Lake Milton, which collects behind a dam just north of I-76. West of Lake Milton, a sewer pipe company operated a pit in the mid-1900s in Pennsylvanian strata at Diamond, just north of the interstate. In the late 1800s shaft mines tapped Sharon coal at Palmyra and Atwater. As this coal was exhausted, miners turned their attention to younger coal seams nearby. Between Diamond and Atwater, particularly west of Deerfield on U.S. 224, strip mines and reclaimed land are evident.

The routes enter a major meltwater channel at Randolph and Rootstown, through which the Grand River and Killbuck lobes shed torrents of meltwater. This is a hilly area with lots of kettles, kettle lakes, kames, and kame terraces. Sand and gravel operations tap the vast deposits of outwash. Near the I-77 and U.S. 224 junction in downtown Akron, Illinoian sediments appear.

Youngstown's Natural Jewel

On the southern edge of Youngstown's metropolitan area, in Boardman, U.S. 224 crosses Mill Creek and passes entrances to Mill Creek Park, which covers some 2,600 acres. The park was established in 1891, but it preserves much older features.

Mill Creek now flows north into the Mahoning River just south of downtown Youngstown. From Canfield Road (U.S. 62) to near its mouth, the creek drops 23 feet at Lanterman's Falls and then flows through a narrow rock-walled gorge with rapids. The rock is Pennsylvanian-age Massillon sandstone, Quakertown coal, and Meadville shale. The sandstone forms the lip of the falls, which became the site of several gristmills and sawmills as early as 1800. Lanterman's Mill, built between 1845 and 1846, ground flour at this site until 1888. In 1892 the abandoned structure became part of the new park, and today the restored mill is a highlight of the park.

Below the sandstone cap of the falls there is a recessed area called Cave of the Winds, which developed in the softer Meadville shale. The East Gorge Walk leads downstream from the restored mill through the narrowest part of Mill Creek gorge. Pleistocene meltwater once gushed through here, heading south onto the Allegheny Plateau. In places the meltwater flowed across loose sediment, but here it encountered hard rock and carved the deep notch into the bedrock. Along the trail is Umbrella Rock, another overhanging ledge of Massillon sandstone, and also Sulphur Springs, which miners created when drilling for coal in 1884. At 107 feet, sulfur water flowed up the hole and the would-be coal mine became an artesian well. In 1900, crowds flocked to the spring to imbibe its supposed health-giving waters. The spring continues to flow, one of many springs in the park. Look for other springs where the sandstone and underlying shale meet at Umbrella Rock and Shelter Rock on the West Gorge Trail.

Iron smelting was also part of the industry of the Mill Creek area. In 1835 workers erected Trumbull Furnace on a hillside along Mill Creek. As with many other furnaces along the plateau, all the necessities were nearby: Pennsylvanian-age iron ore and limestone for flux; timber for making charcoal; and later, coal for fuel. Later the furnace operators shipped iron ore in from Lisbon and the limestone arrived from Lowellville. The foundation of this furnace still exists near the old Heaton woolen mill, now called the Pioneer Pavilion, at the north end of Lake Cohasset.

Lanterman's Mill and Falls in Youngstown's Mill Creek Park. The Pennsylvanian-age Massillon sandstone forms the lip of the falls.

Another interesting geologic site in the park is Bears Den, near the park's western boundary. Large blocks of Massillon sandstone lie scattered along Bear Creek, a tributary of Mill Creek, having broken loose along joints that formed in the sandstone before slowly sliding downhill. Quarries periodically operated here beginning in the 1890s, removing building stone. They were short-lived, for the stone proved to be poorly suited for construction. The park, however, found the stone suitable for various bridges, dams, and walls. A more successful quarry worked the Massillon sandstone near the Ford Nature Center in the 1880s.

Selenite Crystals near Ellsworth

Southeast of Ellsworth in the bank of West Branch Meander Creek, clear, colorless crystals of selenite, a variety of gypsum (hydrous calcium sulfate), occur in gray clays. This site, now closed to collecting, yielded spectacular twinned crystals, which settlers first described in the 1820s. Twinned crystals involve the intergrowth of two crystals of the same mineral; in this case the selenite formed doubly terminated crystals.

Before the glaciers came, northeastern Ohio streams had carved deeply into the landscape, forming an integrated network of valleys. As the ice sheets fluctuated across the region, east-west-trending valleys often became lakes as outwash built up in north-south-trending valleys, effectively blocking their connection with tributaries. Lake muds and silts filled these former lakes. Much later, mineral-rich groundwater percolating through the sediments was drawn to the surface as it evaporated, precipitating gypsum, which formed selenite crystals. West Branch Meander Creek then cut into the former lake bed, exposing the crystals as they weather out of the banks. Sometimes people find selenite with present-day plant roots, suggesting these crystals are still forming.

Although the Ellsworth site is the most famous site in Ohio, people have found selenite along the Cuyahoga River in Portage and Summit Counties, near Lake Milton west of Youngstown, and along the reservoir in West Branch State Park east of Ravenna.

Kames and Kettles Galore—the Kent Moraine

From Brimfield on I-76 and Suffield on U.S. 224, Ohio 43 leads to downtown Kent. In Kent, if you turn west on Main Street, you will be afforded views of the Cuyahoga River's deep cut through the Kent moraine. If you head east, you will be rewarded with views across the moraine. The Kent moraine ranges from 5 to 15 miles wide and formed in the interlobate area between the Killbuck and Grand River lobes. Unlike many end moraines, the Kent moraine does not possess a well-defined, narrow ice margin, which would have formed where the ice remained at relative equilibrium, neither advancing nor retreating; instead the ice stagnated over a wide belt of ground, up to 15 miles across. This area filled with a jumbled arrangement of blocks of ice and meltwater streams. Outwash and till were deposited in a complex mass around and over these blocks and in crevasses, eventually forming a rolling landscape of kames and kettles once the ice blocks melted away. This all happened about 20,000 years ago.

Twin Lakes and Brady Lake, excellent examples of ice block–derived kettle lakes, are north of Kent. Lake Rockwell is an artificial impoundment of the

Cuyahoga River. The Kent State University campus still provides some open views of the glacially deposited countryside.

East Canton and Vicinity Coal and Clay Industries

Settlers discovered coal in the hills of Stark County as early as 1806. By the 1830s, coal banks around Osnaburg, now East Canton, supplied most of Canton's coal needs. These early drift mines tapped the Brookville and Lower and Middle Kittanning seams. Because railroads connected these mines to larger markets, mining flourished from 1900 until World War I, especially along the Conotton Valley Railroad in the Indian Run valley.

Two ghost towns—Browntown and Redtown—grew up around mines of the Steiner Coal Company south of East Canton. Strip mines replaced the underground mines after World War I and continue operating in the area today. The northernmost location of the Blackband ore, an iron carbonate layer that

CANTON AREA

Brick and ceramic plants as well as clay and shale quarrying operations have operated in the Canton area for hundreds of years. A number of plants have been reclaimed, and their sites are no longer recognizable; others stand as industrial ruins; and yet others continue to operate today. The following table is a small chunk of a longer history.

Town	Industry
Alliance	In 1867 Alliance Fire Clay, one of the first clay firms in Stark County, incorporated here. The company also produced pottery, including yellowware, Rockingham ware, stoneware; drain tile; sewer pipe; and all kinds of fireclay products. Coal to fuel the kilns came from mines on the property.
Carrollton	Home to a brick factory as early as the late 1820s. Deckman-Duty Brick Company operated a plant from the 1890s to the 1940s. Operations mined Middle and Lower Kittanning clays.
East Canton (formerly Osnaburg)	National Fireproofing Company operated a plant. Stark Ceramics began in 1908 as a brick-making operation, eventually becoming the world's number one supplier of ceramic tile.
East Sparta	Operations mined Middle and Lower Kittanning clays.
Magnolia	National Fireproofing Company operated a brick plant.
Malvern	Five brick plants worked the Middle Kittanning clay beginning in 1886–87. Deckman-Duty Brick Company operated a plant from the 1890s to the 1940s.
Mapleton, Minerva, North Industry, Pekin, Robertsville, Sandyville	Operations mined Middle and Lower Kittanning clays.
Waynesburg	National Fireproofing Company operated a brick plant.

Eroded spoil bank in Pennsylvanian-age Allegheny group coals just east of East Canton on U.S. 30.

appears brownish black and has color banding, was east of East Canton. Here the ore is associated with the Upper Freeport coal of the Allegheny group. The Grafton Iron Company mined iron here until the mid-1880s and shipped it to their furnace in Leetonia, north of Lisbon. Small iron mines also worked near Robertsville.

The Pennsylvanian rocks in the area also contained the ubiquitous clays and shales people used at other locales to manufacture bricks and tiles. Early brick firms evolved into a number of important companies in the early 1900s. Stark Ceramics is an example; today the company is a major supplier of structural walls for all types of public and commercial buildings.

An oil boom hit the area in 1966 when a vast pool was discovered on the Stark Ceramics property in East Canton. The oil came from the Silurian-age Clinton horizon. Drillers sank over eight hundred wells with nary a dry hole. Oil rigs, still operating, are visible along the drive into the plant.

Canton to Crestline and Akron to Willard

Near the Richville exit west of Canton, U.S. 30 enters Wisconsinan-age sediments of the Killbuck lobe; from here to Crestline, it crosses various deposits laid down by this lobe in late Wisconsinan time. The highway crosses another outwash-filled meltwater channel south of Massillon, which is now occupied by the Tuscarawas River

The year 1842 marks the birth of E. Houghton & Company, which was a stoneware pottery at Dalton. The company used local underclay and shale underlying the Pennsylvanian-age Bedford coal. One of the company's shale pits was ½ mile south of U.S. 30, but it has long since been reclaimed. The prominent valley of Sugar Creek, just east of Riceland, contains eroded remnants

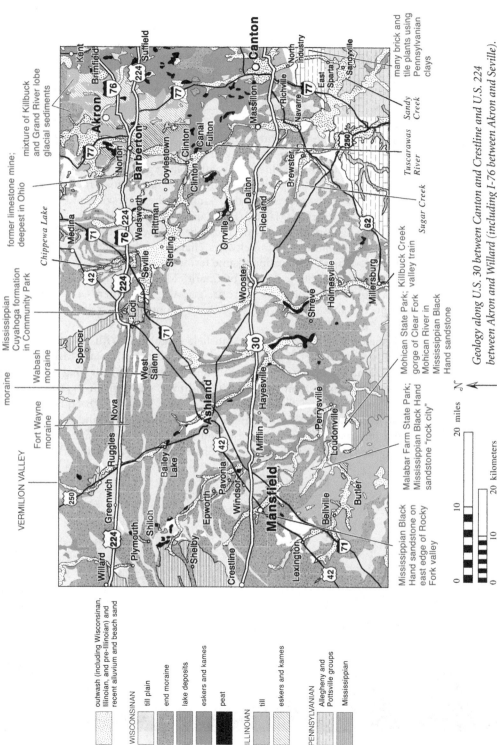

Geology along U.S. 30 between Canton and Crestline and U.S. 224 between Akron and Willard (including I-76 between Akron and Seville).

mixture of Killbuck and Grand River lobe glacial sediments

former limestone mine; deepest in Ohio

fossiliferous Mississippian Cuyahoga formation in Community Park

many brick and tile plants using Pennsylvanian clays

Sandy Creek

Tuscarawas River

Sugar Creek

Mohican State Park; gorge of Clear Fork Mohican River in Mississippian Black Hand sandstone

Killbuck Creek; valley train

Malabar Farm State Park; Mississippian Black Hand sandstone "rock city"

Mississippian Black Hand sandstone on east edge of Rocky Fork valley

Defiance moraine

Wabash moraine

Fort Wayne moraine

VERMILION VALLEY

N

0   10   20 miles

0   10   20 kilometers

WISCONSINAN

outwash (including Wisconsinan, Illinoian, and pre-Illinoian) and recent alluvium and beach sand

till plain

end moraine

lake deposits

eskers and kames

peat

ILLINOIAN

till

eskers and kames

PENNSYLVANIAN

Allegheny and Pottsville groups

Mississippian

of Wisconsinan outwash and kame terraces. Pennsylvanian bedrock disappears about 6 miles west of Dalton and is replaced by sandstones and shales of Mississippian age. Any Pennsylvanian strata that existed west of here was eventually eroded away along the east side of the Findlay Arch. Brick factories used Cuyahoga formation shale at Orrville and Wooster.

The Wooster bypass crosses the wide, flat-bottomed valley of Killbuck Creek, a major meltwater channel. The once-meandering course has since been straightened, but meander scars, often still filled with water, are visible just west of the Ohio 3 intersection on both sides of U.S. 30. U.S. 30 crosses Muddy and Jerome Forks, streams running in similar meltwater channels near the Wayne and Ashland County line. The gently rolling ground moraine near these waterways marks the transition from younger, Pennsylvanian rocks onto older, Mississippian sandstones and shales. South of the Hayesville and Mifflin exits there are two parks—Malabar Farm and Mohican State Parks—that exhibit scenic Mississippian sandstone rock shelters, waterfalls, and cliffs. Mifflin lies on the edge of the deep meltwater channel of the Black Fork. The valley has well-developed Wisconsinan kame terraces along its walls, but they are not visible since the valley is flooded by Charles Mill Lake south of U.S. 30. Near Pavonia, north of the highway, there are glacial lake sediments and peat that were deposited in a lake that formed behind the meltwater sediments of the Black Fork channel.

At Mansfield, U.S. 30 crosses the Mansfield Highland. With elevations between 1,450 and 1,510 feet, it is the second highest point in the state. Mississippian Black Hand sandstone caps these hills and is a major cliff-forming rock in the region. For some reason this area of sandstone was difficult to erode. The ice sheet split into the Killbuck and Scioto lobes at Mansfield. Eventually, Wisconsinan ice topped the Mansfield Highland and deposited end and ground moraine and outwash sediments, but it encountered too much resistance to encroach very far onto the Allegheny Plateau. Shales of the Cuyahoga formation

U.S. 30 cuts through Mississippian-age Black Hand sandstone
on the east edge of Mansfield. Other cuts occur on I-71.

are exposed in the area's valleys. The Black Hand sandstone borders the highway as it descends into the valley now occupied by Rocky Fork. Approximately 14 miles north of U.S. 30, Ohio's roughly southwest-northeast-trending drainage divide between Lake Erie and the Ohio River crosses the area; streams near the Richland and Huron County line flow north, while all the streams around Mansfield carve into the Allegheny Plateau and flow south. U.S. 30 crosses remnants of the Johnstown, Powell, and Broadway end moraines between Lexington Spring Mill Road west of Mansfield and about 1 mile east of downtown Crestline. Between Mansfield and Epworth, north of U.S. 30, brick manufacturing plants took advantage of the shales of the Cuyahoga formation.

On the west side of Akron, U.S. 224 and I-76 join and head west to near Seville. Pennsylvanian-age coal seams lie hidden below glacial sediments along this stretch. Between Barberton and Norton, Wisconsinan till once again becomes the roadbed as the route leaves the Illinoian till of Akron behind. The first factory built in Barberton was National Sewer Pipe in 1891, the largest sewer pipe factory in the world at that time. Its daily capacity was 200 tons of pipe. The clay the factory used came from south of town and was shipped to the factory by railroad. A company that bought National Sewer Pipe closed the plant in the 1940s, and the factory site is now a shopping center. The Diamond Brick and Tile Company, which was adjacent to National Sewer Pipe's clay pits, provided most of the brick for the buildings and streets of downtown Barberton.

The deepest limestone shaft mine in Ohio is in Norton. It is now inundated by Lake Dorothy. The shaft, begun in 1942, eventually reached a depth of 2,300 feet and its underground workings covered a 540-acre site. The tunnels of this mine are huge, big enough to drive large dump trucks and loaders through. Two shafts led to the underground levels. The mine produced crushed limestone from the Devonian-age Columbus formation to be used in soda ash and cement until closing in 1976.

The River Styx, a misfit stream, seems out of place in the wide meltwater channel just west of Wadsworth. At Seville the route crosses the Wabash moraine of the Killbuck lobe. Chippewa Lake lies in another wide meltwater channel north of U.S. 224. It is a remnant of a much larger lake that formed behind the Wabash moraine as a stranded block of ice slowly melted. I-76 ends at the junction with I-71, and U.S. 224 continues west.

Between Seville and Willard, U.S. 224 follows a series of closely spaced end moraines; they rise 20 to 60 feet on both sides of the highway. Streams along this stretch cut deep valleys into the glacial deposits and underlying soft shales deposited in Pennsylvanian time.

At Lodi U.S. 224 crosses another meltwater valley that was once occupied by a glacial lake that stretched between Burbank and Penfield. Today only the flat bottomland underlain by lake silts remains, with the East Branch Black River passing through it. Mississippian Cuyahoga formation siltstone, limestone, and shale form the bed of East Branch Black River in Lodi. A good place to observe these strata is in Community Park. The siltstone and limestone contain abundant brachiopod and netlike bryozoan fossils and scattered molds of clams and snails; over one hundred species have been found here. Lodi's stone arch railroad bridge was made of Devonian-age Berea sandstone shipped in from Plymouth to the west. One mile west of Ruggles, the route crosses the Vermilion Valley. There are prominent kame terraces north and south of the highway. The Vermilion River begins in the terminal moraine to the south near Bailey Lake and cuts through more end moraines northward to Vermilion and Lake Erie.

Massillon—Clay, Coal, Iron, and Sandstone

Beginning in the 1860s, quarriers began pulling thick sandstone overlying the Pennsylvanian-age Quakertown coal from the ground in the Tuscarawas River valley northwest of Massillon. The Massillon sandstone, as it was called, found ready use as a building stone, attested to by the numerous stone structures in Massillon. The Worthorst Quarry exposed 65 to 70 feet of the stone at Massillon. One hundred men worked here, shipping 300 to 400 carloads of building stone and 1,500 to 2,000 tons of grindstones per year. Most of this went to Pittsburgh, Philadelphia, and Baltimore. The Paul Quarry upriver at Canal Fulton sold crushed sandstone to Pittsburgh glass companies. The market for Massillon sandstone dwindled by 1915; however, it is still quarried for building stone in Coshocton County.

The first iron company in Massillon was the Massillon Iron Company, which erected a furnace on the west bank of Sippo Creek on the east edge of town in 1833. Later, during the 1870s, two local iron furnaces used Blackband ore, a carbonate iron ore that was associated with the Upper Freeport coal. The operations ended within five years. Massillon also sat in the middle of an important Sharon coalfield of the Pottsville group. People began mining in the 1840s, and eventually over eighty mines operated between Wadsworth and Barberton and south to Brewster and Navarre. Some 1,500 men worked in six mines around Massillon in the mid-1800s. Companies opened new mines in the early 1880s as miners exhausted the reserves of older mines. Most were shaft or slope mines since the coal only approached the surface along the Tuscarawas River near Canal Fulton and Newman Creek near North Lawrence.

Massillon had a brickyard and pottery by 1813. These meager beginnings led to a thriving clay product industry by the late 1800s. The Massillon Pottery Company, on the east side of town, became known for its stoneware in the 1890s. The Massillon Stone Company, later Massillon Stone and Fire Brick Company, began with a small Massillon sandstone quarry on South Erie Street in 1881. They found it necessary to supplement their stone sales and opened a firebrick plant at Newman Station, northwest of town, in the late 1880s. The company had recurring problems with local clay supplies, but it carved a niche for itself by manufacturing special refractory products. The plant, along with Massillon Refractories, survived into the late 1970s.

Mansfield and Southern Richland County

The Mississippian-age Black Hand sandstone member of the Cuyahoga formation came from a ridge near the former state reformatory in Mansfield in the late 1800s. The most notable product from this site was a variegated sandstone with colorful yellow, red, maroon, brown, and black layers that had many iron nodules and produced swirls of color. No simple explanation of these colorful bands and swirls is known. It probably involved a mobilization of iron ions in the rock due to the downward movement of groundwater in fractures.

Regardless, it certainly is the most attractive building stone ever produced in the state. Unfortunately, not all the colorful stone could be used because much of it was crumbly. The quarrying operations ended around 1912. You can see the stone in a number of churches across the state, including ones in Napoleon, Shelby, and Upper Sandusky. Quarries in the hills north of Windsor and at Bellville also worked the Black Hand sandstone, though it wasn't the beautiful, variegated stone of the reformatory area. Most of this stone went into bridge abutments and piers.

Off Fleming Falls Road between U.S. 30 and U.S. 42 a tributary of the Black Fork Mohican River cascades over a lip of Black Hand sandstone, now the site of a church retreat. The Black Hand is well jointed here and contains many beds of shale. Downstream of the falls there is a small rock shelter and a seasonal waterfall. Fleming Falls was once a popular picnic grounds for the people of nearby Mansfield.

Malabar Farm State Park, west of Perrysville, offers scenic exposures of Black Hand sandstone riddled with joints. Follow Butternut Trail back into the woods about ½ mile to what appears to be a cave in a moss-covered cliff. It's actually a widened joint that connects with another, creating a passageway between the blocks of stone. These are often called "rock cities," or ledges, in northeastern Ohio. They formed as a result of the freezing and thawing of ice near the ground.

Mohican State Park, southwest of Loudonville, is developed around the 300-foot-deep Clear Fork Gorge, a national natural landmark. Mississippian Cuyahoga and Logan sandstones and conglomerates form the rocks here. The glacial boundary runs through the park roughly following the path of the Clear Fork Mohican River—north of the river is Wisconsinan till. The oldest rocks in

Black Hand sandstone forms interesting joint-exploited passageways along a trail at Malabar Farm State Park.

the park occur in Clear Fork Gorge, downstream of the Pleasant Hill Dam spillway. Interbedded sandstone and shale of the Portsmouth shale member of the Cuyahoga formation occur at normal water level. The Black Hand sandstone member rests on this, and you can readily examine it in a cliff near the parking lot for Pleasant Hill Dam. Cross-bedding is prominent at the west end of the cliff, indicative of flowing or turbulent water in the past, and beds are also tilted here due to the tendency of the more resistant sandstone to slide downslope on the less resistant shale underneath. The rock also shows characteristic honeycomb weathering caused by differences in the cement strength between grains. The Logan formation sits on top of the Black Hand sandstone, but only the lower two members—the Berne and Byer sandstones—are exposed in the higher hills of the park.

Hiking the Lyons Falls Trail along the western rim of the gorge will lead you to rock shelters in the Black Hand sandstone and Little and Big Lyons Falls along tributaries of the Clear Fork-Mohican River. Both falls drop off hanging valleys that were left stranded high as meltwater flow enhanced the downcutting of Clear Fork Mohican River in Wisconsinan time. Near Butler and outside of the park, Hemlock Falls tumbles 50 feet over more Mississippian sandstone.

Wadsworth-Rittman Area
A search for oil at Wadsworth, west of Akron, ended in failure in 1889, but prospectors encountered a thick bed of rock salt in the Silurian-age Salina group 2,400 to 2,700 feet down. The Wadsworth Salt Company was organized

Big Lyons Falls cascades from the roof of a rock shelter in Mississippian-age Black Hand sandstone at Mohican State Park near Loudonville.

in 1891. Hot water injected down wells dissolved the salt, which rose to the surface as brine. Workers placed the brine in steam-heated vacuum evaporating pans where halite, or sodium chloride, crystallized. They then dried the halite in huge warehouses for at least a month before screening it to sort it by grain size. In the early 1900s the plant produced 1,200 barrels daily and employed 150 people.

The Ohio Salt Company, incorporated in 1898, tapped similar Silurian-age rock salt beneath Rittman, south of Wadsworth. Its workers obtained the brine and produced halite in a similar fashion. By the early 1900s this plant produced 2,000 barrels daily. The packing and dairy industries used the processed salt for ice cream, cheese, and butter refrigeration. The halite was also used as table salt and fertilizer. The Morton Salt Well Field continues to produce artificial brine on the southeast side of Rittman.

An upland capped by Pennsylvanian-age Sharon conglomerate, about 4 miles northeast of Rittman, supplied building stone in the late 1800s. The Wayne County Courthouse in Wooster contains this stone. The Clinton, Doylestown, and Hametown area was once dotted with over one hundred mines in the Sharon coal. Fifty-one mines operated in Rogues Hollow, off Clinton Road southeast of town, over a span of 105 years beginning in 1840. Mule-drawn railcars hauled the coal to Clinton, where workers loaded it onto canal packets for shipment to Cleveland on the Ohio & Erie Canal. It was said that a miner could walk the 7 miles between Clinton and Hametown entirely underground.

U.S. 33

Columbus—Pomeroy

101 miles

For about 17 miles south of I-70 in Columbus, U.S. 33 crosses Wisconsinan-age Scioto lobe sediments before passing onto an Illinoian valley train south of Carroll and a narrow stretch of Illinoian till at Lancaster. The route follows the Hocking Valley through Mississippian- and Pennsylvanian-age rocks (Pottsville, Allegheny, and Conemaugh groups) of the Allegheny Plateau between Lancaster and Athens. Between Athens and Pomeroy, U.S. 33 cuts across uplands capped by rocks of the Monongahela group.

Columbus to Lancaster

U.S. 33 heads southeast from Columbus along the Alum Creek valley. Wisconsinan kames rise above the ground moraine west of Groveport and north of Canal Winchester. Somerset Road between Pickerington and Baltimore follows an esker. Just east of Canal Winchester, north of Walnut Creek valley, the highway crosses hills marking the Canal Winchester end moraine. Kettles and kames form Pickerington Ponds State Nature Preserve east of U.S. 33, 2 miles west of Pickerington. Kames rise out of the landscape west of the highway at Carroll in a tributary valley of Walnut Creek near the Johnstown end moraine. Gravel pits mark the kames, and other sand and gravel operations dig in outwash terraces south of town. This streamless valley was

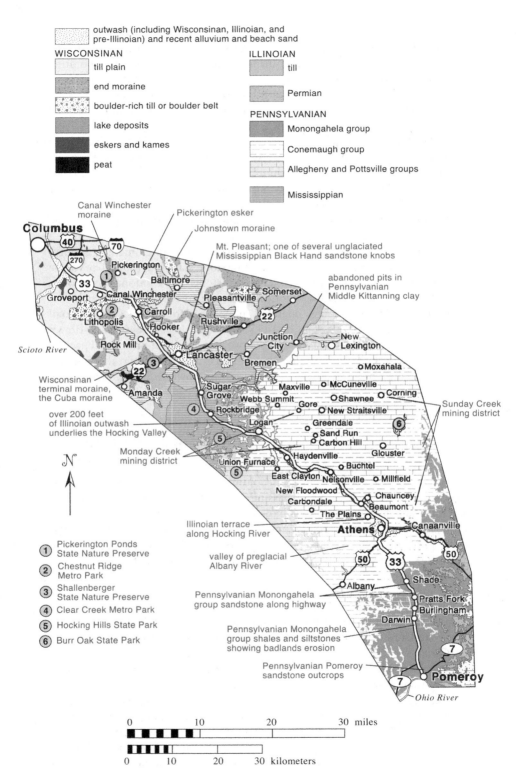

Legend:

outwash (including Wisconsinan, Illinoian, and pre-Illinoian) and recent alluvium and beach sand

WISCONSINAN
- till plain
- end moraine
- boulder-rich till or boulder belt
- lake deposits
- eskers and kames
- peat

ILLINOIAN
- till
- Permian

PENNSYLVANIAN
- Monongahela group
- Conemaugh group
- Allegheny and Pottsville groups
- Mississippian

Canal Winchester moraine
Pickerington esker
Johnstown moraine
Mt. Pleasant; one of several unglaciated Mississippian Black Hand sandstone knobs
abandoned pits in Pennsylvanian Middle Kittanning clay

Columbus
40 70
270
Pickerington
33 Baltimore
Groveport Canal Winchester Pleasantville Somerset
Lithopolis Carroll 22
Hooker Rushville Junction New
Rock Mill City Lexington
Scioto River Lancaster Moxahala
Wisconsinan terminal moraine, the Cuba moraine Bremen
Amanda 22 Sugar Maxville McCuneville
Grove Webb Summit Shawnee Corning
over 200 feet of Illinoian outwash underlies the Hocking Valley Rockbridge Gore New Straitsville
Logan Greendale Sunday Creek mining district
Monday Creek mining district Sand Run
Union Furnace Carbon Hill
Haydenville Glouster
East Clayton Buchtel
Nelsonville Millfield
New Floodwood Chauncey
Carbondale Beaumont
The Plains
Athens Canaanville
Illinoian terrace along Hocking River
valley of preglacial Albany River 50 33 50
Albany Shade
Pennsylvanian Monongahela group sandstone along highway Pratts Fork
Burlingham
Darwin
Pennsylvanian Monongahela group shales and siltstones showing badlands erosion
Pennsylvanian Pomeroy sandstone outcrops 7
7 Pomeroy
Ohio River

1. Pickerington Ponds State Nature Preserve
2. Chestnut Ridge Metro Park
3. Shallenberger State Nature Preserve
4. Clear Creek Metro Park
5. Hocking Hills State Park
6. Burr Oak State Park

N

0   10   20   30 miles
0   10   20   30 kilometers

Geology along U.S. 33 between Columbus and Pomeroy.

obviously the path of a much larger stream at one time. It also served as the route of the now-abandoned Hocking Canal. The broad, flat area east of the highway, north of the airport, marks a glacial lake that formed when Wisconsinan outwash piled up in the streamless valley, temporarily preventing tributaries from draining into the outwash channel. Many east-west tributaries in Fairfield County experienced similar damming and ponding; areas of lake sediments—broad, flat valley bottoms—abound to the east and west of the highway.

At Hooker the Hocking River enters the valley from the west and parallels U.S. 33 to Athens. At Lancaster, U.S. 33 intersects the Wisconsinan terminal moraine, the Cuba moraine, and crosses a narrow band of Illinoian till. The unglaciated Allegheny Plateau lies about 4 miles south of the city. Note the rounded knobs to the southwest and prominent Mt. Pleasant to the east. These bedrock knobs are underlain by Cuyahoga formation and represent portions of the plateau that erosion has isolated. This may be due to a particular resistance the rocks had to erosion or to the quirks of meandering streams that simply flowed around them. Just south of downtown, another wide, streamless valley stretches to the east. The preglacial Groveport River, part of the preglacial Teays drainage, flowed through here.

Lithopolis-Carroll Area

Mississippian-age Black Hand sandstone holds up Chestnut Ridge, which lies about 1½ miles east of Lithopolis and terminates to the north on the south side of the Walnut Creek valley. Walnut Creek is bordered by Devonian-age Ohio shale to the north. The buried valley of the preglacial Newark River underlies this portion of the Walnut Creek valley, passes under Canal Winchester, and heads north to Reynoldsburg.

Siltstone, below the Black Hand sandstone of the Cuyahoga formation, occurs along the edge of Chestnut Ridge. Lithopolis freestone, the quarry workers' name for this stone that split easily into slabs, was an important product of quarries along Chestnut Ridge in the late 1800s. Quarry men pried into the thin, easily broken layers of siltstone and marketed it as tombstones, sills, capstones, and flagstone. From a 78-foot quarry face carved into the ridge, 4 feet of stone was usable as building stone. Workers sawed the stone at the quarry in early years, and later it was shipped on the Hocking Canal from Canal Winchester to Columbus-area stonecutting mills. The Lithopolis quarries closed around 1910. Chestnut Ridge Metro Park, east of Lithopolis, has trails that pass by small, overgrown quarries. Other quarries southwest of Carroll and at Rock Mill once provided Black Hand sandstone as well.

Rock Mill, about 6 miles south of Lithopolis, sits above a deep, narrow gorge cut by the headwaters of the Hocking River. A waterfall continues to erode the sandstone next to a still-extant gristmill and covered bridge. Note the curving splash pool below the falls. Native Americans called this river the *Hock Hocking*, which meant "gourd," because this circular splash pool looked like the bowl of a gourd, with the narrow upstream gorge representing the neck. Settlers gradually shortened its name to the Hocking River. The pool below the falls is eroded to a depth of nearly 50 feet. The gorge above the falls is 20 to 40 feet deep and 4 to 40 feet wide. Quarried blocks of Black Hand sandstone line

Quarried blocks of Mississippian-age Black Hand sandstone lie near Rock Mill. Many of these discarded blocks have tooled surfaces.

the road by the old mill. Many flowing springs occur where the Black Hand sandstone meets underlying shale in the area. The most famous spring in the area was at Jefferson, north of the Chestnut Ridge Metro Park. Water bottled here was sold in Columbus and was in great demand during the typhoid scare around 1900.

Lancaster Area

Mississippian-age building stone also came from central and southern Fairfield County. The Black Hand sandstone underlies till in the hills and mounds surrounding Lancaster. Several higher mounds, including Mt. Pleasant (2,000 feet in elevation), escaped glaciation. Hike to the top of Mt. Pleasant in Rising Park for a great view of the Hocking Valley and surrounding countryside and a close look at the sandstone. Downtown Lancaster is built on a high Wisconsinan-age terrace; the stone courthouse and city hall sit on a remnant of the Wisconsinan terminal moraine that rises above the terrace gravels. South of downtown, Wisconsinan till gives way to rugged Illinoian ground moraine and high Illinoian terraces that stretch as far as Horns Mill, but Wisconsinan outwash continues alongside U.S. 33 to Sugar Grove.

Sandstone quarrying was centered east and south of downtown. Most quarries exposed the Byer and Berne sandstone members of the Logan formation, and the Black Hand sandstone of the Cuyahoga formation. The Black Hand, forming 25- to 75-foot high walls, was the main stone quarriers were after, but the younger units saw some local use. The Baumaster Quarry on the southwest edge of Mt. Pleasant was one of the earliest quarries in the area, operating in the early 1840s. East of town there were a number of quarries, and Quarry

Mt. Pleasant, or Standing Stone, looms above the Fairfield County Fairgrounds in Lancaster. Many bedrock mounds like this occur along the edge of the unglaciated portion of the Allegheny Plateau. Mississippian-age Black Hand sandstone forms the crest.

Road hints at their existence. Lancaster's city hall, built between 1896 and 1898, is constructed of stone from the Allegheny Quarry, located at a railroad station east of town called Sandstone. Other quarries stretched from downtown Lancaster to Logan. Each shipped Hocking Valley Black Hand sandstone on the Hocking Canal, opened in 1841, and later the Columbus & Hocking Valley Railroad into the early 1900s.

Lancaster firms also made brick and tile products using local glacial tills and Mississippian shales. In 1889, Lancaster claimed the deepest (1,948 feet) cased gas well in the world, located at the corner of Maple Street and Fair Avenue on the south edge of Mt. Pleasant. A cased well was one that had tubing inserted into the borehole to prevent collapse of the walls and movement of unwanted fluids into the well. Just down the street, the county fairgrounds offered the first nighttime horse racing, which was illuminated by gas from another copious well.

A Side Trip on U.S. 22—Clay, Fossils, and Fuel

U.S. 22, east or west of Lancaster, leads to some interesting geologic sites. Two miles east of Rushville, well-preserved plant fossils were discovered in a shale layer within the Quakertown coal in the 1860s. Seventeen species of swamp-dwelling plants new to science were identified, including *Cardiocarpon* (Cordaites group); lycopods *Lepidodendron* and *Lepidophloios*; progymnosperm *Archaeopteris*; seed ferns *Alethopteris, Eremopteris, Megalopteris,* and *Orthogoniopteris*; and the scouring rush *Asterophyllites*. The strata, formerly exposed in a roadside ditch, are pretty much buried today. Southeast of Rushville near Junction City, plant fossils were found in the clay pits of the Rush Creek Clay Company in the Middle Kittanning coal horizon. The abandoned and built-over pits are northeast of the junction of Ohio 37 and 668.

Sharp Quarry near Sugar Grove was typical of the operations that removed building stone from the Mississippian bedrock. The uppermost part of the quarry wall is thinner-bedded Logan formation; below that is an excellent building stone horizon of the Black Hand sandstone member of the Cuyahoga formation. —Ohio Department of Natural Resources, Division of Geological Survey photo

Drillers tapped a major pool of petroleum and gas in the Silurian-age Clinton sands around Bremen, Junction City, Pleasantville, and Rushville between 1907 and 1909. The Rush Creek Oil & Gas Company started the exploration in 1896 when a Bremen-area well encountered a strong gas play at 1,790 feet. For a year and a half, Bremen, Rushville, and West Rushville enjoyed the advantages of local natural gas. Unfortunately, saltwater ended this supply. As gas was released from the Silurian strata, water and brines from surrounding strata moved into the pores that had held the gas. It was left to the newly organized Bremen Gas & Oil Company to initiate the first economical oil production in the area in 1907, which was also the first yield in the area. A well drilled along the Rush Creek valley, northeast of Bremen, initially produced 140 barrels per day; in 1910 it was still producing 20 barrels. More wells followed, all producing from the Silurian Clinton sands, around 2,500 feet down. Some initially yielded 300 barrels per day. Besides Bremen Oil & Gas Company, many other companies met with success. Drilling spread in all directions, but the best yields were south of Bremen. The Rush Creek valley was soon peppered with holes and prospectors gave little regard to the proper spacing of derricks; six wells may have existed where only one should have been drilled. Failures soon followed, but a few wells still pump today.

Farther east U.S. 22 also passes through the Somerset gas and oil fields that tapped the Devonian-age Berea sandstone and Silurian Clinton sands, 1,800 and 3,000 feet down respectively. A brick plant operated northeast of Somerset until the 1970s, utilizing Pennsylvanian Quakertown and Massillon shales. A workforce of fifty in the mid-1940s produced up to 65,000 face bricks per day.

Coal and clay mining and products dominated the economy of New Lexington and Junction City. Companies began shaft mining in the late 1800s and had switched to strip-mining by the early 1950s. Local tile and brick plants used the Clarion, Massillon, Middle Mercer, and Quakertown shales and Flint Ridge and Middle Mercer clays during the early and mid-1900s. The Junction City Sewer Pipe Company used a mixture of Middle Mercer shale and Flint Ridge and Middle Mercer clays in sewer pipe and clay specialty products, like fireproofing, refractories, and conduits. The scars of a shale pit are visible off Ohio 37 in Junction City. West of New Lexington, on Tile Plant Road, lie the abandoned quarries of the Ludowici-Celadon Company, a roofing tile plant that employed three hundred people in the 1950s. The plant was thoroughly mechanized, with a conveyor belt system that carried Clarion shale from the adjacent quarry to the plant. In the 1920s, five firms processed Lower and Upper Freeport sandstones into molding sand along Little Rush Creek, east of New Lexington. By the 1950s, one company, near Gosline, remained in business.

Heading west of Lancaster on U.S. 22, Allen Knob towers 250 feet above the highway. It's the home of Shallenberger State Nature Preserve, a good place to get a feel for a common plant and animal habitat of the southeastern hills—a sandstone ridge or mound. The tops of these unglaciated knobs have a thin sandy soil that retains little moisture. Chestnut oak (*Quercus prinus*), mountain laurel (*Kalmia latifolia*), polypody fern (*Polypodium virginianum*), and several other ferns are the dominant plants. The Mississippian Black Hand sandstone caps this erosion-resistant hill, or outlier, that rises some 240 feet above the parking area.

Some 330 million years ago during Mississippian time, this area was part of a shallow sea bottom that was being covered with sand and silt carried in by

Oil wells at Bremen in 1909, marking the initial significant production from the Silurian-age Clinton horizon. —John A. Bownocker photo, Ohio Department of Natural Resources, Division of Geological Survey

Black Hand sandstone in the wall of Stonewall Cemetery. Note the expert craftsmanship involved in constructing this mortarless structure.

rivers from the worn-down Acadian Mountains. After another 30 million years the area became land as the sea withdrew westward and eastern North America rose with the uplift of the Appalachian Mountains. More sand and mud accumulated on top of the Mississippian delta sands, which had since hardened to sandstone, now called the Black Hand sandstone (a member of the Cuyahoga formation). Through Mesozoic and Cenozoic time the forces of wind, ice, rain, and running water cut down from the surface, eventually exposing the Mississippian strata along the west side of the Allegheny Plateau. Where streams left the sandstone-dominated plateau to carve into less-resistant limestones and shales of earlier Paleozoic time, or to join the larger rivers of the plateau like the Hocking River, they carved more deeply into the plateau and increased in number. The landscape became riddled with stream channels flowing around more-resistant patches of sandstone. The rock left between these many channels formed knobs (mounds) or ridges, like those that occur around this preserve.

A trail leads to the top and an old sandstone quarry. Note the rough surface of the Black Hand sandstone at spots, something geologists call *honeycomb weathering*. Geologists think that this reticulate pattern of raised dark brown ridges is caused by the concentration of iron oxides in certain parts of the rock, making these parts more resistant to weathering than the in-between areas. The rocks also exhibit cross-bedding, a pattern of intersecting layers caused by the cutting and filling of channels in ancient deltas.

Wisconsinan ice came to a halt here, surrounding the knob but not covering it while laying down its terminal moraine. Look carefully and you can find evidence of the ice's presence in small rounded pieces of igneous and metamorphic rocks that were mixed in with the local sandstone. It's fairly easy to distinguish these long-traveled glacial erratics from the rocks that came from nearby. They are not the typical tan or gray colors of local rocks. Many are speckled with red, pink, white, and black spots, some are dark green or black, and others are specked gray and white. These rocks, some carried here from as far away as central Canada, contain a wider assemblage of minerals than Ohio's native sedimentary rocks. Kames form hills to the south of the highway and on the northeast side of Allen Knob.

South of U.S. 22 and east of Allen Knob on Stonewall Cemetery Road is the cemetery of the Nathaniel Wilson family, who settled here in 1799. What is interesting is that skilled stonemasons enclosed the plot with a twelve-sided wall of sandstone blocks in 1838. Emulating the construction techniques used to build Solomon's Temple, the Black Hand sandstone blocks were dressed at the quarry site on top of Allen Knob and designed to fit perfectly together without the usual hammering and shaping at the job site. The wall continues to be one of the finest examples of mortarless masonry in the state.

Lancaster to Athens

Black Hand sandstone appears along U.S. 33 just north of Stump Hollow Road as the highway passes the glacial boundary at Sugar Grove. Sand and gravel excavations tap the outwash valley fill. The rocky uplands of the Hocking Hills dominate the landscape from Rockbridge to Logan. Steep-walled gorges, rock shelters, waterfalls, and natural bridges in Black Hand sandstone make this one of the more visited natural areas in the state. Fossiliferous Maxville limestone of Mississippian age was quarried just north of Maxville on Ohio 668

Pennsylvanian-age rocks appear in the uplands east, south, and west of Logan. Burr Oak State Park, north of U.S. 33 and northeast of Glouster, has trails that lead to rock shelters in Buffalo sandstone like Buckeye Cave and erosional remnants like Table Rock; both occur because of the weathering of the underlying shale. Besides scenery, the Pennsylvanian strata yield fossils and mineral resources. Shale exposures in Brush Creek along Ohio 56 just west of Athens are noted for containing molds of clams and snails. U.S. 33 passes an operating tile plant on the south edge of Logan; others lie abandoned at Greendale, Nelsonville, and New Straitsville. Sand and gravel are dredged in ponds along the Hocking River south of Logan, and other pits operate at Nelsonville. At a rest stop between

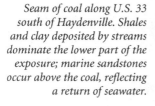

Seam of coal along U.S. 33 south of Haydenville. Shales and clay deposited by streams dominate the lower part of the exposure; marine sandstones occur above the coal, reflecting a return of seawater.

Haydenville and Nelsonville, Lock number 17 of the Hocking Canal, which connected Carroll and Athens, has been preserved. It's built of local sandstone. Allegheny group strata form the roadsides on the south edge of Nelsonville. The hills are capped by Conemaugh-age strata; sandstones form vertical cliffs, and shales and clays form the slopes.

A high Illinoian-age terrace is the site of The Plains, an appropriately named community underlain by 110 feet of outwash that buries a preglacial valley—a tributary of the Albany River, which was a tributary of the Teays River. U.S. 33 leaves the Hocking Valley and follows this terrace to the west side of a mound of Pennsylvanian strata. The abandoned workings of the Hocking Valley Mining Company's No. 1 Mine occurs within this mound and spread east and west of U.S. 33. Middle Kittanning coal was mined from 275 feet below the hills until the mine closed in 1949. The river flows around the east side of the mound and rejoins U.S. 33 at Athens. Pottsville group strata outcrop at road level to just north of Athens. From the junction with Ohio 683 to the south edge of Athens, U.S. 33 crosses younger Conemaugh group rocks. Founding fathers established Athens at a large bend in the Hocking River where Illinoian and Wisconsinan outwash forms terraces above the floodplain. The city has since spread onto the hillsides.

Clear Creek Metro Park Area

South of Lancaster, Mississippian-age Cuyahoga and Logan formation sandstones provide some interesting scenery—the Hocking Hills. Following roads up into the knobs and hills bordering the Hocking Valley offers many geological views. Wahkeena Nature Preserve, west of Sugar Grove, exhibits more Black

The erosion of a softer zone in the Black Hand sandstone causes overlying, more resistant layers to slump in the valley of Clear Creek, a tributary of the Hocking River.

Hand sandstone that remained just beyond the reach of Ice Age ice sheets. Clear Creek Metro Park, between Sugar Grove and Rockbridge, offers a view of a typical small tributary valley of the Hocking River. County Road 116 (Clear Creek Road) provides access.

Note the wideness of the mouth of the Clear Creek valley where a small lake existed during Wisconsinan time. The lake developed when outwash in the Hocking Valley dammed Clear Creek along with many other tributaries. Just west of U.S. 33 on County Road 116, a sandstone cliff with tilted slump blocks is close to the road. Leaning Lena, a house-sized block of Black Hand sandstone liberated from the main cliff, makes it a tight squeeze for vehicles on the road and forms an informal east entrance to the park. Most of the tributaries of Clear Creek begin on ridgetops in Logan formation sandstones, cut downward, and cascade over the harder, erosion-resistant upper portion of the Black Hand sandstone into the Clear Creek valley, which is bottomed by Mississippian shale. About 5 miles from U.S. 33 at the western edge of the park there is a colorful undercut cliff of sandstone called Written Rock.

As with other colorful sandstones, the coloring agents were probably iron, manganese, and sulfur. Due to changing chemical conditions within the pore spaces of the rock, iron, manganese, and sulfur ions have alternately been dissolved and precipitated. During the dissolved stage, ions become mobile and can move up and down or laterally within the strata. As conditions change and certain ions precipitate, they may form color swirls, concentric rings, or nodules. Although rocks of this nature are attractive, the process often weakens certain parts of the strata, causing the sandstone to crumble easily.

Written Rock, a particularly colorful part of the cross-bedded Black Hand sandstone in Clear Creek Metro Park. Much of this sandstone has reddish hues because of iron and manganese staining and oxidation. Note that thin iron-impregnated layers stand out because of their relative hardness.

Rockbridge

Ohio's largest natural bridge, located 1½ miles southeast of Rockbridge off Township Road 503, began forming as a rock shelter where a Hocking River tributary and its glacial ancestors cascaded over a cliff of Black Hand sandstone. The roof of the shelter was resistant sandstone, while the shelter opening underneath developed in underlying softer shale. Ever-present vertical joints caused blocks of the ceiling of the shelter—about 20 feet from the cliff's edge—to collapse. The cascading water immediately shifted to this hole in the roof, and the remaining ceiling rock of the cliff edge became the natural bridge.

The bridge measures 4½ to 20 feet wide, 5 feet thick, and over 100 feet long. It stands about 40 feet above the opening the water created. It has been a popular destination since the late 1800s and is now a state nature preserve. It's best to visit this feature when vegetation dies back because it's in the midst of a well-wooded area, making it difficult to see or take pictures of.

Two miles south of the town of Rockbridge a quarry operation exposed about 60 feet of Mississippian-age sandstone, which was crushed and sent to glass plants, steel mills, and brick plants. A fire destroyed the plant in 1915, and it never reopened.

A Side Trip to the Hocking Hills

South of Rockbridge on Ohio 374 there are six areas of geologic interest—Ash Cave, Cantwell Cliffs, Cedar Falls, Conkles Hollow, Old Man's Cave, and Rock House—carved into the rocks of Hocking Hills State Park. The hilltops in this region are capped by the Pennsylvanian-age Pottsville group, valley walls are Mississippian-age Logan formation and Black Hand sandstone of the

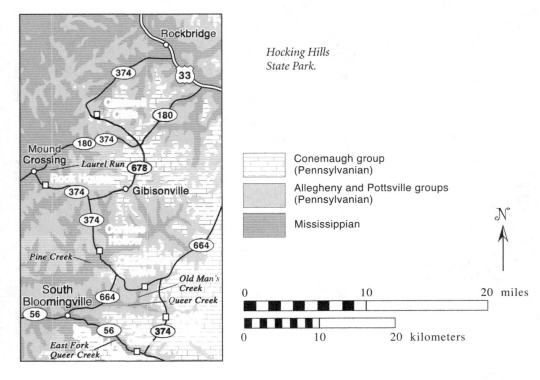

Hocking Hills State Park.

Cuyahoga formation, and deeper valleys expose shales of the Cuyahoga formation. The rocks of the plateau dip eastward into the Appalachian Basin, so the oldest rocks occur at the surface in the westernmost features of this park.

The erosional effects of streams, groundwater, gravity, and the differing resistances of rock layers within the Black Hand sandstone to weathering are responsible for the geologic features of Hocking Hills State Park. The ice sheets of pre-Illinoian, Illinoian, and Wisconsinan time apparently stopped short of the park area, so no till is present. Terrace remnants of lake sediments, however, have been identified both north and west of South Bloomingville in Pine and Queer Creeks, and there is Wisconsinan outwash in the Queer Creek valley as far east as South Bloomingville. So streams in the area, other than undergoing remarkable changes in water flow dependent on whether glaciers were advancing or melting back, have been free to carve into the Logan County landscape for at least a couple million years or so.

Since the Cuyahoga formation, like many Mississippian and Pennsylvanian rocks, is a mixture of layers of hard sandstone, siltstone, and limestone and softer shale and mudstone, downward-cutting streams have had either a difficult or easy time eroding streambeds and underlying bedrock. Groundwater soaking down through sandstone layers has aided this ongoing process by either dissolving or hardening the cement holding the grains of various rocks together. The harder strata became waterfalls that continue to move upstream with time as the underlying softer rock erodes back, undercutting the harder layers. Eventually the overlying rock will break off along joints, causing the waterfall to adjust slightly upstream and creating a rock-walled gorge downstream of the falls. Repetition of this process over thousands of years has created the rocky gorges that characterize this park. Rock shelters form along the walls of the gorges or underneath waterfalls; where the lower walls are more susceptible to weathering and erosion, the upper parts of the walls eventually greatly overhang them.

This kind of scenery has always attracted people. The Hocking Hills region was a destination of many city and country folk who enjoyed a picnic or a hike to scenic vistas. The park began developing when the state purchased 146 acres surrounding Old Man's Cave in 1924; within five years all areas were under state control.

Ohio 374 leads to all six areas. Cantwell Cliffs occurs first, southwest of Rockbridge and north of Ohio 374. Trails lead into the gorge of Buck Run or along its rim. The gorge ends at a 150-foot-high rock shelter that water carved in the softer parts of the Black Hand sandstone. After rains and in spring a small waterfall cascades from the edge of the top of the shelter. Vertical joints are particularly well developed in this gorge, forming a number of narrow clefts that extend from the rim to the cliff base. Slump blocks occur where the jointing resulted in slippage.

Rock House is about 7 miles south of Cantwell Cliffs on Ohio 374. This unique erosional feature above Laurel Run is a striking example of the erosion that can occur along perpendicular joints. Rock House, a 25-foot-high by 20- to 30-foot-deep passage, was carved along a northeast-trending joint halfway up a 115-foot cliff of Black Hand sandstone. It parallels the cliff face for 200 feet. A steep trail and steps lead up to the feature and continue through it. The passage of Rock House is open at both ends, and it also has five openings, or "windows," carved along northwest-trending joints. This feature formed as water percolated down through the joints and selectively hollowed out the middle, weaker part of the Black Hand sandstone. Then blocks of stone pulled away from the cliff and came to rest, jutting out from the cliff face. The passage formed behind them. Most of the excavation of Rock House took place during wetter glacial times. On rainy days groundwater seeps from the rear wall near the floor of Rock House, where it encounters the harder, lower portion of the Black Hand sandstone. It eventually falls to the base of the cliff through the windows.

Continuing south on Ohio 374 will lead you past Conkles Hollow, home of the highest cliffs in Hocking Hills State Park. At the mouth of the hollow, which is a tributary of Pine Creek, 200 feet of Black Hand sandstone and overlying Logan formation tower above the hollow's floor. Trails pass along the rim of the hollow or at stream level. Conkles Hollow is one of many rock gorges in this region that were carved along northwest-trending joints where they intersect Pine Creek valley. The ½-mile-long gorge narrows to 300 feet at its upper end, which is a rock shelter. The gorge has been cut downstream of the rock shelter as this waterfall has retreated—½ mile from its original position—upstream.

The Old Man's Cave area south of Conkles Hollow is the largest, most diverse, and most popular attraction in Hocking Hills State Park. Trails at this site follow Old Man's Creek, along which are some interesting geologic features. At road level the creek is in a wide valley carved in soft Logan formation sandstones, and the harder Berne conglomerate member underlies the flat stretch of

Typical cross-bedded Black Hand sandstone near Old Man's Cave in the Hocking Hills. A small fault offsets the layers.

Swirling torrents of water and sediment continue to carve the Devil's Bathtub along one of the trails near Old Man's Cave.

the creek. The creek then plunges 30 feet over a cliff of the uppermost portion of the Black Hand sandstone at Upper Falls and follows a northeast-trending joint—now a gorge—to Queer Creek. Downstream of Upper Falls is Devil's Bathtub, an oval hole, or pothole, that swirling water, sand, and pebbles wore into the streambed. Old Man's Cave, farther downstream, is a rock shelter carved in the softer middle part of the Black Hand sandstone. It's one of the larger ones in the Hocking Hills park system, measuring 200 feet long, 50 feet high, and 75 feet deep along the wall of the gorge. Old Man's Creek flows over the resistant lower part of the Black Hand sandstone at Lower Falls, downstream of the cave. Behind Lower Falls there is another large rock shelter. The floor of this shelter is soft shale of the Cuyahoga formation.

East of Ohio 374 and southwest of Old Man's Cave, Queer Creek flows in a wide, joint-controlled valley where bedrock is Logan formation of Mississippian age. Just west of the road, the creek plummets 50 feet over a cliff of upper Black Hand sandstone. This is Cedar Falls. Note the rock shelter behind the sizable plunge pool.

Another joint-controlled valley features the most impressive rock shelter in Hocking Hills State Park—Ash Cave. A tributary of East Fork Queer Creek plunges over upper Black Hand sandstone and cascades 90 feet to the shelter floor. The shelter, behind the waterfall, is 100 feet deep and nearly 500 feet long.

Ash Cave is one of many rock shelters in the Hocking Hills. Note that the waterfall is receding along a joint.

Sandstone blocks broken loose along joints high above the shelter and strewn at its base and piles of sand attest to the great amount of erosion that formed the feature.

Hocking River—Ice Age Changes

In preglacial time the Logan-Lancaster River, a Teays River tributary, flowed to the northwest from a source near Haydenville, which marked a drainage divide. Other streams headed southeast from Haydenville. During Illinoian time, ice advanced south and stopped just north of Sugar Grove. Torrents of meltwater eventually breached the divide at Haydenville, creating the present-day southeast drainage of the Hocking River to the Ohio River. During this time the ancestral Hocking River's valley was wide from south of Rockbridge to Logan and beyond; meltwater deposited over 200 feet of outwash in this valley, which now underlies U.S. 33. After Illinoian ice retreated, the ancestral river began cutting into the deposits, and by Wisconsinan time it had begun cutting a narrow, rocky gorge east of present-day U.S. 33. Then the river rejoined its old valley just north of Logan. Wisconsinan meltwater did little, however, to alter the valley.

Salt, Iron, and Clay in the Hocking Valley

The Hocking Hills gorges show the beauty of Mississippian sandstones, but the overlying limestones, shales, underclays, and coals of Pennsylvanian strata provided much of the region's early economic base. Salt was one of the earliest geologic resources that locals tapped. Salt came from Salina, later called Beaumont, from 1820 to 1877. Wells tapped brine in the Mississippian Black Hand sandstone of the Cuyahoga formation about 570 feet below the surface. Chauncey developed near these saltworks in 1839. A saltworks also operated 3 miles west of Athens, and in the early 1870s brine came from wells sunk into the Mississippian Cuyahoga formation at Nelsonville.

Iron smelting was another early industry of the Hocking Valley. Iron ore came from the Block ore associated with the Lower Mercer limestone, the Ferriferous ore above the Vanport limestone, and the Blackband ore from the Upper Freeport coal horizon. These were calcareous iron deposits, or siderite, that were easily detected where oxidation caused rusty stains in the mixed Pennsylvanian layers.

The first charcoal-fired furnace in the area, the Logan Furnace, was erected at Logan in 1852. Logan Iron Company once sat south of Hunter Street on the north bank of the Hocking Canal, most likely close to present-day Furnace Street. In 1853 the Union Furnace opened just southwest of New Cadiz, the town now called Union Furnace, about 0.7 miles south of town. Ore came from company-owned land 1 to 2 miles east of town. The furnace produced 11½ tons of iron ore daily in the late 1860s. Peter Hayden, a Columbus financier, purchased iron-bearing land in Hocking County in 1852 and had a furnace at Hanging Rock, along the Ohio River, dismantled in 1856 and shipped up the Hocking Canal to a site that became Haydenville. The Hocking Furnace, as it was called, sat against the hill on the east side of Main Street. The first furnace using Hocking Valley coal went into blast in 1875, halfway between Maxville

The National Fireproofing Company during its peak years of the 1920s in Haydenville. Little trace of this huge operation remains. —Wilber Stout photo, Ohio Department of Natural Resources, Division of Geological Survey

and McCuneville. Fourteen more furnaces were operating in the area by 1881 (*see also* the Ohio Valley chapter for more information about furnaces of the Hanging Rock Iron Region).

The major iron-bearing units in the valley were the Blackband, Block, and Baird (Ferriferous limestone) ores. The field was mined in the Hocking Valley, Monday Creek valley, Sunday Creek valley, and the Shawnee–New Straitsville area. After the Civil War, demand for local pig iron declined as both Missouri and Lake Superior iron made great inroads into the market; by 1885 the Hocking Valley iron boom was ending. Little trace remains of these furnaces today.

Where there was iron, there was clay. The Hocking Valley had many brick plants, sewer pipe plants, and potteries that took advantage of local clays that had been deposited in an ancient sea. Peter Hayden's pig iron business prospered into the late 1870s, but increased competition with bigger mills in the Cleveland to Youngstown belt and Pittsburgh caused him to shift to ceramics. He named his new operation the Haydenville Mining and Manufacturing Company. Construction of the clay plant began in 1883. Since there were not enough houses in the isolated location, Hayden began constructing a company town around the plant. Hayden died in 1888, and the town fell under the control of National Fireproofing Company from Pittsburgh in 1907.

Initially, the Haydenville plants produced a wide variety of paving and building bricks and tile when the change of ownership occurred, and later electrical conduit was a major product. Strip and drift mines supplied Lower and Middle Kittanning clay, and the company briefly used Flint Ridge clay for fireproofing materials. The company had its own coal mine that tapped Middle Kittanning coal, which fueled its kilns until 1934. The deep clay mines closed in 1957. In 1964, the Haydenville operation closed its doors forever. Although the plants and company store are gone, many of the company houses remain. The church contains samples of all major brick and tile styles the Haydenville Mining and Manufacturing Company produced.

Nelsonville was the site of the first large commercial brick plant in the Hocking Valley. It opened around 1870 on the east side of town, getting Lower Kittanning clay from the adjacent hills. By 1876 four other plants were in operation. Nelsonville Sewer Pipe Company opened in 1887, followed by Nelsonville Brick Company in 1903. The two companies merged in 1907 and concentrated on producing brick pavers using Lower Kittanning clay. It shipped several of its products—Nelsonville Block, Hallwood Block, and a famous sidewalk paver embossed with a star—across the Midwest. The last brick plant in town, Athena Brick, operated into the late 1940s.

The Hocking Valley Coalfield

East of the Hocking Valley up tributaries like Federal, Monday, and Sunday Creeks, thick seams of coal brought a coal boom to the area that continues still. This region of coal became known as the Hocking Valley Coalfield, and it stretched from Logan, Nelsonville, and Athens east to Shawnee, Corning, and Glouster by the 1880s and 1890s. The Middle Kittanning coal, averaging 5 to 7 feet of marketable coal, was the main seam that was mined, but miners also dug into the Lower Kittanning and Lower Freeport seams. Starting in the late 1870s, most of the coal went to fire iron smelters.

Two beehive kilns of the Nelsonville Brick Company remain as a tribute to this important Nelsonville industry. Nelsonville pavers surround a historical plaque.

The largest section of this coalfield opened in 1880 in the Sunday Creek valley, which runs northeast of Beaumont, upon completion of the Ohio Central Railway. The railroad's purchase of 12,000 acres of coal land initiated production. The subsidiary Ohio Central Coal Company handled the mining of the Middle Kittanning and Upper Freeport coals. The Middle Kittanning horizon averaged 11 feet thick but at places reached 14 feet in the valley. Miners worked the seams with shaft and drift mines in the 1880s, removing 7 to 9 feet of marketable coal.

With the cheap supply of fuel came iron furnaces. Four furnaces produced pig iron in Shawnee; others did so at Buchtel, Floodwood (now New Floodwood), Moxahala, and New Straitsville. The Troy Mines at New Straitsville yielded the first fossil reptile tracks recorded in Ohio. Iron Point, just northeast of Shawnee, marks the first discovery of Blackband ore, a carbonate iron ore associated with the Upper Freeport coal. Miners found that this ore, known locally as the Shawnee ore, extended beyond Iron Point in the subsurface, having been found in coal mine shafts south of Shawnee and to the north around Moxahala.

The heart of the Hocking Valley field was up Monday Creek. Many mines in this valley worked 6- to 8-foot-thick seams of Middle Kittanning coal. The Baird's Furnace Mines at Gore supplied the first furnace coal of the Hocking Valley in 1874. The Baird ore, associated with the Lower Kittanning coal, is an iron-rich limestone, which led entrepreneurs to erect thirteen furnaces in the Monday Creek valley.

CLAY MANUFACTURING IN THE HOCKING VALLEY

Town	Company (years of operation)/Products/Strata
East Clayton	The East Clayton Brick Company, located 3 miles west of Nelsonville, was in operation by 1870. It made pavers. The plant burned down in 1892 and was never rebuilt. Ruins still mark the site on the south side of the Hocking River.
Glouster	The Wassall Brick Company had an annual output of 10 million pavers around 1910 and used Brush Creek shale of the Conemaugh group.
Greendale	The Hocking Valley Products Company, built in 1904, was once recognized as the largest brick plant of its kind in the world. The plant made standard construction bricks from the Conemaugh group Brush Creek shale. The bricks had rough outer surfaces that resembled thick carpet, resulting in their marketed name—the Greendale Rug Brick. They were used in buildings across Ohio and as far away as Montreal, Quebec, and Vancouver, British Columbia. The huge plant closed in 1929; only ruins and a few company-built houses remain in the woods.
Logan	Logan Fireclay (1890), Logan Granite Clay, and Hocking Clay Manufacturing Company (1885) produced various bricks and sewer pipe using Brookville and Tionesta clays from a ridge southeast of town. Logan Pottery Company (1903–1964) employed some seventy people in the early 1900s, making a wide variety of stoneware items using Tionesta clay from pits along Clay Bank Road.
Nelsonville (Diamond)	The Diamond Brick & Clay Company, established in 1910, originally made fireproof tiles; later it manufactured pavers and standard building bricks. Through a number of corporate changes, the plant survived into the mid-1990s. It utilized the Lower Kittanning clay of the Allegheny group.
New Straitsville	The Straitsville Impervious Brick Company, built in 1904, initially used Flint Ridge and Middle Mercer clay of the Pottsville group to make vitrified face brick (a brick with a glassy surface) and later used Tionesta clay and shale of the Pottsville group to manufacture tiles for furnace flues. The company had clay and coal mines along Rock Run Road north of New Straitsville.
Shawnee	The Ironclay Brick Company, west of town, mined Lower Kittanning clay to make building brick. It closed in the early 1950s. Claycraft Mining and Brick Company had a shaft mine in Brookville clay. The plant started out making firebrick and then shifted to tile manufacturing.
Trimble	The Trimble Brick Company, which operated in the early 1900s, mined the Brush Creek shale to make pavers.
Union Furnace	The Columbus Brick & Terra Cotta Company (1881–1921) specialized in building brick using Middle Mercer and Flint Ridge clays of the Pottsville group.

Although the first coal mining in the Hocking Valley began around 1820, Nelsonville became a mining center with the arrival of the Hocking Canal in 1840, which stretched from Athens to a connection with the Ohio & Erie Canal at Carroll. The opening of the Columbus & Hocking Valley Railroad in 1869 led to a boom: the railroad shipped 200,000 tons of coal in 1871, and another 75,000 tons were shipped by canal that year. By the 1870s at least seven mines were operating in Nelsonville. The hills bordering Nelsonville contained little coal by 1875, and that is when many operations stretched north into the Monday Creek valley. Twelve coke ovens were built in Happy Hollow near Nelsonville because of the Nelsonville boom. The production of the Hocking Valley Coalfield peaked between 1895 and 1924.

The coalfield made history in 1884 when disgruntled workers set fires at mines in Carbon Hill, New Straitsville, Sand Run, and Shawnee in the midst of a six-month-long strike. (On the hill above the south side of New Straitsville there is a rock shelter, Robinson's Cave, where coal miners met and started the United Mine Workers in the 1880s.) The arson led to underground fires that burned for decades, destroying millions of dollars of coal. New Straitsville even made *Ripley's Believe It or Not*, a popular radio program (1930–1948), because of the fires. Tourists flocked to see the smoking vents and feel the hot earth. Visit the History Museum in New Straitsville to learn more and be sure to check out the model coal mine in the basement.

The coalfield made news again November 5, 1930, when sparks from a trolley wire ignited methane gas, causing a huge explosion in the former Poston No. 6 mine at Millfield. Eighty-two people died, making it the worst mine disaster in the state. Coincidentally, the catastrophe happened while officials were touring the mine to see new safety features that the new owners, Sunday Creek Coal Company, had recently added. The mine reopened and operated until 1945. A historical plaque marks the site among its crumbling ruins.

Mine 266 of Ohio Collieries in the Middle Kittanning seam in the Hocking coal region. This mine near Glouster operated between 1888 and 1926. Photo dates to around 1907.

*An abandoned drift mine in the side of a hill along Ohio 13
near Millfield. Stone and concrete block the entrance.*

Acid mine drainage plagues the coal district today. Note the orange and yellowish stains along certain stream courses. Acid mine drainage is the result of coal mining and releases sulfides into surface water and groundwater, which react to form acids that are extremely detrimental to native animals and plants. Metals, particularly iron and aluminum, eventually precipitate in the streams and color them reddish orange or milky white. For all practical purposes these streams are dead. Many local streams are also choked with sediment as a result of improper mining practices. Mining companies have learned a lot from their errors and lack of concern for the surrounding environment. Examples of reclamation efforts are becoming more prevalent. At Carbondale, for example, a series of artificial wetlands traps mine waters so that bacteria have time to render the sulfides harmless before they form acids. Cleanup of rock or gob piles (piles of coal waste) at abandoned mines is also important in reducing the amount of pollution reaching nearby streams. Some mines have been gated to provide bats places to hibernate.

Athens Area

At Athens the Hocking River heads east from U.S. 33, roughly paralleling U.S. 50 and then Ohio 144 to the Ohio River at Hockingport. Radiating out southwest of Athens is the valley of the preglacial Albany River, a tributary of the ancient Teays River system. The headwaters of this preglacial river was in present-day southern Perry County. These smaller tributaries flowed south, generally underlying the present Monday Creek and Sunday Creek valleys, through Athens and to the Hebbardsville area, where the two streams combined to form the Albany River. From Hebbardsville the river flowed southwestward to join the preglacial Marietta River under present-day Rio Grande. Pennsylvanian-age strata of the Conemaugh group form the bedrock—shales, sandstones, and limestones—of this old valley.

Athens became a center of learning in early nineteenth century Ohio. Ohio University was established in 1804. All the resources a growing community needed were nearby. Its mineral resources were Pennsylvanian in age. The first brick homes appeared as early as 1803, and a small commercial brick plant was operating by 1823. In the 1870s workers dug clay from the river valley for making bricks, and later they got clay from Conemaugh group shales in the hills overlooking Athens. The Athens Brick Company used shales above and below the Ames limestone to produce Athens Block pavers in the early 1900s. In its heyday the company had thirty-one kilns that produced 15 million pavers annually. Miners extracted Middle Kittanning coal from a 200-foot-deep shaft in downtown Athens in 1878. Most of the coal mines that came later were outside the city limits. Coal mines at Canaanville supplied the Athens Brick Company with fuel for its kilns. The Brush Creek limestone thickens to an astounding 18 feet southwest of Athens; it's normally 1 to 2 feet thick. Stone quarries operated near Albany, and one continues to supply crushed stone today. A number of other short-lived limestone quarries operated east of Athens.

Athens to Pomeroy

Sandstones, shales, and siltstones—Monongahela group strata of Pennsylvanian age—flank the road between the U.S. 50 junction in Athens and the Ohio River. South of Athens U.S. 33 passes through a rural, hilly landscape. Sandstone borders the road at Shade. The town sits at a wide spot in the Shade River valley. Locals found fossilized Pennsylvanian-age tree trunks along the river here in the 1870s. Locals used flint from local Cambridge limestone as doorsteps to their homes. More sandstone outcrops at Pratts Fork. Oil wells appear near the highway in Burlingham. Wells now tap sandstone in the Cow Run and Maxton sandstones of the Conemaugh group, but in 1924 oil came from a deeper sandstone below the Devonian-age Berea sandstone, the Cussewago sandstone. Maroon and greenish gray Monongahela group shales and siltstones laid down on ancient floodplains appear on the edges of a hill near the rest area south of Darwin and just north of Township Road 19. The soft strata are deeply cut by small channels, what geologists call *badlands erosion*. Massive Pomeroy sandstone, also of the Monongahela group, lies below these redbeds on the hills above Pomeroy.

*The 1913 Hocking River flood at Athens. Beehive kilns of the
Athens Brick Company are in the foreground.*

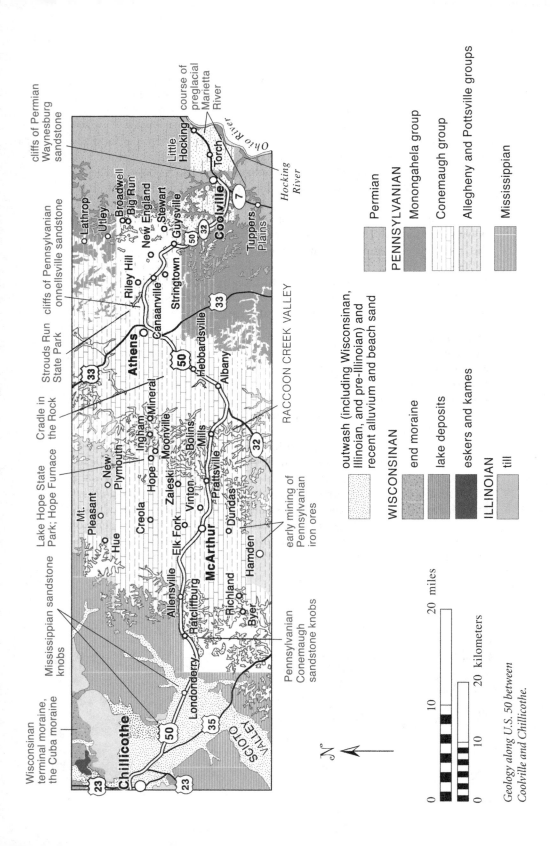

Geology along U.S. 50 between Coolville and Chillicothe.

cliffs of Permian Waynesburg sandstone

course of preglacial Marietta River

Ohio River

Little Hocking

Torch

Hocking River

cliffs of Pennsylvanian onnellsville sandstone

Lathrop

Utley

Broadwell

Big Run

New England

Stewart

Guysville

Coolville

Tuppers Plains

Strouds Run State Park

Riley Hill

Canaanville

Stringtown

Hebbardsville

Albany

RACCOON CREEK VALLEY

Cradle in the Rock

Athens

Mineral

Lake Hope State Park; Hope Furnace

Mt. Pleasant

New Plymouth

Ingham

Hope

Moonville

Bolins

Mills

Prattsville

Dundas

Mississippian sandstone knobs

Hue

Creola

Zaleski

Vinton

Elk Fork

Allensville

Ratcliffburg

Richland

Byer

Hamden

McArthur

early mining of Pennsylvanian iron ores

Pennsylvanian Conemaugh sandstone knobs

Londonderry

Chillicothe

SCIOTO VALLEY

Wisconsinan terminal moraine, the Cuba moraine

N

WISCONSINAN

outwash (including Wisconsinan, Illinoian, and pre-Illinoian) and recent alluvium and beach sand

end moraine

lake deposits

eskers and kames

ILLINOIAN

till

Permian

PENNSYLVANIAN

Monongahela group

Conemaugh group

Allegheny and Pottsville groups

Mississippian

20 miles

10

0

20 kilometers

10

0

Coolville—Chillicothe
76 miles

U.S. 50 stretches across the unglaciated Allegheny Plateau, from Permian-age sandstones and shales at Coolville, near the mouth of the Hocking River, to Mississippian-age strata around Londonderry. Between Londonderry and Chillicothe the route follows Illinoian tills and outwash, paralleling the glacial boundary. Since the strata of the plateau dip eastward, the rocks get older as you travel west on U.S. 50.

Coolville lies about 100 feet above the Hocking River valley on Permian strata where rock shelters develop below prominent cliffs of erosion-resistant Waynesburg sandstone. Pennsylvanian-age Monongahela strata form the river bluffs, and Wisconsinan and Illinoian outwash form terraces in the bottomland. The preglacial Marietta River once flowed southwest from Little Hocking and Torch (formerly Torch Hill), through Coolville and Tuppers Plains and into West Virginia. This drainage now follows the Ohio River, which developed after glaciation. Ohio 7 follows this ancient valley from Coolville to the southwest, but otherwise it is disguised well under outwash and lake sediments. U.S. 50 winds through the Permian uplands west and northwest of Coolville. At Guysville the highway returns to the Hocking Valley, and Pennsylvanian rocks of the Monongahela group form the bedrock. Up the Federal Creek valley from Stewart, the Pittsburgh coal forms an important field around Broadwell. The coal is extensive, of high quality, and at places over 7 feet thick. Workers made coke in 125 ovens at Utley, but a lack of sufficient coal led the owners to close the ovens around 1900. Black Diamond Coal & Coke Company ran a mine at Lathrop beginning around 1890. Fifty coke ovens at this mine were used sporadically. The town of Big Run had a shaft to the Pomeroy seam in the 1870s.

North of Stewart and west of Broadwell, a shale associated with the Waynesburg coal, the youngest coal of Pennsylvanian age, yields well-preserved compression fossils of ferns, horsetails, lycopods, and seed ferns—plants that began establishing themselves as ancient seas receded from Ohio. The age of this layer is viewed as Permian by some geologists, Pennsylvanian by others.

Near Canaanville, Conemaugh group strata appear in the valley bottoms. This small mining community had the deepest mine in the state in 1911. A shaft reached the Lower Kittanning coal at a depth of 440 feet. The main coal this town extracted was the Middle Kittanning. Gas wells pepper the countryside around Canaanville, New England, Riley Hill, and Stringtown. Most of the wells were drilled after 1928 and tap Devonian-age Cussewago sandstone. Fifty-foot cliffs of Pennsylvanian-age Connellsville sandstone flank the Hocking Valley east of Athens. Note that the river flows in a straightened channel east of the U.S. 33 junction. Strouds Run State Park lies north of U.S. 50. The completion of a dam in 1960 formed Dow Lake, the focal point of the park. Trails in the park lead to sandstone-capped ridges of second-growth timber. Linscott Spring near the park office constantly replenishes Strouds Run.

Southeast of Athens, U.S. 50 parallels the wide valley of Margaret Creek, the former path of the preglacial Albany River. From Hebbardsville to Albany the valley is flat and wide with no significant stream evident; the valley is clearly the work of an ancient predecessor. North and south of the highway and just west of Albany there are limestone quarries in Pennsylvanian Brush Creek limestone. The limestone layers thicken to 6 feet here but usually they average 1½ feet. North of Albany along Fox Lake, Buffalo sandstone of the Conemaugh group forms the cliffs of minor rock shelters in Brush Creek limestone at Cradle in the Rock. The area is part of the Fish Lake Wildlife Area. At Bolins Mills the route crosses the deeply incised valley of Raccoon Creek. This creek follows an erratic path on its way to the Ohio River, often doubling back on itself, shifting from one preglacial valley to the next. Middle Fork Salt Creek occupies a wide valley between Allensville and Ratcliffburg, underlain by another preglacial valley of a Teays River tributary. U.S. 50 passes two large sandstone erosional knobs southwest of Ratcliffburg where the floodplain widens to 0.7 miles. Near Londonderry another knob lies north of the highway along Salt Creek. Rattlesnake Knob, another prominent erosional knob, is about 3 miles west on the edge of the Scioto Valley. This is the edge of the Allegheny Plateau, where the Scioto River and its tributaries and ancient drainage systems before them have been carving away the western part, exposing the underlying Devonian rocks. The tributaries flowing into the Scioto Valley experience an increase in their downcutting ability as their slopes steepen to reach the level of the Scioto floodplain. In the process, knobs of Mississippian strata are sometimes isolated from the main plateau. The Scioto Valley between Londonderry and Chillicothe is filled with Illinoian and Wisconsinan outwash forming steplike terraces above the current floodplain.

Hope Furnace and Other Vinton County Mineral Resources

U.S. 50 transects the northern part of the Hanging Rock Iron Region. Six charcoal-fired furnaces, the Eagle, Cincinnati (later called the Richland),

Hope Furnace in Lake Hope State Park.

Hamden, Hope, Vinton, and Zaleski, began operating between 1852 and 1858 in Vinton County. The Ferriferous ore of the Allegheny group formed the backbone of the early iron industry, and though it was more plentiful in the southern portion of the Hanging Rock region in Lawrence County, the layer was still thick enough in Vinton County to supply these furnaces. The region between Dundas and Hamden was a prominent mining area. Besides the Ferriferous ore, a 7-inch layer of iron carbonate associated with the older Pennsylvanian-age Upper Mercer limestone was used widely in these furnaces. Another siderite ore, associated with the Lower Mercer limestone, fed the Cincinnati (Richland) Furnace east of Richland. Creola was a major shipping point for Sand Block ore (blocky, siliceous rock associated with the Upper Mercer coal) and Upper Mercer ore from 1880 to 1888. During this period, twenty-five railroad cars a day left Creola for the smelter at New Straitsville.

Perhaps the best known of the early furnaces is Hope Furnace, now viewed by thousands in Lake Hope State Park, off Ohio 278 north of Zaleski. Hope Furnace, built in 1854, produced 15 tons of cast iron daily in 1870. Its ore came from near Vinton. Blocks of Lower Freeport sandstone, quarried from nearby cliffs, compose the furnace stack, which was lined with the older Clarion sandstone. Although the stack is well preserved, most of the plant is gone. (To see a reconstruction of a typical furnace operation, visit Buckeye Furnace State Memorial southeast of Wellston.) Around 1875 Hope Furnace was one of the first furnaces in the area to close; others, like Cincinnati and Vinton, lasted into the late 1880s because they were modified to burn coal or coke. Ultimately, the increased availability of higher-grade iron ore from the Lake Superior region and southwestern Missouri led to their abandonment.

Along Elk Fork about 1.7 miles southwest of Prattsville lies the remains of the Vinton Furnace. The furnace went into blast in 1854, fueled by charcoal

An abandoned drift mine along Ohio 278 north of Hope.

made from the surrounding forests. In 1868 the company sank a shaft down to the Quakertown coal, but later they found that it was unsuitable for furnace use. Next the company built twenty-four brick ovens and made coke from nearby Clarion coal. Unfortunately, the high sulfur content of the coal made this uneconomical as well. Finally, the company turned to coal from the Hocking Valley and operated a few years until the depression of 1873–1879 caused it to abandon the operation. The stone foundation of the stack, some building foundations, and the ovens buried in vegetation remain, on private property.

Local Pennsylvanian-age strata contained numerous other useful materials besides iron ore. Paleo-Indians found the black Zaleski flint of the Allegheny group in valley bottoms perfect for arrowheads and spear points. There is also Allegheny group Vanport flint in the area. Some limestones, like the Vanport, formed in nearshore areas where water conditions allowed the precipitation of silica compounds within the bottom sediments and by various organisms, including sponges and microorganisms called *radiolarians*. Sometimes quartz

SOME VINTON COUNTY CLAY AND COAL MINING TOWNS

Town	Mining History
Dundas	Mines in Clarion coal.
Elk Fork	Mines in Quakertown coal.
Hamden	Puritan Brick Company, established east of town in 1909, used Clarion shale and clay. The company town of Puritan was established. Mines in Clarion coal.
Hope Station	Mines in Middle Kittanning coal.
Hue	Potters used Flint Ridge and Middle Mercer clays from 1845 to 1903.
Ingham	Mines in Middle Kittanning coal.
Kings Switch	Mines in Middle Kittanning coal.
McArthur	McArthur Brick Company made face brick from various local clays and shales from 1905 to 1961. Mines in Upper Mercer and Clarion coal.
Mineral	Mines in Middle Kittanning coal.
Moonville	Mines in Middle Kittanning coal.
Mt. Pleasant	Mines in Clarion coal.
New Plymouth	Mines in Clarion coal.
Prattsville	Mines in Brookville coal.
Vinton	Mines in Quakertown coal.
Zaleski	Drift mines in Clarion clay produced clay for firebrick manufacturing. Mines in Upper Mercer, Clarion, and Middle Kittanning coal. Mines in Brookville coal south of town along Wheelabout Creek. The bulk of Middle Kittanning coal production came from two mines at Zaleski.

(silicon dioxide) crystals accreted around siliceous skeletons on the sea bottom, forming nodules; at other times layers of gelatinous silica coated the sea bottom. The presence of quartz mixed with the ever-present calcite made the Vanport a much harder stone. It was too porous for good points, but settlers noticed its similarity to the hard French stones that were fashioned into buhrstones for water-powered mills. McArthur started as a center of buhrstone manufacturing around 1805, marketing local siliceous Vanport limestone as "Raccoon Millstone" for some thirty years. Nonsiliceous parts of the Vanport limestone served as flux for iron smelting, agricultural lime, concrete aggregate, and road gravel. Old quarries remain at the town of Elk Fork; Oreton, a ghost town; and elsewhere in the hills.

Sandusky—The Ohio River
162 miles

U.S. 250 traverses 20 miles of Huron-Erie lake plain between Sandusky to just south of Norwalk. Then for 73 miles it crosses rolling till plain and end moraines of the Wisconsinan-age Killbuck lobe from Norwalk to the Stark and Tuscarawas County line near Wilmot. From there to the West Virginia line the route cuts through the Allegheny Plateau to the Ohio River valley.

Sandusky to Mt. Eaton

Devonian-age bedrock lies close to the surface around Sandusky, and limestone quarries are common in the area. One of the larger quarries is visible west of U.S. 250, stretching about 1½ miles between Strub Road and Ohio 2. Soils around Sandusky are waterworn moraine material and sand associated with postglacial lakes.

Glacial lake sand is especially prominent in the Huron River valley north and south of Milan. The Huron River meanders across this easily eroded material, creating a wide floodplain with many branching, or dendritic, tributaries. Sand pits abound in this area. Norwalk rests atop the Glacial Lake Whittlesey beach ridge and underlying Devonian sandstone. Berea sandstone for building stone was taken from small quarries northeast and southeast of town as early as 1860. Berlin Heights was also the site of a quarry by 1875, and West Clarksfield had a quarry too. These were the westernmost Berea sandstone quarries in the state.

Between downtown Norwalk and U.S. 20, U.S. 250 crosses sandy ridges that were deposited by Glacial Lake Maumee. U.S. 250 climbs 200 feet in the next 8 miles south of Ohio 2 onto till plain, and eventually onto the Defiance end moraine between Olena and Fitchville. An esker lies east of Olena, just west of West Clarksfield.

U.S. 250 bridges the deep valley carved by the Vermilion River through the Defiance moraine at Fitchville. Legend says the river got its name from reddish Mississippian-age shales that yielded a red pigment used by Native Americans.

A small glacial lake once bordered the south edge of the moraine here. Sand and gravel deposited by meltwater form terraces along the river north and south of town. Between Fitchville and New London, kettle depressions, now filled with peat and marsh plants, dot the countryside. Drainage in the area is poor.

Beginning at the junction with U.S. 224 and continuing to near Ashland, about 12 miles, the highway crosses a series of Wisconsinan moraines that are sandwiched between the present-day lakeshore and the Allegheny Plateau of

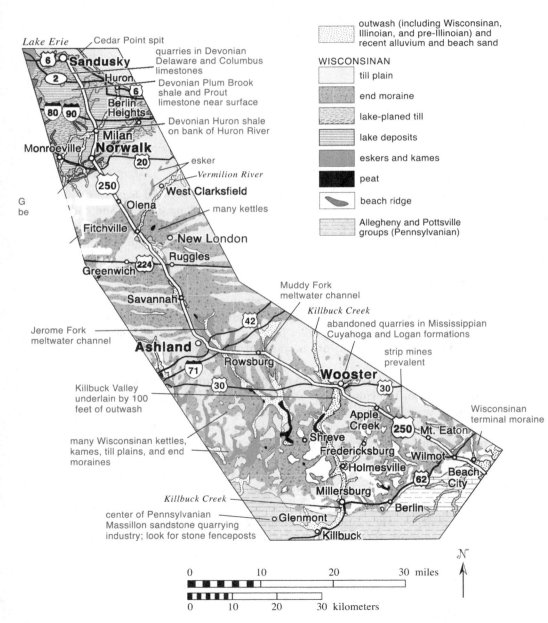

Geology along U.S. 250 between Sandusky and Beach City.

southeastern Ohio. All the end moraines that are spread out between Toledo and Cincinnati, on the west side of the state, occur here. The tills made good bricks. Locals made bricks as early as the 1820s in Savannah, and other brick and tile makers followed. Their craftsmanship is still evident in the brick buildings of the area.

East of Ashland U.S. 250 crosses the outwash-filled valley of Jerome Fork, which is now modified by the Mohicanville Dam farther downstream. This Mohican River tributary served as a major meltwater channel when ice lay at the various end moraines to the north. Sand and gravel terraces dotted by an occasional peat-filled kettle and sandy hill or kame stretch along the channel. Rowsburg sits on the west bluff of a similar meltwater channel, which is now occupied by Muddy Fork. South of town there are small, active oil fields.

On the west side of Wooster U.S. 250 passes over an impressive meltwater channel. Mississippian-age bedrock lies 100 feet below the valley bottom. Killbuck Creek appears minuscule compared to its wide floodplain, a sure sign that the valley originated during the melting of ice sheets when tremendous torrents of water gushed through the area. The creek in this channel flowed north before an ice sheet blocked its path. For a time water backed up, forming a lake in this old valley. Eventually, the lake drained through a low spot in the divide between the former Erigan and Ohio River drainages, establishing the present southward trek of Killbuck Creek. This creek also has long stretches of channel that have been straightened. Note the abandoned stretches of channel that now form narrow, sinuous ponds. Apple Creek, on the southeast side of Wooster, has a similar history; it lies in a bedrock valley, which is buried under at least 185 feet of outwash, and appears to be too small for such a wide valley. Obviously preglacial drainage was much different—much larger. Mississippian-age sandstone and shale and oil and gas are local mineral resources.

Mississippian-age Cuyahoga formation shales in an abandoned quarry of the Medal Brick & Tile Company in Wooster.

The Wooster area had a few operations that took advantage of its Mississippian resources. The Medal Brick & Tile Company quarried shales along Mechanicsburg Road west of town beginning in 1901. By 1920 it was mainly producing paving bricks. The abandoned quarry exposes shale and sandstone of the Cuyahoga formation overlain by conglomerate and sandstone of the Logan formation. Over fifty species of fossils, including brachiopods, bryozoans, snails, clams, and crinoids, were found at this site, especially in the concretionary layers of the shale. Ripple marks and cross-beds dominate the sandstone layers. The plant site across the road is now a recycling business. North of West Wayne Street in Wooster there are abandoned quarries, major suppliers of Logan formation sandstone from the 1870s to 1920s. The quarry area, long since reclaimed, is now a gated residential area.

As you pass through Wooster, try to imagine the U.S. 30 and Ohio 3 interchange completely submerged under water. That is what happened on July 4, 1969. Almost 11 inches of rain in twenty-four hours and a continual deluge for two days was more than Killbuck Creek and its tributaries downstream could handle. Devastation exceeded that of the 1913 Flood, which affected all the lowlands of the Wooster area. A similar amount of rain fell during that flood, but it fell over five days. Twenty-two people drowned and damage exceeded $20 million in the 1969 flood.

Southeast of the village of Apple Creek, strip mines appear, indicating that Pennsylvanian-age coal and associated rocks form the bedrock. Builders used colorful Cuyahoga formation sandstone of Mississippian age from Fredericksburg quarries in several local structures. At Mt. Eaton the Pennsylvanian Putnam Hill limestone was quarried to make lime and aggregate. Many farmers

Quarries in towns south of Wooster, in Holmes and Coshocton Counties, cut building blocks from different brown and tan color phases of the Pennsylvanian-age Massillon sandstone of the Pottsville group.

Gambier	Many Kenyon College buildings are built of this stone.
Glenmont	The Briar Hill Stone Company, established in 1917 by a stonecutter from Amherst, took over a sandstone quarry that had been established in 1857 east of town. The company currently operates several quarries throughout Coshocton, Holmes, and Knox Counties that are dug into different color phases of the sandstone. The cutting and fabricating of the stone is done at the main mill in Glenmont. The company is the sole remaining supplier of Massillon sandstone as a building stone.
Killbuck	Hills west of the town were a source of this colorful sandstone.
Millersburg	Builders constructed the 1885 courthouse in Millersburg with this stone. From 1909 to 1912 a glass factory used local Massillon sandstone, but it closed because it lacked a good gas supply.

in the area dug up the limestone and coal on their farms, piled it 6 to 7 feet high in alternating layers, and then burned it for several days. The end product was a fine lime, which they spread on their fields.

Mt. Eaton to the Ohio River

Strip mines—both active and reclaimed—in Middle and Lower Kittanning coals and other seams cap the hills from southeastern Wayne County to the Sugar Creek valley. The Stark and Tuscarawas County line near Wilmot marks the Wisconsinan terminal moraine and northern edge of the unglaciated Allegheny Plateau. South of Beach City Lake, which formed when Sugar Creek was dammed, U.S. 250 follows the Sugar Creek valley to Dover. Dover was the home of a number of clay plants, and a sewer pipe factory built in 1880 now makes firebricks on the north edge of town. All these plants used local Pennsylvanian-age shales and underclays.

Strip mines blanket the hills between New Philadelphia and Uhrichsville. Brick and tile plants operated at Uhrichsville and Dennison. From Midvale to Uhrichsville, U.S. 250 follows the wide flats of Stillwater Creek and then heads east along Little Stillwater Creek, a misfit stream that occupies only a small part of a wide valley bottom that was carved by meltwater. Tappan Lake, a reservoir of the Muskingum Watershed Conservancy District, floods this valley farther upstream. Fourteen plants, employing 1,500 men, worked Pennsylvanian-age clays around Uhrichsville and Dennison in the 1920s. Area businesspeople promoted the area as the "clay capital of the world." The area's industry began developing in 1883, when Mazurie Brothers started a drain tile factory on the west edge of town. It was followed by a rapid progression of plants, including

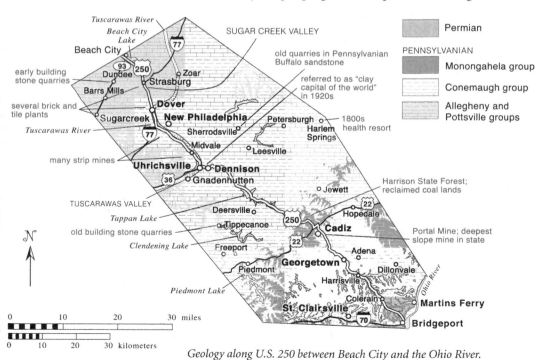

Geology along U.S. 250 between Beach City and the Ohio River.

those of Belden Brick, Buckeye Fire Clay, Evans Pipe, Ross Clay, and Wolf-Lanning Clay, culminating with the impressive forty-three-kiln Robinson Clay Products north of Uhrichsville. The Depression brought consolidation, and by 1995 only a few solitary sewer pipe and brick companies remained.

North and south of U.S. 250 and east of Uhrichsville, quarriers made extensive use of local sandstone. The Buffalo sandstone is one of the thickest and most widespread Conemaugh group sandstones in Ohio. It found extensive use as bridge stone, for grindstones, and even served as glass sand. Many quarries operated before 1900; the last one was the Craig Stone Company, east of Sherrodsville. A 30-by-500-foot scar marked the hillside off Ohio 39 where the stone was quarried. There was another quarry west of town, which closed around 1906.

Builders used slump blocks of Buffalo sandstone that occurred along Stillwater Creek, about 14 miles south of Uhrichsville, as foundation stone in the 1880s. The sandstone came from a 30-foot ledge above the creek. Three quarries opened between 1890 and 1900 near Tippecanoe, sawing the stone into curbs and dimension stone and marketing it for grindstones and flagging. Stone interests from Cleveland operated a quarry in Tippecanoe until 1907. Since then nature has reclaimed the quarries, but a keen observer can still discern vertical grooves on the former quarry walls—signs of the old channelers quarriers used.

Strip mines and reclaimed land surround Cadiz and dominate the landscape for the next 12 miles. Reclaimed mine land is part of Harrison State Forest north of town. The Cadiz Portal Mine is the deepest slope mine in the state. Pittsburgh-age coal was extensively mined in the Cadiz area with both slope and shaft techniques. Visit the Harrison County History of Coal Museum in the Puskarich Public Library on East Market Street in Cadiz and the Harrison County Coal & Reclamation Historical Park on Ohio 519 south of Cadiz to see preserved mining artifacts, equipment, and for more historical information on local coal mining. Just east of Cadiz there is a nice view of the hilly reclaimed countryside. This whole area was strip-mined at one time. U.S. 250 rises, falls, and twists, eventually descending to the Ohio River. The Allegheny Plateau becomes very rugged

COAL MINING OF THE CADIZ AREA

Town	Coal/Mining History
Adena	Miners began stripping the Pittsburgh seam of the Monongahela group around 1916. The coal averages 5½ feet thick in this area.
Deersville	Anderson coal of the Conemaugh group.
Dillonvale	Pittsburgh seam averages 5½ feet thick.
Freeport	Anderson coal.
Harlem Springs	Harlem coal of the Conemaugh group came from six mines in the late 1800s.
Perrysville	Barton coal of the Conemaugh group.
Petersburg	Upper Freeport coal of the Allegheny group.

close to the Ohio Valley because all its tributaries carve away at the Pennsylvanian bedrock because they strive to reach the Ohio River floodplain. At Georgetown, Pennsylvanian-age rocks outcrop on the hill at the north edge of town. Towns, like Harrisville and Colerain, mainly rest on the ridgetops in this region.

Ice Age Mammals of Killbuck Valley

Killbuck Creek and its tributaries buried more than outwash when they served as a major meltwater channel during the recession of the Killbuck lobe. The ice lobe in the main valley often dammed tributaries and turned them into temporary glacial lakes. Scattered bones and teeth of Pleistocene mammals have a history of appearing in drainage ditches and farm ponds dug into these former lake beds. In 1890 the skeleton of a large ground sloth (*Megalonyx jeffersoni*), was excavated near Berlin. It is on prominent display in Ohio State University's Orton Geological Museum.

On Martins Creek, northwest of Berlin, more bones were unearthed in 1993 near the site of 1928 mastodon and 1890 ground sloth discoveries. Researchers with the Killbuck Valley Museum found some flint flakes and bones of a mastodon (*Mammut americanum*), beaver (*Castor canadensis*), deer (*Odocoileus virginianus*), muskrat (*Ondrata zibethicus*), and shrew (*Blarina brevicauda*). Careful laboratory analysis of the flint showed traces of elephant and deer blood, indicating that early Ohioans used the flint to butcher the animals. To learn more and see the bones, visit the Killbuck Valley Museum in Killbuck.

Sugar Creek Valley

Like many other Allegheny Plateau locales, the Sugar Creek valley had industries that developed around its Pennsylvanian-age resources. Wilmot, Sugarcreek, and Strasburg had brick and tile plants that utilized Brookville, Lower Kittanning, and Tionesta clays. The Sugarcreek clay industry began in 1883 with the

This quarry near Tippecanoe supplied Pennsylvanian-age Buffalo sandstone building stone circa 1910.
—D. Dale Condit photo, Ohio Department of Natural Resources, Division of Geological Survey

opening of a drain tile plant at the current site of the Moomaw pit of the Belden Brick Company. Originally the clay came from the top of the Main Street hill, but by 1902 it came from a farm southeast of town. The first face brick plant, the Sugarcreek Clay Products Company, began production in 1912 along Dover Road (Ohio 39). Bricks from a predecessor of this company went into Ohio's only completely brick bridge, built about 1926, which carries Ohio 39 over South Fork Sugar Creek on the east edge of town.

In nearby Shanesville, the Shanesville Coal Company closed its mine after battling groundwater problems for some time. The underground workings stretch for about 1 mile west of town. In 1910 the Finzer Brothers Clay Company formed in order to buy the old mine with the hope of reactivating it. The arrival of electrical power made it possible for the company to pump out the water and obtain the remaining coal reserves. Finzer Brothers also purchased nearby land and began manufacturing drain tile. By 1938 the mine extended some 3 miles southwest of town. After World War I, brick was in demand for a building boom, and the plant expanded to meet the need, as did other plants of the Sugarcreek area. Today Belden Brick operates the plant. Reclaimed land and strip mines lie north of Ohio 39, north of Sugarcreek's business district.

In the late 1800s, Dundee Rocks drew many visitors who climbed and picnicked among the strange rock shapes. Dundee Rocks are erosional remnants and slump blocks that developed along a jointed 25-foot cliff of Pennsylvanian Massillon sandstone along South Fork Sugar Creek in Dundee. Quarries at Barrs Mills and Dundee initially produced dimension stone from this formation. A number of brownstone buildings in New Philadelphia are made of this stone. By 1890 they turned to crushed sandstone and marketed it as molding sand and for other industrial uses. The sandstone averaged 50 feet thick between Sugarcreek and Beach City. As of 2004, the Lower Springs Quarry at Dundee continued to crush the Massillon sandstone for industrial sand. The quarry also supplied some building stone.

OHIO 11
Ashtabula—East Liverpool
96 miles

Ohio 11 stretches, roughly north-south, from Lake Erie to the Ohio River at East Liverpool. It crosses 3 miles of lake plain before crossing about 79 miles of Wisconsinan till plain and end moraine sediments deposited on the northern fringe of the Allegheny Plateau. A short stretch of Illinoian till underlies the route south of the Lisbon exit. The highway joins U.S. 30 at West Point and continues to East Liverpool on unglaciated plateau underlain by Pennsylvanian-age strata.

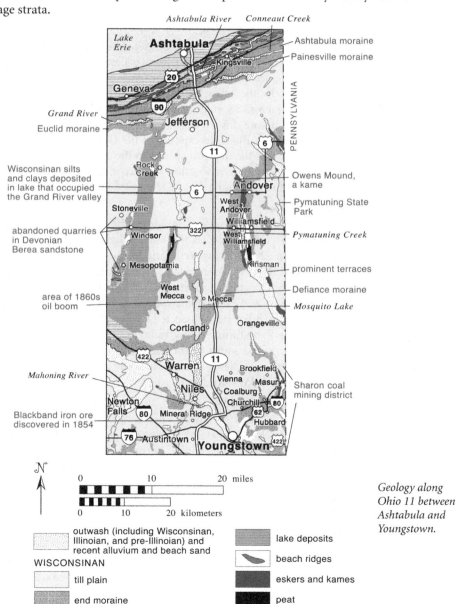

Geology along Ohio 11 between Ashtabula and Youngstown.

Ashtabula to Youngstown

Ohio 11 starts at Lake Road (Ohio 531) east of the Ashtabula River on Pleistocene lake sediments. For 3 miles the route crosses lake plain and beach ridges squeezed between the northern edge of the glaciated Allegheny Plateau and Lake Erie. The low sandy ridges of Glacial Lakes Lundy, Wayne, Warren, and Whittlesey parallel U.S. 20 north and south of it. East 21st Street marks the Glacial Lake Warren ridge, which was deposited around 12,800 years ago when the edge of Wisconsinan ice was just north of present-day Lake Huron in Ontario. South Ridge Road (Ohio 84) follows Glacial Lake Whittlesey's shoreline. Ohio 11 bridges the deep, narrow tree-lined valley of the Ashtabula River that passes through the Wisconsinan Ashtabula and Painesville moraines at the border of the lake plain and the Allegheny Plateau.

At I-90, Ohio 11 crosses sediments that were deposited in a glacial lake that once filled the north-south-trending Grand River valley to the west. The lake formed between the Defiance and Euclid end moraines as the Wisconsinan ice sheet retreated. Marshes dot the valley bottom today. After crossing the Euclid moraine, Ohio 11 heads across 28 miles of relatively flat till plain to the Ohio 88 overpass near Mecca. Before this overpass, at the Ohio 87 junction near the north end of Mosquito Lake, the wide expanse of the Grand River valley and resistant sandstone ridges on the horizon appear to the west. Just north of the interchange with Ohio 5, Ohio 11 crosses the Defiance moraine. A wide stretch of till plain underlies the route from here to a few miles south of Austintown.

Pymatuning State Park

A major reservoir, Pymatuning Reservoir, straddles the Ohio and Pennsylvania border east of Ohio 11. U.S. 6 skirts the north edge of the reservoir, and U.S. 322 skirts the south. Pymatuning Reservoir formed when a swamp, the source of the Shenango River of Pennsylvania, was dammed in 1933. Much of the shoreline is part of Pymatuning State Park. Between Andover and West Andover and Williamsfield and West Williamsfield, U.S. 6 and U.S. 322 cross the Defiance moraine, which has been dissected by Pymatuning Creek and its glacial predecessors. The Pymatuning Creek valley is filled with outwash, and terraces are evident north and south of the highways, evidence that a lot of meltwater passed through here. Between the routes the valley is quite marshy. Owens Mound, just north of West Andover, is a prominent kame.

Sandstone Quarrying around Windsor

North and south of U.S. 322 near Windsor, Phelps and Indian Creeks, as well as others, cut deep channels into the Defiance end moraine on the west side of the Grand River valley, exposing Devonian Berea sandstone. Windsor Mills, 1½ miles west of Windsor, is the site of early sandstone quarries. The Berea sandstone forms the walls of Warners Hollow, which Phelps Creek flows through, where early nineteenth-century settlers pried it loose for local use. The Windsor Stone Company opened a quarry and stonecutting mill in 1890. At that time the quarry exposed 20 feet of Sunbury shale of Mississippian age and 60 feet of Berea sandstone. The community of Stoneville developed 1½ miles north of Windsor Mills around a quarry in this sandstone. Little evidence of Stoneville remains except for a water-filled quarry and the Stoneville Road.

A quarry operated 1 mile southwest of Mesopotamia in the valley of Andrews Creek from the 1860s to 1890. Nearly 50 feet of Berea sandstone provided a good supply of stone for buildings, bridge piers, and foundations in the area. A smaller quarry downstream closed in 1889.

The Mecca Pool

Early settlers of the Mecca area lacked good water wells; the wells had oily films on the surface. The oil was considered a nuisance by some and medicinally valuable to others until Colonel Edwin L. Drake drilled the first successful oil well at Titusville, Pennsylvania, in 1859, starting the nation's oil boom. If there was that much oil on the surface of their wells, Mecca residents wondered, then what was farther down? In 1860 a well at West Mecca, now on the west side of the dammed Mosquito Lake, struck oil 50 feet down in the Devonian-age Berea sandstone. By 1865 the Mecca oil boom had peaked.

Within twenty-five years of that first well, two thousand wells tapped oil horizons some 40 to 60 feet below the surface of central Trumbull County, producing vast quantities of good lubricating oil. The shallowness of the oil explains why there were so many oil seeps in the area—hence the oily water—and the virtual absence of natural gas. The Mecca pool was the pioneer in eastern Ohio's Berea sandstone oil production, but the usable oil disappeared quickly. There's still oil in the pool, but it's so thick that it's nearly impossible to pump out of the rocks.

Mecca and West Mecca survive as towns, but many of the commercial buildings and homes were carted off to Cortland and Warren to the south. Area businesses now focus on the recreational activities at Mosquito Lake.

Salt in the Western Reserve

For many years, Native Americans boiled water from mineral springs in the Mahoning Valley on the south side of present-day Niles and distilled salt. *Mahoning* is a Native American word for "salt lick." In the late eighteenth century, New Englanders exploring the new lands of the Connecticut Western Reserve—lands in northeastern Ohio that Connecticut retained ownership of when it ceded its claims to western lands to the U.S. government—were naturally attracted to these same springs. Salt was a necessary preservative for meat and other winter provisions in the early days. It was also used for medicinal purposes and to make lye soap. In 1788 a saltworks operated for about a year south of present-day Warren in the so-called Salt Spring Tract, selling salt for $6 a barrel, but it closed after the premature death of the owner. The waters were not highly saline, and it's doubtful that the enterprise would have met with much success.

By 1810, salt boilers across the state had discovered that by drilling a hundred feet or so below salt springs they could obtain stronger brines. Although this essentially ended the business of precipitating salt by boiling water from surface springs, the health conscious were still attracted to salt springs for their medicinal purposes. In 1903 a hotel was built in Niles that catered to those who wished to bathe in the salty water. Three years later the construction of the Baltimore & Ohio Railroad through Niles buried all but one of the springs under its roadbed. The remaining spring lies on private property along Salt Spring

Road, south of the railroad tracks and north of Meander Creek Reservoir. Carson Salt Springs Road, between Newton Falls and Niles, and Salt Spring Road (also called Youngstown Road), between Niles and Youngstown, hint at this area's salt distilling history.

Youngstown-Area Coal and Iron

On the northwest side of Youngstown, Ohio 11 heads south, separating from I-80, which it had joined at Girard. Just north of this junction, at Ohltown, settlers began digging coal in 1835 on the west side of a bedrock ridge that stretches between Canfield and Mineral Ridge. This was the oldest coal in the state—the Sharon coal of the Allegheny group. The community of Mineral Ridge formed as Welsh coal miners immigrated to the region. The first mines were small, generally not deeper than 10 feet. The Cambria Mine, the first major one, opened in 1850. It eventually spread underground over 132 acres. At least another dozen larger mines followed at Mineral Ridge. Until 1857 the coal was hauled in wagons to Niles, where it was transported to Cleveland, Warren, and other northeastern Ohio towns on barges on the Mahoning River or Pennsylvania & Ohio Canal, which connected Akron to New Castle, Pennsylvania. By 1864 railroads replaced the waterways. By the late 1880s, these mines were running out of coal; most were abandoned by 1889.

Iron ore was also mined at Mineral Ridge. Although iron ore occurred about 3 feet above the Sharon coal, underlying the coal there was an even better grade of ore. Miners didn't start removing the lower iron ore, the Blackband ore, until 1854, however, when an immigrant miner recognized that the rock was similar to an English iron ore he knew of and convinced the mining company to investigate its worth. The ore rested between layers of the Sharon coal. The ore was only 6 to 10 inches thick, forming a black band in a shaley part of the coal seam. The Blackband ore was siderite, an iron carbonate that formed in a lake or bay within the ever-changing delta complex that covered northeastern Ohio during early Pennsylvanian time. Pennsylvanian iron ores had to be heated and mixed with other ores to make pig iron. In 1859 the Ashland Furnace opened in Mineral Ridge to smelt the ore. It took metallurgists a while to make the best of the Blackband ore, but by 1868 mixtures of the Blackband and ore from the Lake Superior region were some of the finest in the region. The Blackband ore was a late addition to the local ores miners had already been mining for years, which by the 1850s were already being replaced by richer ores from upper New York and the Lake Superior district. The iron ore of Mineral Ridge was pretty much exhausted by 1878. Except for orange-colored streams, polluted by mine drainage, flowing into Meander Creek Reservoir, little remains on the surface to tell of Mineral Ridge's coal and iron mining past.

At least 225 mines once tapped coal seams underlying parts of Mahoning and southern Trumbull Counties. Austintown, Brookfield Center, Canfield, Churchill, Coalburg, Hubbard, Mineral Ridge, Vienna, and Youngstown all became mining centers. The first coal mining of record in the Mahoning Valley took place around 1826 at Crab Creek, a small village long since swallowed up by Youngstown. A local hotel became the first to use coal as a domestic fuel around 1829. Mining spread to Brier Hill, now also a part of northern Youngstown, where the coal seams were more extensive. This was the Sharon coal, or Block coal, which slowly replaced wood and charcoal as the favored heating fuel. By 1847, the Brier Hill mines produced 100 tons per day.

These coals were the first in the nation to be used for cokeless iron smelting and rapidly became the standard fuel for the industry. Shafts and slope mines in the region reached the coal, 50 to 180 feet down. The Mahoning (at Lowell) and the Eagle (south of Brier Hill), both completed in 1846, were the first iron-producing furnaces in Ohio designed to burn these coals. The Quakertown coal, a younger coal closer to the surface, averages only 1 foot thick in this

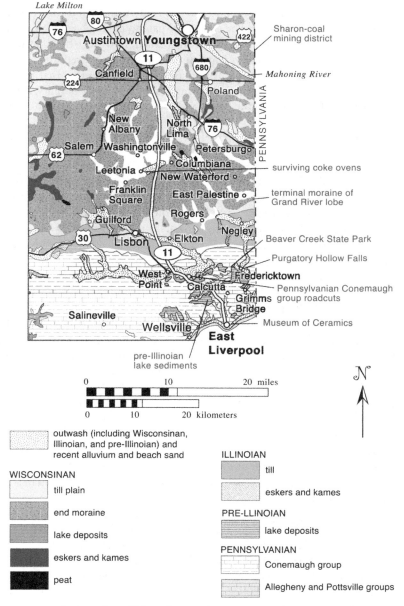

Geology along Ohio 11 between Youngstown and East Liverpool.

region, but iron deposits associated with it at Brookfield Center, Hubbard, and Youngstown made it an important local resource. From 1836 to around 1841, a charcoal-fired furnace at Brookfield Center smelted local iron ore that was turned into small iron implements, pots, and pans. The furnace occurred along Ohio 7 near old Ohio 82 but no evidence of it remains.

Mining spread outward across the Youngstown area, reaching its peak in the 1870s with some seventeen major companies and numerous small mines spread across the Mahoning Valley from Niles to the Pennsylvania border. Daily mine totals for Sharon coal were 4,000 tons in the early 1870s. In 1875 the total coal production of valley mines was 3.5 million tons, and together they employed four thousand men. The coals were patchy, having been deposited in local basins, and thus many mining ventures failed; however, the Sharon coal in the valley was of exceptional quality. Since it was mainly used for domestic heating, the Sharon coal's low sulfur and ash (inorganic residue that remains after burning the coal) content was very important to its marketability. Mining companies rapidly exhausted the fields, and most mines closed by 1900.

Youngstown to East Liverpool

Through Youngstown, Ohio 11 continues to traverse Pennsylvanian-age coal-bearing strata that are buried under glacial material. Numerous abandoned mines snake through the subsurface south of Youngstown, often below the highway. Most towns in this area had one or more slope or shaft coal mines. Mine subsidence is now a huge issue in the Youngstown area, which like most Ohio cities is growing and expanding into areas that were not always part of the urban landscape. Building over coal mines cannot be avoided in this area; if tunnels and stopes (large mined-out rooms) are close to the surface, the added weight of buildings or vehicles on a highway can cause them to collapse. For example, in 1977 a 115-foot-deep hole opened up under a garage in a residential neighborhood of Youngstown when a shaft from an 1884 mine collapsed.

Miners extracted the Lower Mercer coal, 100 to 150 feet above the Sharon coal, at Canfield and in the Austintown area. More valuable than the coal, however, was the overlying 2 to 20 feet of limestone, the most extensive in the area. A quarry at Canfield shipped tons of it to Leetonia to be used as furnace flux. Brookville coal, the uppermost, youngest coal in the area, was another important local seam. It was 3 to 6 feet thick and was mostly mined at Canfield. The Canfield area was home to four companies, formed in 1858 and 1859, that converted local coal, through a rather expensive distillation process, to coal oil. The largest of the companies produced 75 barrels of this highly desired illuminant per week. The successful drilling of petroleum in 1859 spelled the end for these companies. South of the exit for Lisbon and Rogers, Pennsylvanian strata form high banks along the highway.

Between I-76 and Lisbon there is a complex of till plains, end moraines, kames, terraces, and valley trains marking the edge of the Wisconsinan Grand River lobe. This is the Kent moraine, as much as 12 miles wide along this stretch. The oldest till in this deposit dates to about 40,000 years ago when an early Wisconsinan ice sheet spread south of the present-day Erie Basin and deep into the Allegheny Plateau. The last ice advance this far south in northeastern Ohio covered the region with a thin till around 23,000 years ago. The most recent sediments date from 23,000 to 12,000 years ago, marking the final advances and retreats of the Grand River lobe.

Between the Middle Fork Beaver Creek bridge, east of Lisbon, to 1 mile north of West Fork Beaver Creek, Ohio 11 traverses eroded Illinoian till. Strip mines begin to appear because the coal-bearing strata isn't covered by so much till. West Fork Beaver Creek cuts into Wisconsinan valley train deposits near West Point, penetrating into the unglaciated Allegheny Plateau. In low-lying areas between here and East Liverpool, clay settled out of pre-Illinoian lakes, forming the Minford clay. Pennsylvanian strata of the Conemaugh group flank the road just south of its junction with U.S. 30 on to East Liverpool.

Ohio 11 passes through California Hollow as it descends into East Liverpool. Fifty feet of Pennsylvanian-age Lower Freeport sandstone occurs on the west wall of this valley, which workers quarried and used as pulp stone into the 1920s. Pulp stone is used in the papermaking process. The quarry shipped the stone to paper-processing plants in the Elyria area. Apparently its texture was more suitable than Elyria's local Berea sandstone of Devonian age. The Lower Freeport sandstone had also been used as a local building stone since the founding of East Liverpool. Quarries also operated north of town at Fredericktown and Grimms Bridge, where builders used the stone for foundations and bridge piers.

A New Iron Town—Leetonia

The opening of the Ohio & Pennsylvania Railroad through Columbiana County in 1851 led to the exploitation of rich coal and iron ore deposits east of Salem. The Lower Kittanning coal, up to 3 feet thick in areas, was widely mined around Leetonia in the late 1800s. The seam extended north to Greenford, east to Columbiana, west to Salem, and south toward Lisbon; it was thickest around Leetonia and Washingtonville. It made a high-quality coke when heated and a good coal for fuel in rolling (steel) mills. Some of the coal was mined for firing railroad steam locomotives. The Lower Kittanning formed the backbone of Leetonia's iron industry and coke business. Initially, the iron ore came from a blackband-type ore in the overlying Lower Kittanning shale around Leetonia and from nodular ore dredged from the gravels lying in the bed of Middle Run Little Beaver Creek east of Lisbon. Later a mixture of the native ore was mixed with ore shipped in by rail from the Lake Superior district. The older Clarion coal was also worked around Leetonia; though it was a thicker seam, the coal had a lot of impurities, including a lot of sulfur, and it produced a large number of clinkers after being burned.

The Leetonia Iron & Coal Company formed in 1865, established the town of Leetonia, and started the Leetonia Furnace (later known as the Cherry Valley Furnace) in 1867. The Grafton Iron Company came to town in 1867, and the area's iron boom was on. From nothing but farm fields in 1865, Leetonia was a thriving village of 1,800 people by 1870 and was destined to become one of the leading iron producers in Ohio. The state government reported that in 1873 Leetonia industry produced 33,901 tons of pig iron; 4,487 tons of iron rods, bars, and nails; 1,000 tons of iron stoves and hollow ware; 611 tons of miscellaneous iron castings; 165 steam engines; and 129 boilers.

The economic depression of the early 1870s caused the Leetonia Iron & Coal Company to reorganize in the hopes of becoming more competitive, but things didn't greatly improve until Pittsburgh interests took control of the company in 1900. The Cherry Valley Iron Company—as it became known—upgraded its facilities, and by 1905 it had two hundred coke ovens in operation, burning impurities from coal to make coke. Grafton Iron expanded its operations on the opposite side of town during the depression and weathered the financial panic. The furnaces and foundries of the Cherry Valley Iron Company disappeared in the mid-1900s, and only the coke ovens survived—an impressive industrial remnant to say the least! If you want to see the ovens, on the east side of Leetonia follow the signs for the Cherry Valley Coke Ovens and Arboretum, where you can walk paths past the uncovered brick ovens in a developing arboretum.

Along Little Beaver Creek

South of West Point, Ohio 7 leads north to Beaver Creek State Park. The park consists of several parcels, but the majority of it lies between Elkton and Fredericktown. Lower Freeport sandstone forms prominent cliffs in the park. Little Beaver Creek, a State Scenic River and a National Wild and Scenic River, deeply dissects Pennsylvanian-age strata, which perhaps is best viewed at the overlook on Sprucevale Road near the park's southern boundary. Several locks of the Sandy & Beaver Canal lie within the park's boundaries, all masterly constructed of Lower Freeport sandstone. The Dogwood Trail near the north end of the park leads past old stone quarries where men extracted the stone to build the canal locks.

Purgatory Hollow Falls, where Little Beaver Creek plunges 45 feet across a lip of Lower Freeport sandstone, occurs in Fredericktown. Potholes dot the rocky flats above the falls and a large boulder contains the fossilized imprint of a Pennsylvanian scale tree (*Lepidodendron* species).

Abandoned coke ovens of the Cherry Valley Iron Company at Leetonia.

A charcoal-fired furnace operated along Hazel Creek near Calcutta and the Sandy & Beaver Canal, which carried the pig iron the furnace produced to Pittsburgh. Pennsylvanian iron ore from nearby fields fed the furnace for about a year and a half. The failure of the canal in 1853 led the furnace's owners to abandon it. The canal failed because of competition from railroads and the catastrophic failure of the dam at Cold Run Reservoir near Lisbon in 1852.

Calcutta and Fredericktown became the centers of a short-lived oil boom in 1866. Some fifteen wells produced 100 barrels daily for several years, but by 1876 the wells were going dry. The oil was a thicker lubricating type. About the same time, investors lost their shirts in a well-drilling foray in nearby West Point. A number of wells struck gas, but workers burned it off because they considered it useless.

America's Crockery City

Ohio 11 ends in the Ohio River town of East Liverpool—often proclaimed America's Crockery City. East Liverpool was an excellent location for a pottery. It had ready transportation (the river), vast clay deposits, and dense forests for fuel. The Pennsylvanian-age Clarion and Lower Kittanning clays provided the raw material for numerous clay and ceramic industries.

The first pottery opened around 1839, and others followed in rapid succession. Yellowware, a type of bright yellow pottery, was first fired in East Liverpool in 1844 using Clarion clay. Most of the early potteries in East Liverpool were powered by horses, and their kilns burned wood or charcoal. These potteries struggled through the Civil War years, the major problem being that workers

MAHONING AND COLUMBIANA COUNTY COAL MINES

Mahoning and Columbiana Counties had hundreds of operating coal mines in the nineteenth and early twentieth centuries. Many of them developed along newly opened rail lines. Miners worked the Allegheny and/or Pottsville groups, mainly the Sharon, Clarion, and Lower Kittanning seams.

Austintown, Coleman, Columbiana, East Fairfield, East Palestine	Upper Freeport cannel coal east of Ohio 11 on Ohio 46 was in local use by the 1840s. It burned with a smokeless flame; local factories desired it. By 1860 larger-scale mining was in process.
Elkton, Franklin Square, Ginger Hill, Guilford, Jim Town, Lisbon, Millville, Moultrie, Negley	Mines worked the Upper Freeport and Mahoning coals.
New Albany, New Waterford, North Georgetown, North Lima, Petersburg, Rogers, Salem, Teegarden, Washingtonville, West Point	Mines produced Upper Freeport coal, which averaged 4 feet thick in the area; some worked the thinner Mahoning seam.
Woodworth	Mines, which closed in 1904, supplied Lower and Middle Kittanning coal for the Cherry Valley Iron Company at Leetonia.

weren't available. Ohio River floods also plagued the businesses. The industry rebounded, however, and by 1875 some twenty potteries were operating in East Liverpool. Coal replaced charcoal as kiln fuel. This proved economical since the clay seams were overlain by coal and both resources could be mined through one opening. Many U.S. potteries of the late 1800s began importing clays from other areas to satisfy the exotic tastes of some customers. East Liverpool became famous for producing ironstone, queensware, and white granite; before this time, people in the United States could only obtain white ware from Europe. With prosperity and increasing consumer demand, the plants grew larger around 1900 and hired ceramic chemists; soon all types of fancy china with American trademarks became available. Smaller potteries merged and larger corporations like Hall China Company—now the largest producer of specialty chinaware—appeared. Local Lower Kittanning clay became a favorite material in the 1920s, but most East Liverpool companies imported clays from elsewhere.

Seventeen potteries flourished in 1923, employing seven thousand people, but by this time East Liverpool was only the center of a large pottery district. Because the city was situated on the floodplain, there was little room for potteries to expand, so many firms began to leave the area. By 1940 only six dinnerware firms remained in East Liverpool.

Visit the Ohio Historical Society's Museum of Ceramics on East Fifth Street to learn more about the history of East Liverpool's historic potteries. The still-operating plants and the museum of the Homer Laughlin China Company, which had its beginnings in East Liverpool, are located across the Ohio River in Newell, West Virginia. The Hall China Company and the restored Goodwin-Baggott pottery are in East Liverpool.

East Liverpool companies also used local clays to manufacture items other than pottery. The N. U. Walker Company used Clarion clay to make terra-cotta and sewer pipes in the 1860s and 1870s. East Liverpool Brick Manufacturing Company used Clarion clay and imported clay from Clearfield, Pennsylvania, to manufacture firebrick; it used local shales to make construction brick. The company produced ten thousand bricks daily in the 1920s. American Vitrified Products used Clarion clay to manufacture sewer pipes in the 1920s.

Drillers first struck gas in East Liverpool in 1859 while prospecting for salt 450 feet down in the Devonian-age Berea sandstone. Well drillers seeking fuel oil in the 1860s also tapped gas. In 1874 they piped it into their homes and used it for light and heat. To them it seemed like the supply was endless, and as a result a tremendous amount of gas was wasted. By the late 1870s, gas lines ran through East Liverpool. Street lights were lit twenty-four hours a day and a vast supply of gas was available to the twenty or so potteries that prospered because of this steady source of heat. The local gas supply lasted into the 1880s, and then gas was piped in from neighboring West Virginia. Gas was still cheap, and many glass plants appeared in the area. Wasteful behavior continued; pressure-release standpipes flared day and night. By the 1890s gas supplies dwindled, and local potteries started burning coal again. The local gas companies found other sources in other states, however, and replenished the supply. Some potteries returned to burning gas.

Many pottery ovens like this one once dotted East Liverpool.

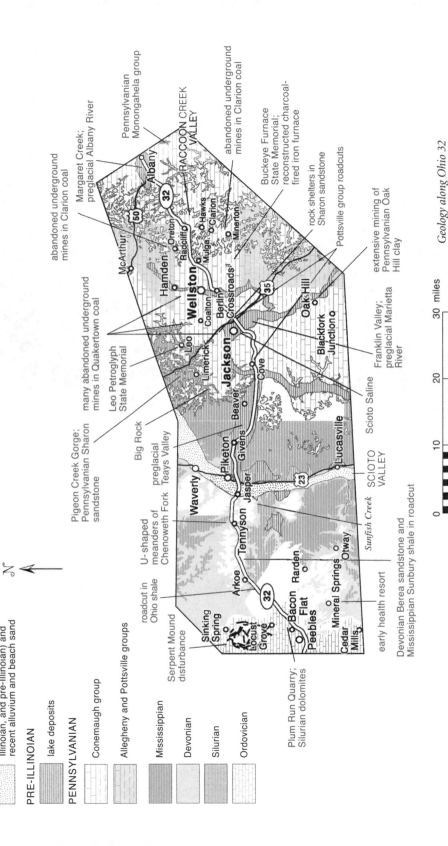

*Geology along Ohio 32
between Albany and Peebles.*

OHIO 32
Albany—Peebles
80 miles

Ohio 32 and U.S. 50 are one and the same between Coolville and Albany; the geology along that stretch of road is covered in the U.S. 50 road guide. Ohio 32 crosses 70 miles of unglaciated Allegheny Plateau underlain by Mississippian- and Pennsylvanian-age strata from Albany to near Locust Grove just northeast of Peebles. These strata were deposited in ancient deltas spreading westward from the rising Appalachians. As with other east-west-trending roads in Ohio, the bedrock exposed at the surface gets older as one travels west.

Albany to Jackson

Ohio 32 follows Margaret Creek, which lies in the old valley of the preglacial Albany River northeast of Albany. Ohio 32 leaves the preglacial valley south-west of Albany; the preglacial Albany River flowed to the south from here to Rio Grande. Conemaugh group sandstones cap the hilltops around Albany. Allegheny group strata appear in the Raccoon Creek valley, west of Albany, and dominate the landscape to Jackson. Strip-mined land borders the highway southeast of Wellston where miners removed Lower Kittanning coal. The main coal mined in the Wellston area, however, was the Quakertown coal. All the shaft mines that reached this deeper coal have been long abandoned, and essentially the coal was mined out in the area. The junction of Ohio 32 and U.S. 35 at Jackson overlies the Globe Iron Company No. 4 mine in the Sharon coal. It closed in 1946.

The highway leaves the hilly uplands and enters the wide, flat Franklin Valley, the former path of the preglacial Marietta River, at Jackson. Little Salt Creek, which runs in a northerly direction, is a misfit stream and is dwarfed by this valley. The buried valley of the Marietta River underlies downtown Jackson north of Ohio 32 and then heads west to Sharon. Little Salt Creek leaves the ancient valley on the northwest side of Jackson, but the ancient Marietta River valley, now only drained by short creeks, is still evident as a broad, curving valley extending west of Jackson and southwest to Cove, where it once joined the preglacial Teays River. The bedrock floor of the buried valley is Mississippian Logan sandstone, and the valley is filled with pre-Illinoian Minford clay. High sandstone cliffs along the road are Pennsylvanian Sharon conglomerate. Following U.S. 35 north from Jackson a few miles takes you by some roadcuts in the Pennsylvanian Pottsville group strata, and a number of rock shelters are visible from the highway, especially during the winter months when vegetation does not hide them. The highway also leads to side roads to Canter's Cave 4-H Camp and Leo Petroglyph State Memorial.

Jackson-Area Landmarks

- **Big Rock:** This is an example of the many isolated mounds, or knobs, of cross-bedded sandstone, often 100 to 200 feet high, that exist in this area. It occurs near the town of Big Rock near the Jackson and Pike County line.

- **Boone Rock:** A 57-foot-high, south-facing cliff of cross-bedded Pennsylvanian Sharon conglomerate occurs on the northwest edge of

The Pennsylvanian-age Sharon conglomerate, here mainly a sandstone, is well exposed in a gorge at Leo Petroglyph State Memorial.

Jackson at the sewage treatment plant. At its base a small rock shelter undercuts the rock face. The shelter is 11 feet high at the tallest point and occurs in a weakly cemented part of the sandstone. An earlier meander of Little Salt Creek, now an oxbow lake, lies just south of the sewage facility. Since this feature occurs near salt wells, many thought that it likely was a site frequented by early native people. Although the site had been highly disturbed by souvenir hunters, archaeologists uncovered three hearths, stone and flint tools, pottery of the Fort Ancient culture, human and animal bones, and mussel shells, which were probably used as tools. Boone Rock was also the site of sandstone quarrying, which removed both the eastern and western ends of this cliff.

- **Buzzard Rock:** A similar cliff of Sharon conglomerate occurs about 2 miles down Little Salt Creek from Boone Rock and has two shelters. The larger one is about 30 feet above stream level and measures 75 feet along the cliff face. It's 11 feet high at its highest point and extends 24 feet back into the cliff. Above it is a smaller one, about half the size of the lower shelter. Archaeologists have found numerous animal bones, artifacts, and three human burials at the Buzzard Rock shelters, mainly in the larger one. All are related to the Fort Ancient people.

- **Canter's Cave:** Another rock shelter lies at the base of a 60-foot-high cliff of Sharon conglomerate about 5 miles north of Jackson on Caves Road, east off U.S. 35. Native Americans were the first to use the salty brines seeping from the cliff face. Later, settlers made saltpeter—a general name for potassium or sodium nitrate—for gunpowder at this location. The salts were dissolved from underlying strata and rose into the Sharon conglomerate since they were lighter than the groundwater surrounding them. Since the Sharon formation had plenty of open pore space, the salt was able to move laterally as well and seeped out as springs where streams had cut into the Pennsylvanian rocks. Softer strata under the resistant Sharon conglomerate eroded back into the base of the cliff, as is typical of many similar shelters throughout this region. The shelter, along with associated waterfalls and narrow rocky gorges, is now part of Canter's Cave 4-H Camp.

- **Leo Petroglyph State Memorial:** 5 miles north of Jackson, off U.S. 35 near Leo, Fort Ancient people carved caricatures of animals, humans, and footprints in Sharon sandstone/conglomerate between AD 1,000 and 1,650. A trail leads into a typical rock gorge of this region, where 20- to 65-foot-high cliffs of Pennsylvanian-age Pottsville group strata are exposed on the way to salt springs. The strata show cross-bedding, honeycomb weathering, and rock shelters.

- **Katharine State Nature Preserve:** Sharon conglomerate forms the cliffs of this preserve off County Road 76 (State Street) just northwest of Jackson. Trails offer close-up views of sedimentary features (including cross-bedding, ripple marks, iron banding and nodules, and honeycomb weathering), rock shelters, and abandoned drift mines. The lake formed from the damming of Rock Run in 1946.

The Scioto Saline

The flat bottomland of Little Salt Creek at what became Jackson played an important role in early Ohio history. By 1725 pioneers knew this area was a source of brine, or saltwater, that was coming from underlying rocks. Ohioans came to know this area as the Scioto Saline, Scioto Salt Springs, or Scioto Salt Licks. Native Americans and early European settlers dug pits into the Pennsylvanian-age Sharon conglomerate in Little Salt Creek's streambed at times of low water, allowing brine to slowly flow into the pits. They then boiled the solution to precipitate salt. Later, folks discovered that stronger brines flowed into deeper pits, and that these brines also occurred in the deep valley fill below the creek. As a result of the Treaty of Greenville, the U.S. government removed the area's Native Americans and created the Scioto Salt Reserve in 1796, the largest and most important salt reserve in Ohio. Upon statehood in 1803, Ohio's legislature controlled the development of this 23,040-acre plat even though the U.S. government retained ownership.

The heyday of the Scioto Saline occurred between 1806 and 1808, when thirty wells and twenty salt furnaces produced 50 to 70 bushels of salt weekly. The center of production was in Jackson, between what is now Broadway Street and Harding Avenue, but other saltworks stretched 4 miles along the valley and up and down its tributaries. Two spring sites along Little Salt Creek were popular with salt makers and appear on early maps—one at Boone Rock and another at the foot of Broadway. The discovery of more-concentrated brines deep below Kentucky around 1810 led drillers in Jackson to deepen the wells in the Scioto reserve, but they had little success. By 1818 the Ohio legislature decided to sell the reserve; however, inaction on the part of Congress delayed the breakup of the salt lands until 1825. Today, little trace remains of the industry that gave birth to the city of Jackson; portions of the original bottomlands lie below tons of man-made fill in the downtown area.

The reason brines are concentrated at Jackson may be related to movement that occurred in deep-seated Precambrian rocks, which led to fractures in the overlying Paleozoic strata and conduits for the brines originating in Silurian strata. Most of the evidence to date relates to a curious bend in the preglacial Marietta River valley at Jackson, which a portion of Little Salt Creek flows through; local streams flow away from this bend in all directions. Some geologists think that this might be a dome, an upward bulge in the underlying Paleozoic strata of the Jackson area. It could have been caused by the upward movement of salt. This is commonplace along the Gulf Coast where fractures allowed underlying rock salt of Jurassic age to rise upward, forming salt domes and pushing the overlying rocks into a local dome. The preglacial Marietta River may have flowed around this slightly higher region, accounting for the northward loop of its buried valley at Jackson. Future investigative drilling and geophysics should solve this puzzle.

Jackson and Vinton County Mineral Resources

The salt boom in Jackson County ended in the early 1820s, only to be followed by iron smelting and coal mining. Between 1834 and 1856, a dozen charcoal-fired furnaces went into blast in Jackson County using mainly Upper Mercer and Ferriferous ores of the Pottsville group. Many coal- and coke-burning furnaces appeared between 1864 and 1875, which locals fed Jackson Sand Block, Ferriferous, and Upper Mercer ores to make pig iron. Depending on local availability, furnace operators also used the Sharon, Boggs, Sand Block, and Zaleski Flint horizon ores.

Jackson and Wellston were centers of iron smelting, but other furnaces worked local ores at Berlin Crossroads, Coalton, Hamden (in Vinton County), Keystone Furnace, Monroe Furnace, Oak Hill, Oreton, and Rempel. Jefferson Furnace in Oak Hill was the last charcoal-fired furnace to operate in Ohio; it

Typical Pennsylvanian-age Pottsville group strata on U.S. 35 on the north side of Jackson.

MINERAL INDUSTRIES OF
JACKSON AND VINTON COUNTIES

Town	Strata and Products
Altoona, Berlin Crossroads, Buffalo, Chapman, Clarion, Coalton, Comet, Davisville, Garfield, Glen Nell, Glen Roy, Goldsboro, Hawks, Iron Valley, Jackson, Jonestown, Keystone, Minerton, Mulga, Oreton, Petrea, Radcliff, Ratchford, Tom Corwin, and Wellston	Mines in Lower Kittanning, Clarion, Quakertown, and Sharon coals were centered at Jackson and Wellston and scattered around other towns.
Coalton and Jackson	Quarries in the Sharon conglomerate existed near these towns, providing sand for foundries and other general uses; some also went to a Jackson glass plant. This is the only part of Jackson County where the Sharon conglomerate is mainly a sandstone and thus usable for industrial sand. The Jackson Sand Mining Company quarried over 70 feet of sandstone north of Coalton in the early 1900s.
Oak Hill	Between the Lower and Middle Kittanning coals are the Lower Kittanning and Oak Hill clays, which served as raw material for six clay plants around Oak Hill in 1920. What makes the area unique is the high-quality flint and plastic clays within the 3-foot-thick Oak Hill clay. A flint clay can resist high temperatures without becoming glassy and does not become plastic when ground up and mixed with water. A plastic clay becomes plasticlike when mixed with water. People used the Oak Hill clay as oven and kiln linings and to manufacture firebricks. This was the only spot in Ohio where it was used so extensively, beginning in 1873 with the opening of Aetna Fire Brick and Oak Hill Fire Brick Companies in Oak Hill. Furnace builders favored the Clarion sandstone, quarried around Oak Hill, for relining the stacks of charcoal furnaces. Today, the quarries lie abandoned.
Wellston	This town became the home of two portland cement plants around 1900. Workers quarried the Vanport limestone, exposed at the site of Cornelia (or Lincoln) Furnace east of Wellston, and shipped it to the Alma Cement plant in Wellston. Later the company obtained limestone from Kitchen Station (now Kitchen) and limestone and coal from Oreton. Wellston Portland Cement, which later became Lehigh Portland Cement also used Vanport limestone. Both plants were abandoned by 1920.

was extinguished in December 1916. Oreton became a ghost town in the 1950s, and little trace remains of the fifty or so buildings that housed and served the eight to nine hundred people who once lived there. It was the site of Eagle Furnace from the 1850s to 1880s. The area is now part of the state forest. Most of the coke furnaces were extinguished by World War I because it was no longer economical for them to operate in the area; a few lasted into the 1920s. Better-quality ores had been depleted in the area and ores from other regions were available. Only a solitary brick stack remains in Wellston's Veterans Memorial Park, marking the site of the city's 1875 Wellston Twin Furnaces. Buckeye Furnace State Memorial, where the Ohio Historical Society's reconstruction of a typical charcoal-fired iron furnace community provides a glimpse of this time period, is southeast of Wellston. The Madison Furnace, still standing near Rempel, is unique because its builders cut the lower part of it directly into the adjacent cliff of Clarion sandstone. Abandoned furnaces also remain near Berlin Crossroads, Blackfork Junction, Byer, Clay, Jackson, Mulga, and Oak Hill.

Coal mining in Jackson County dates as far back as the early 1800s, beginning in the Quakertown coal near Chapman Station (now Chapman). The construction of Jackson Furnace near Oak Hill in 1836, the first charcoal-fired furnace in the county, later led its owners to mine Brookville coal on company property as fuel for their boilers. The Marietta & Cincinnati Railroad, which passed through Jackson and Petrea, was the first railroad to provide a means of transporting the area's mineral resources to markets. Large-scale coal mining progressed slowly until other rail lines were laid through the county during the 1870s and 1880s. Coal mining in Wellston began with the 1873 organization of the Wellston Coal & Iron Company, which sank a 50-foot shaft to the Quakertown coal and erected the Wellston Twin Furnaces. The Sharon, Clarion, Lower Kittanning, and other minor coals came from numerous drift, slope, and shaft mines in the area.

For fifty years, beginning around 1870, coal mining ruled the hills of Jackson County as smelting tapered off. After World War I many mines closed as the seemingly endless coal reserves were exhausted. Today, abandoned railroad grades, crumbling foundations, cinders, and crushed coal are stark reminders of the industry's past importance to the local economy. Scattered strip mines continue to remove the Clarion, Lower Kittanning, Middle Kittanning, and Upper Freeport coals primarily from the hills east of Wellston and Oak Hill. At least twenty-three mines lie abandoned below Wellston at an average depth of 80 feet, over which the city is built on 60 feet of glacial lake sediment that fills an ancient buried valley. Mine subsidence continues to be a concern.

Cove to Peebles

Ohio 32 enters the wide preglacial Teays Valley at Cove and follows it to the Scioto Valley just south of Piketon. Devonian-age Berea sandstone and Mississippian Cuyahoga formation form the uplands between Cove and Piketon, and pre-Illinoian Minford clay that was deposited in a glacial lake fills the valley. Devonian Ohio shale forms the bed of No Name Creek and Chenoweth Fork, the path of Ohio 32 from Jasper to 1½ miles west of Arkoe; Berea sandstone and Cuyahoga formation rocks form the hills. Note the U-shaped meanders at Tennyson, where Chenoweth Fork enters the valley of Sunfish Creek.

The streams in this region were not affected by meltwater during the Ice Age since the ice sheets did not advance far south of Chillicothe, which lies some 12 to 20 miles north of Ohio 32. The streams have comparatively wide floodplains underlain by mud and sand derived from the erosion of the surrounding hills. The lower parts of the hills are composed of Devonian Ohio shale, which is easily cut away by meandering streams. The stream erosion undercuts the more resistant strata above, eventually causing its collapse to the valley floor. Streams encountering larger blocks of rock and coarser sediment seek new routes around the material, leading to their winding courses. Meanders may also be a result of the gentle slope of streams in this area, which have been carving into the plateau for tens to hundreds of thousands of years. The streams are adjusted to moving a certain volume of sediment, and if this changes, perhaps due to landslides, deforestation related to forest fires, or more recently, by mining activity, readjustments will be made to suit the new stream conditions. Where streams flow into one another, such as at Tennyson, excess sediment may be deposited, at least temporarily, creating a blockage and more meander adjustments.

LAKESHORE OHIO

Lake Erie forms the north edge of Ohio from the Pennsylvania border to Toledo, some 210 miles. It certainly represents the most dynamic aspect of present-day geology in Ohio. A quick glance at historical maps of the shoreline shows great changes in geography that have happened in just one hundred years or so. Striking changes in beach width and sediment type occur seasonally. Even a strong storm may cause overnight changes. This is geology in motion! All of the bedrock you encounter in this chapter I discuss in more detail in preceding chapters, including their origins and composition.

History of Lake Erie
Lake Erie is one of the youngest geologic features of Ohio, and its predecessors are no more than 16,000 years old—youngsters in the context of geologic time. The lake, as we recognize it, is only about 5,000 years old. Its predecessors developed as Wisconsinan ice relinquished its grip on the Ohio landscape.

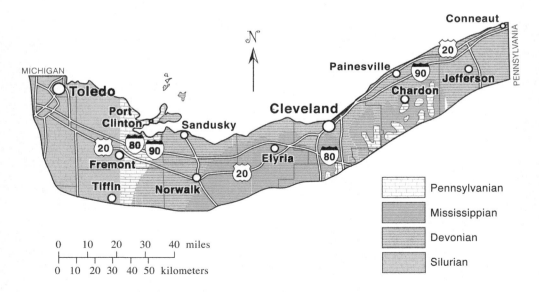

Highways and bedrock geology of the Lake Erie coast. —Modified from Ohio Department of Natural Resources, Division of Geological Survey publication

Before Pleistocene time, over 2 million years ago, northern Ohio was indistinguishable from southern Ontario. It was a rolling terrain of hills and stream valleys—part of the drainage basin of the preglacial Erigan River—that was destined to be a lake. The Erigan River was a major east-flowing river that many geologists feel carved the valley that is now covered by Lake Erie. Some think it may have been the downstream part of the preglacial Teays River. Other geologists feel the Teays flowed to the west. Unfortunately, the depth of burial and glacial erosion do not permit definite conclusions about buried river channels in northern Ohio.

The Wisconsinan ice sheet alone couldn't have cut a basin as deep as the Erie Basin, so the basin must have been a lowland before glaciation even though the thickening of ice over this low spot would have increased the ice sheet's erosional capacity. North-flowing streams discharged into the Erigan system from the Allegheny Plateau, the uplands of eastern Ohio, and groundwater seeped from limestone karst areas of northwestern Ohio, feeding the Erigan River. All of this is speculation, of course, since the area was covered by several major ice sheets during Pleistocene time. Each ice advance and retreat removed evidence of previous ice movement. Thus the only events that geologists can clearly understand are those related to the last, or Wisconsinan, ice sheet. It's possible that pre-Wisconsinan glacial lakes existed where Lake Erie is, but no physical evidence remains.

About 17,000 years ago rising temperatures caused the ice lobes covering Ohio to retreat back to the present confines of the Lake Erie Basin. This retreat was not continuous but consisted of many stops and starts; readvances were as common as retreats as the climate fluctuated. Just minor changes in temperature, precipitation, and the configuration of the underlying landscape translated into different responses by the ice. The land was somewhat lower near the ice margin due to the weight of the ice sheet, leading to piles of deposited sediment and ponds of meltwater. It was in such a low spot that Lake Erie was born.

Proglacial lakes, those that existed along the ice sheet's margins, constantly evolved, changing shape and depth. The north shore of these lakes lapped against the ice front, while the rest of the shoreline spread across the Ohio landscape. Advances and retreats of the ice caused lake level to rise or fall. Water sought the lowest spots, creating outlets that regularly fluctuated until the ice moved far north of the Erie Basin. Geologists have names for the series of proglacial lakes that covered northern Ohio from 16,000 years ago to present. They recognize them by low, arcuate sand piles, or beach ridges, that mark their ancient shores; layered silts and clays that were deposited on the lake bottoms; fossils of freshwater clams, snails, ostracods, and aquatic plants; and eroded tills, planed down by the movement of waves along their shores. The ridges are often difficult to decipher because of erosion and destruction by farming and development. Along the present-day Lake Erie coast, beach ridges are squeezed between the present shore and the edge of the Allegheny Plateau and on top of the northwestern till plains. Geologists are still trying to unravel clues about Lake Erie's past that remain hidden in these deposits.

Evidence of Glacial Lake Leverett, the oldest known lake of latest Wisconsinan time in northwestern Ohio, is scanty. It existed around 16,000 years ago, after the initial retreat of the ice sheet into Ontario and before the last advance of the ice sheet back into central Ohio. The general fine-grained nature of the lower part of the till laid down during the last Wisconsinan ice advance into Ohio, about 15,000 years ago, is evidence of this lake's existence. The Union City (Miami lobe) and Powell moraines (Scioto lobe) formed as a result of this final Wisconsinan incursion into Ohio, which didn't last long. The ice sheet was back in Canada within one thousand years. Meltwater again ponded between the ice sheet's margin in Ontario and the higher land in Ohio that marked the drainage divide between present-day Lake Erie and the Mississippi River. This ponded water became Glacial Lake Maumee I about 14,000 years ago. After the ice retreated, Glacial Lake Maumee I spread to an elevation about 800 feet above today's sea level. The lake extended nearly to Fort Wayne, Indiana, and drained through the Wabash River in Indiana. The best-developed beach ridges from this time are near Bryan and West Unity.

As ice continued to recede, the lake dropped to 760 feet and drained through a lower outlet in central Michigan. This lake level is called Glacial Lake Maumee II. From the Cleveland area to Berlin Heights, the Maumee II lakeshore closely paralleled the present coast, though it was some 2 to 12 miles inland. The ridges it left behind are low and hard to distinguish. A readvance of the ice sheet submerged the Maumee II beaches as the lake was forced inland, and its outlet may have shifted back to the Wabash River. Glacial Lake Maumee III, as it is known, assumed a 780-foot level. Beaches of this age stretch along the entire North Coast. They are among the more prominent beach ridges in Ohio. To see examples, follow Ohio 12 from Findlay toward Delphos or Ohio 613 from Van Buren toward McComb or Fostoria in northwestern Ohio. As the fitful ice once again receded northward, the lake may have again drained through the central Michigan outlet.

The next lake was Glacial Lake Arkona, which fluctuated between 695 and 710 feet around 13,600 years ago. Its indistinct sand ridges occur sporadically along the North Coast. Arkona beach ridges are visible west of Portage along Powell Road and southwest of Rudolph along Mermill Road and Ohio 281 in northwestern Ohio. A readvance of the ice sheet pushed lake level to 735 feet, forming Glacial Lake Whittlesey about 13,000 years ago. The lake drained through central Michigan. Prominent sand ridges, reaching heights of 5 to 10 feet near Ashtabula, mark this shoreline from the Pennsylvania border into northwestern Ohio. Between Painesville and Ashtabula, Ohio 84 is built on this sand dune–topped ridge, which reaches widths of 1,000 feet and a height of 10 feet. Lake Whittlesey beach ridges also appear southwest and east of North Baltimore in northwestern Ohio.

Geologists think that Lake Whittlesey came to a sudden end when the ice sheet receded far enough north to open a low outlet in the Niagara Gorge area of New York and Ontario. Lake level is thought to have dropped dramatically, probably below present-day Lake Erie's level of 569 feet. As the ice readvanced and then fluctuated back and forth between 13,000 and 12,000 years ago, Glacial Lake Warren (670 to 686 feet above present sea level) formed. There were three levels of this lake related to continued downcutting of the outlet in central Michigan. U.S. 20 follows a Lake Warren beach between the Pennsylvania line and Lakewood. Farther west, where Lake Warren was shallower, the former shoreline sand is spread into low dunes. Some broad and linear patches of sand may represent spits, deltas, or offshore bars rather than the typical beach deposit. Glacial Lake Warren beach ridges are visible along Sand Ridge Road between Bowling Green and Weston in northwestern Ohio.

Between the last two levels of Lake Warren, glacial geologists refer to a short-lived low-lake-level stage, which drained east across New York, as Glacial Lake Wayne. Lake Wayne (660 feet above sea level) beach ridges are low and discontinuous since they were reworked when the New York outlet was blocked by an ice readvance and water returned to the higher Lake Warren levels; drainage returned to central Michigan. As the ice sheet began to recede again, Lake Warren drained to the east through an outlet near Buffalo. As Lake Warren water levels lowered, it became what geologists refer to as Glacial Lake Grassmere. Geologists believe that Lake Grassmere (640 feet above sea level) drained east through New York but are not certain. It formed sometime between 13,000 and 12,500 years ago. The Grassmere beach underlies Lake Road between Conneaut and Madison-on-the-Lake in northeastern Ohio as discontinuous sandy ridges about 5 feet high.

Glacial Lake Lundy, the next lake level, fluctuated between 590 and 640 feet. Geologists think that indistinct sandy areas associated with this lake are offshore bars or dunes. Early Lake Erie began forming after 12,500 years ago. It has a history of water level changes, related to the slow rise of terrain that had long been weighted down by heavy ice, and changing outlets, related to this rise and the melting of ice. As ice retreated farther north, the outlet through Niagara Gorge allowed Lake Erie to drain to such a degree that there was only water in the deeper eastern basin and perhaps part of the central basin (Lake Erie has three distinct basins, which are discussed in the following section). As the Niagara region rose after being released from its yoke of ice, water level slowly rose in the Erie Basin. Water level has been rising in the last 2,600 years. Currently, Lake Erie lies at 569 feet above sea level.

About a 2- to 12-mile strip of lower land lies between the Allegheny Plateau and the present Lake Erie shoreline. This strip of land between Lorain and the Pennsylvania border becomes lower in a steplike progression. Geologists speculate that the three prominent, gently sloping terraces, separated by steeper areas, represent wave-cut terraces and cliffs in older Wisconsinan till that formed in a lake pre-dating the postglacial lake sequence just discussed. Where Devonian bedrock is exposed at the lakeshore, the features slope much more steeply toward the lake. The terraces and wave-cut cliffs formed 35,000 to 23,000 years ago during an ice-free time. The readvance of Wisconsinan ice sheets about 23,000 years ago eliminated this lake, but the ice was not extremely deep in northeastern Ohio and therefore did not destroy the earlier erosional features; instead it coated them with a thin layer of younger till. On each terrace, except along the bedrock portion of the shoreline, there are two to six more-recent beach ridges. The beach ridges on the terrace at the highest elevation, farthest inland, formed during the time of Glacial Lake Maumee; the middle cliff beach ridges are associated with Glacial Lake Whittlesey; and the discontinuous ridges on the lower terrace formed during the time of Glacial Lake Warren.

Today, a similar wave-cut cliff and wave-cut terrace is forming along Lake Erie's shore.

It's likely that more pieces of the puzzle that is the complex geologic history of Lake Erie and its glacial lake predecessors will be discovered as geologists continue to apply new technologies to the glacial cover and bedrock surface of northern Ohio.

Lake Erie Basins

Due to the gentle arching of the bedrock underlying Ohio, which is related to the Findlay Arch, different strata underlie Lake Erie from Conneaut to Toledo. The Pleistocene ice sheets and preglacial streams found it easier to hollow out the eastern and central part of the lake, underlain by Devonian- and Mississippian-age shale, and more difficult to erode the western part, underlain by older Silurian- and Devonian-age limestone and dolomite. As a result, Lake Erie has three distinct basins: the eastern basin east of the Ohio and Pennsylvania border, the central basin from the Ohio and Pennsylvania border to Sandusky, and the western basin from Sandusky to Toledo.

The central basin is relatively flat bottomed and averages 60 feet in depth. The western basin is the smallest and averages 24 feet in depth. The Bass Islands, Kelleys Island, and Pelee Island separate the central and western basins. Rocky cliffs and slumping shale bluffs characterize the central basin shores, while low sandy beaches border the western basin.

Shoreline Dynamics

The narrowness of Lake Erie and the fact that its east-west axis parallels regional winds and storm tracks makes it susceptible to high storm waves and a sloshing action called a *seiche*. Sometimes storm waves more than 12 feet high pound the shores with great force, often undermining cliffs and causing the shoreline to collapse and slump. Wind can blow water from the shallow west end of the lake toward Buffalo, New York, setting the stage for it to swash back, much like water in a bathtub. As much as a 15-foot change in water level may develop during a seiche, which is a significant erosional force. Huge blocks of lake ice can add to the destruction during winter storms. Even thunderstorms with winds blowing shoreward can produce 4-foot waves in about thirty minutes.

Although the lake is most dynamic during storms, every day is a continual battle between the forces of erosion and deposition. Waves and currents are the main players in the changing geology of a lake system. Winds blowing across the long stretch of Lake Erie set the surface water in motion and push it toward the shoreline. A chain reaction causes underlying water to move in a circular path as it travels in the direction of the wind. In the open lake in deeper water, bottom waters remain undisturbed, but as the upper section of the water column moves into shallower coastal areas, the entire water column is set in motion. Friction with the bottom slows down the circular motion of the water near the bottom and causes a wave at the surface to grow higher and slow down. As the wave washes onto shore, it grows taller and eventually rolls over before washing back into the lake. A tremendous amount of energy is expended by this bottom friction and wave collapse, stirring up sediments that then travel along the beach. Sometimes successive waves can cut a knick in a beach at the point where they continually collapse. In other places a cliff may form as other sediments are ripped away by constant wave impact. These lake cliffs recede inland over time as they are undercut, eventually collapsing. All these erosional effects are more intense during storms.

Wherever erosion is not the dominant force, deposition probably is. Sandbars form in the shallows where the underwater part of waves just begin to interfere with the bottom. They also develop behind rocks that jut above the water. Since most waves approach the shore at an angle, they generate what are called *longshore currents*, which carry sand and silt parallel to the shore. These currents form narrow stretches of beach. Where bays lead back from the shore, narrow fingers of sand develop across the channel forming spits. Cedar Point near Toledo is an example. If a spit grows across a bay, it is called a *baymouth bar*. The beach at Crane Creek State Park is an example. A sandbar that connects an island or offshore rock to the mainland is called a *tombolo*, such as along the shoreline of Catawba Island.

Ohio has been losing shoreline at a record pace since at least the early 1800s as settlers began to alter it. Before the 1800s, a narrow, sandy beach blanketed the Ohio shoreline. Between 1796 and 1838, 132 feet of it was lost to wave erosion from the Pennsylvania border to Huron. Erosion forced Ohioans to relocate U.S. 6 between Ruggles Beach and Huron in the 1930s. At that point, nearly 300 feet of shoreline had disappeared around Huron since settlement. Remaining segments of the old highway end at the lake cliff along this stretch. Around Sandusky the shoreline lost around 9 feet per year from around 1830 to 1950. During the 1970s an episode of shoreline loss is attributed to a rise in lake level. In less than twenty years wave spray doused houses that had been built several hundred feet from the shore in the 1950s and 1960s.

A sandbar at Crane Creek State Park. Note the groin in the distance, trapping sediment that is moving parallel to the beach face in longshore currents.

A major storm hit Lake Erie and the North Coast on November 13 and 14, 1972. Northeast winds averaging 20 knots blew directly down the long axis of the lake. At times wind speed increased to 60 knots in the western basin. Lake water was pushed from the eastern basin into the shallower central and western basins. Lake level at Toledo reached a maximum of 6 feet above normal November levels. Waves crashed into shoreline structures with great force and shoreline areas were submerged under lake water, causing some $22 million in damage.

The level of Lake Erie has fluctuated through the last 4,000 years. Rises and falls in lake level are partially related to variations in the amount of direct precipitation falling into the lakes and runoff entering the lakes from adjacent land areas across the drainage basin of Lakes Erie, Huron, Michigan, and Superior. Another cause of lake level fluctuations is the slow rise of the land surface north of the Great Lakes. This area was depressed by the weight of thick ice sheets during the Ice Age and is still slowly recovering, like a sofa cushion after a person gets up. Engineers estimate this rebound to be about .37 foot per century.

Part of the reason for growing shoreline loss is due to human attempts to control shoreline processes. As with river floodplains, people like to live along the lake coast. To preserve property and investments, people use many techniques to offset the effects of wave erosion, deposition, and flooding. As early as the 1820s, groins, engineering structures built perpendicular to the shore, appeared along beaches. By blocking the path of longshore currents, these groins allowed sand to accumulate on their up-current side, widening the beach near them, a result residents were looking for. The longer the groin, the wider

Shoreline is disappearing near Huron where waves carve into Wisconsinan lake clays and back-shore forests of hardwoods.

the beach. The cost, however, is that erosion occurs on the down-current side of the groin because the water is sediment starved. Since longshore currents on the down-current side of a groin have less sediment in suspension, more erosion will take place when waves sweep this water onto the beach. Sediment-laden water is less erosive since it must expend energy to keep the sediment in suspension. The beach gains on one side of the groin but loses on the opposite side. To reduce this effect, engineers regularly build groins in a series, but wherever they end erosion takes over.

Other structures engineers use to try to control the lake are jetties, breakwaters, and seawalls. A jetty consists of parallel walls that project into the lake on either side of a channel. These walls keep the channel open, fighting the longshore currents' tendency to build bars across the mouths of channels. Jetties, like groins, trap sediment on the up-current side and cause erosion down the shore. Breakwaters are obstructions built parallel to the shore in shallow water. They reduce the impact waves have on the shoreline, causing sediment to accumulate in the quiet water behind them. They simply cause more-forceful wave impact where the breakwater ends. Seawalls are massive walls of concrete, riprap, or other materials and are meant to withstand wave impact and prevent shoreline undercutting. They require continual maintenance because of the beating they endure and cause greater erosion to occur along the natural shore. Modifying the natural shore tends to concentrate erosion and deposition at certain points, rather than spreading it along the entire shoreline.

Shoreline structures have their pros and cons, and one wonders if Lake Erie residents can live without them? One thing we can be assured of is that the Lake Erie coast will continue to change. Driving the roads of the North Coast will allow you to see the most changeable part of Ohio's landscape.

Groins at Crane Creek State Park trap sediments carried by longshore currents on the near side.

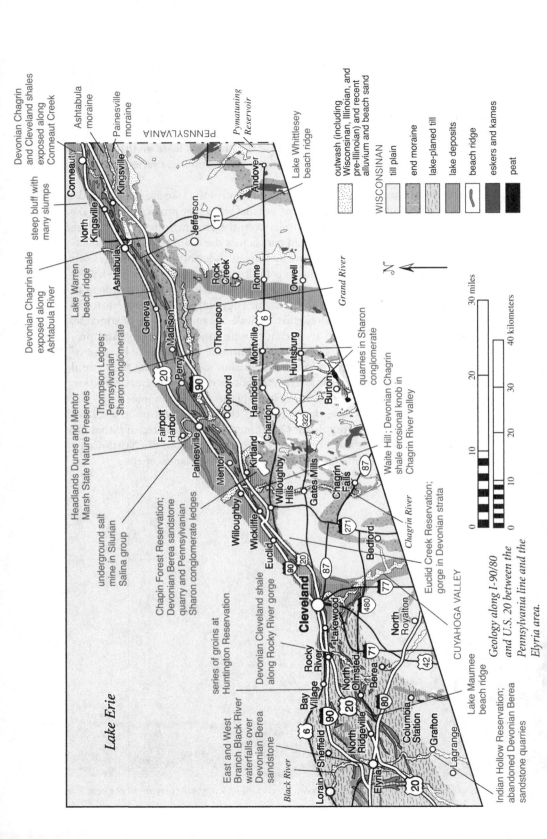

Geology along I-90/80 and U.S. 20 between the Pennsylvania line and the Elyria area.

INTERSTATE 90 AND U.S. 20
(Including I-80 between Elyria and Toledo)
Pennsylvania Line—Toledo
178 miles

Between the Pennsylvania line and Cleveland, I-90 and U.S. 20 roughly parallel the boundary of the Huron-Erie lake plain and the glaciated portion of the Allegheny Plateau. Between Cleveland and Sandusky, the routes skirt the border of the lake plain and western Ohio's till plains. The roads stretch across the flat lake plain from Sandusky to Toledo.

Many rivers and streams drain into Lake Erie, and thus the routes cross over them. Many of the smaller streams between the Pennsylvania line and Cleveland have sources in the Ashtabula and Painesville moraines, between Lake Erie and the edge of the Allegheny Plateau, and empty into the lake a few miles downstream. The end moraines in northeastern Ohio mainly formed from two major ice advances of the Wisconsinan Grand River and Killbuck lobes.

Pennsylvania Line to Painesville

From the Pennsylvania border to Conneaut, I-90 and U.S. 20 traverse lake plain. U.S. 20 follows the Glacial Lake Warren beach ridge, a relatively persistent ridge, while the interstate follows the Painesville moraine. The beach ridges are subdued along this stretch; If you look north to the lake, you'll see that the landscape drops gradually but steadily in steplike fashion to a 40-foot-tall wave-cut cliff at the present lakeshore. These are the wave-cut terraces and cliffs of a lake that existed during the ice-free time that separated major advances of the Grand River lobe.

Lake Road (Ohio 531), an alternate route, parallels the lakeshore atop the cliff and provides public access to a number of beaches. Mass wasting, when large amounts of material move downslope in a relatively short period of time, is the dominant geologic process along this stretch of shoreline. Wisconsinan lake sediment and underlying Wisconsinan till and Devonian-age shale continually slide into the lake. Breakwaters and jetties extend out into the lake on either side of Conneaut Harbor, preventing sediment from clogging the shipping channel. Conneaut Township Park, off Lake Road, is a good spot to view the shoreline. In the park a road climbs down the bluff along a narrow valley that was carved by a stream that originates in nearby beach ridges. The stream twists wildly as it passes onto a flat picnic area.

East of Kingsville, I-90 bridges the deep valley of Conneaut Creek, which begins in a Wisconsinan end moraine near Conneaut Lake, Pennsylvania. From Conneaut Lake the creek flows north to the Painesville and Ashtabula moraines, parallels the end moraines to a low spot in them at Kingsville, and cuts through them and loops back east 10 miles to Conneaut and Lake Erie. A 60- to 80-foot valley in Devonian Chagrin and Cleveland shales marks its course. A plant located along Conneaut Creek in Conneaut made bricks from the Chagrin shale in the early 1900s, but the site is unrecognizable today. A plant in Ashtabula also made bricks using this shale. Off U.S. 20, east of North Kingsville on Poore Road, is North Kingsville Sand Barrens, a preserve of the Cleveland Museum of

High bluffs underlain by Wisconsinan till form the shoreline east of Ashtabula. Note the small earthflows and clumps of sod that are slowly making their way to the lakeshore below.

Natural History. These sandy hills formed some 12,000 years ago on the shores of Glacial Lake Warren.

East of the Ashtabula exit, I-90 crosses the flat-bottomed valley of the Ashtabula River, which was carved in the Painesville and Ashtabula moraines. The Ashtabula River begins on an end moraine near Andover, flows north to the Painesville and Ashtabula moraines, then parallels the end moraines west to near Ashtabula, where it cuts through the moraines. On the north side of the moraines the river flows west again and meanders north to Lake Erie at Ashtabula, Harbor. The Chagrin shale forms its valley walls, which are 100 feet high near Ashtabula. Severe erosion marks the lakeshore around Ashtabula. High bluffs slowly yield to crashing storm waves and ice that gouges them in winter. Breakwaters and jetties protect the harbor, and groins maintain narrow beaches below the wave-cut bluffs. Cottages along Lake Road lie dangerously close to the cliff.

West of Ashtabula you can see the wave-cut terraces and cliffs—dips in the landscape—of a mid-Wisconsinan glacial lake north of the routes. After the final retreat of the Wisconsinan Grand River lobe, Glacial Lake Whittlesey at its highest level stretched south to the base of the Painesville moraine. Ohio 84, sandwiched between I-90 and U.S. 20, follows its beach ridge. U.S. 20 follows the former shoreline of Glacial Lake Warren. South of I-90 the meandering course of the Grand River is nestled within the Painesville and Ashtabula moraines. The Grand River begins about 98 miles south near Champion, flows north to the Painesville moraine through glacial lake deposits, and parallels it west to south of Painesville where it cuts through the moraines and beach ridges and empties into Lake Erie at Fairport Harbor. The Grand River also cuts into Devonian shales, which are visible where U.S. 20 crosses the river in downtown Painesville.

Waterfalls, some over 50 feet high, cascade over harder sandstones and silt-stones that cap the softer Devonian rocks along the heads of short valleys at Madison, Painesville, Chardon, and Kirtland. Gildersleeve (south of Kirtland) and Little Mountains (south of I-90 and southeast of Mentor) are among the northernmost outposts of the sandstone-capped Allegheny Plateau. They are isolated knobs of Pennsylvanian-age Sharon conglomerate that were separated from the main plateau by stream erosion. Quarries at Concord and Kirtland marketed Devonian-age Berea sandstone and Sharon conglomerate for bridge and building construction in the years after the Civil War. A number of quarries in the Berea sandstone operated near Chardon; one in Munson Township sup-plied stone for the Chardon courthouse. Plants made bricks using the Chagrin shale at Wickliffe and Willoughby Hills in the early 1900s. Between Geneva and Painesville, the Lake Warren and Lake Whittlesey beach ridges are readily seen and are known as North and South Ridge, respectively. U.S. 20 continues to fol-low the Glacial Lake Warren shoreline west of Painesville, and Ohio 84 follows the Glacial Lake Whittlesey beach ridge.

Northeast Ohio Earthquakes

A Jesuit priest, Reverend Frederick L. Odenbach, set up the first seismograph in Ohio in Cleveland in 1900. It was one of the first in the United States. Ohio was not a center of earthquake activity, but seven weak to mild shocks had struck Ashtabula, Cuyahoga, Lake, Portage, and Summit Counties in the 1800s. Father Odenbach's pioneering work allowed researchers to record about a dozen mod-erate quakes during the 1900s. The most notable quake in the area, however, occurred much later, on January 31, 1986, southeast of Painesville. It registered about 5 on the Richter scale. Two people suffered minor injuries; items fell off store shelves in Chardon, Mentor, and Painesville; windows and plaster walls cracked; and area well water reportedly changed color and taste. Portable seis-mographs placed around the area the day after the main quake recorded at least thirteen aftershocks as high as 2.4 Richter magnitude. The great interest in this quake was probably related to the fact that a nuclear power plant operates in nearby Perry.

Small faults are common in Devonian-age rocks in northeastern Ohio. Other, more ancient ones lie more than a mile below the surface in Precam-brian rocks. All these faults experience periodic adjustments, caused by the release of stress or pressure in the subsurface, which are recorded as earth-quakes at the surface.

On January 25, 2001, a quake registering 4.6 magnitude rattled the Ashtabula area. This was only one of dozens that had shaken the area since 1987. Injection of waste fluids down a 6,000-foot-deep well into the Cambrian Mt. Simon sandstone from 1987 to 1993 has been linked to this seismic activity. The injected fluids seep along cracks in the sandstone, leading to slips along faults. The faults have been there for hundreds of millions of years and have just come to life in recent years because of artificial lubrication.

A Side Trip to Thompson Ledges

Head south from U.S. 20 or I-90 on Ohio 528 to visit the flat-topped ridges of Pennsylvanian-age Sharon conglomerate known as Thompson Ledges. From the routes Ohio 528 climbs onto the Painesville moraine, crosses the deep valley of the Grand River with its cliffs of Devonian-age Chagrin shale, and climbs the Euclid moraine and Allegheny escarpment, which developed in Pennsylvanian-age rocks. Near Thompson the road surmounts Sharon conglomerate, which caps the uplands along the Grand River valley. This bedrock ridge continues to the south beyond Montville, and only a thin covering of glacial till overlies it. Granite boulders—originally part of Ontario's bedrock—scattered about northern Ohio are evidence that ice sheets once covered this area.

Thompson Ledges forms a ridge that loops around the north side of Thompson. The flat-topped rock outcrops with cavelike crevices are a popular picnicking area; a hotel was even built after the Civil War to cater to visitors. The best place to observe the ledges is 0.3 miles east of the town square on Thompson Road at Thompson Ledges Township Park. A trail winds along the crest of the ridge, permitting close looks at the rock, tight squeezes through widened joints, and scenic vistas of the Grand River valley to the east.

Stone quarrying began north of Thompson in the mid-1800s. The Sharon conglomerate and sandstone found use in many local bridges, walls, and foundations. Quarries also operated near Chardon, Hambden, Huntsburg,

Pennsylvanian-age Sharon conglomerate forms Thompson Ledges, where enlarged joints create a pleasing maze of narrow passageways.

Montville, and South Thompson. By 1910 local entrepreneurs turned their interest to sand and gravel. Beginning in 1932 R. W. Sidley began quarrying the rock on the south side of Thompson along Ohio 528 and crushing it for use in the concrete and construction industry. The industry still contributes to the economy of the Thompson area.

Salt under Fairport Harbor

The rooms and pillars of a salt mine lie 2,000 feet beneath Headlands Beach State Park and extend 2 miles out under Lake Erie. The salt, part of the Silurian-age Salina group, is about 50 to 80 feet thick. It formed as part of a thick accumulation of sediments as the shallow Silurian sea evaporated. Morton Salt opened the mine in 1959 along the Grand River across from Fairport Harbor and mainly produces crushed rock salt for snow and ice control. The company, which still operates the mine, claimed it was the deepest and most modern salt mine at the time. It produced over 2 million tons of rock salt in 2004.

The brine industry along Lake Erie preceded underground mining by about sixty years. Salt wells dotted the terrain around Fairport Harbor, Grand River, and Painesville by the early 1900s. The brines workers boiled down came from Devonian and Silurian strata. Diamond Alkali used the brine to manufacture soda ash, caustic soda, chlorine, carbon tetrachloride, cement, and related products. The closing of the plant in 1976 left a legacy of abandoned chemical plants and polluted ground in the area.

Kirtland and Its Quarries

Just west of Mentor, Ohio 306 south dips impressively from U.S. 20 and I-90 into the valley of East Branch Chagrin River, carved in Devonian-age Bedford and Chagrin shale. On the south side of the valley is the historic community of Kirtland, home of the Kirtland Temple of the Church of Latter-Day Saints.

Devonian-age Berea sandstone taken from the Stannard Quarry near Kirtland was used in the temple of the Church of Latter Day Saints. The quarry is now part of Chapin Forest Reservation. Note the grooves from the channeling machine.

View northwest toward Lake Erie from ledges of Pennsylvanian-age Sharon conglomerate at Chapin Forest Reservation.

Chapin Forest Reservation, a Lake County park south of town, offers visitors the chance to see old quarries and more area ledges. The Stonecutters Trail on the reservation encircles the Stannard Quarry, which was excavated in the Devonian Berea sandstone. Look for ripple marks on the rock surfaces, representing shallow-water conditions in the Devonian sea, and tool marks on the quarry walls as well as loose blocks of stone. Builders used stone from this quarry to construct the Mormon temple in the 1830s.

Trails up Gildersleeve Mountain, also part of the reservation, lead to ledges of Pennsylvanian Sharon conglomerate and striking views across the countryside to Lake Erie. The 100-foot cliff on the north side of the mountain was left from more-recent quarrying, which crushed the pebbly sandstone into aggregate. The actual ledges of the area are off-limits and can only be visited with special permission.

Painesville to Willoughby

The extensive battering of Ohio's northeast coast by waves led residents to put shore-control structures—breakwaters, groins, jetties, and seawalls—in place as early as 1827. The mouth of the Grand River at Fairport Harbor is a prime example of coastal engineering. Sand travels eastward along the shore here, and if left unattended it builds sandbars across the channel, causing navigation problems for boats. Initially, jetties were extended a short distance into the lake to keep the harbor entrance open, but by 1876 they stretched 2,000 feet into the lake. Breakwaters appeared in the early 1900s. These structures altered the shoreline west of Fairport Harbor significantly. Because of sediment buildup, the harbor changed from a nearly east-west orientation in the 1820s to northeast-southwest by 2004. An immense pileup of sand west of the river mouth increased beach width by nearly 1 mile. This area includes Headlands Beach State Park and Headlands Dunes State Nature Preserve.

East of the harbor structures, sediment-starved waters cut away the shore at an equally impressive rate. A good place to observe the destruction of shoreline is Painesville Township Park east of Fairport Harbor. For a 4-mile stretch of lakeshore, beach widths are decreasing. A 60-foot bluff cut in glacial till towers above the narrow beach, and slumps are common. Note the toppled trees on the beach. Piles of rubble on the beach and groins slow the erosive onslaught. Lake Road ends abruptly at an overlook on the east side of the park; it used to continue farther east. This road is threatened in many places, and houses on the north side of it are rapidly losing ground to the lake. Similar loss of land is common at many other points along the northeast coast, including nearby Mentor Headlands, where Headlands Road has been relocated at least three times.

Headlands Dunes State Nature Preserve has an undisturbed stretch of backshore dunes that harbor the unique flora of Atlantic coastal plain plants, including beach grass (*Ammophila breviligulata*), beach pea (*Lathyrus maritimus*), purple sand grass (*Triplasis purpurea*), sea rocket (*Cakile edentula*), and seaside spurge (*Euphorbia polygonifolia*). The plants migrated into the Great Lakes region during the waning stages of Wisconsinan glaciation when the Champlain Sea, a finger of the Atlantic Ocean that extended into the present-day Champlain Valley, stretched to the present-day Lake Ontario basin. Transverse

Headlands Dunes State Nature Preserve near Painesville.
Sand dunes form the upper back-shore to the right.

dunes, ridges of windblown sand perpendicular to wind direction, formed as onshore winds piled sand that waves and longshore currents had spread on beaches. The dunes parallel the shoreline. Depressions among the dunes are blowouts, places where masses of sand were moved elsewhere.

Mentor Marsh State Nature Preserve is just west of Headlands Beach State Park in the abandoned channel of the Grand River. The Grand River once flowed parallel to the coast west of Painesville, entering the lake between the towns of Mentor-on-the-Lake and Mentor Headlands. A combination of river and wave erosion cut through the ridge that separated the river and lake between Painesville and Fairport Harbor, shortening the river by about 5 miles. The former river mouth is dredged and is now part of Mentor Harbor and adjacent lagoons. In the nature preserve the invasive reed *Phragmites australis* has taken over the open wetlands while swamp forest covers other areas of the abandoned channel.

At Willoughby, the Painesville moraine thins where the Chagrin River flows north to Lake Erie. Devonian-age Chagrin shale walls the valley south of U.S. 20 to I-90 and beyond. The town of Waite Hill surrounds an erosional knob of Devonian Chagrin shale lying in the middle of the Chagrin Valley just south of I-90 where the river's east and west branches converge. South of Willoughby at Gates Mills another isolated hill of Chagrin shale on the floodplain marks a change in the course of the Chagrin River. At one time the river flowed around the west side of this hill, but the river abandoned that meander when it cut through its neck and began to flow around the east side. At Pleasant Valley near U.S. 6 the Chagrin River flows through the Euclid and Painesville

Glacial till overlies Devonian-age Chagrin shale where the Chagrin River passes under U.S. 20 at Willoughby. The old bridge pier is composed of Berea sandstone.

moraines and onto lake plain. Devonian sandstone caps the uplands of the plateau. Atop an eroded bank of Chagrin shale, Hach-Otis State Nature Preserve, off U.S. 6 near Willoughby Hills, offers a great view of the Chagrin Valley some 150 feet below. Note the twisting course of the river as it approaches the lake. As is typical of many streams flowing down to the lake from the uplands, the Chagrin River loses much of its downcutting ability and strength to carry its sediment load of sand and mud eroded from the plateau. It is forced to dump much of its load in its channel at the base of the plateau. As it flows across the gently sloping lake plain to Lake Erie, it winds between piles of eroded debris to reach the lakeshore.

Euclid Creek Reservation

Euclid Creek Reservation, a Cleveland Metropark, offers glimpses of flagstone quarries that locals first worked in the 1850s. The reservation encompasses Euclid Creek, which flows into Lake Erie just west of Euclid. Quarrying developed on the east bank of Euclid Creek first before spreading to its west side. The quarries are gone now but an iron ring, probably used to anchor a derrick, still juts out of a sandstone boulder at the "quarry area" parking lot. The Euclid bluestone makes up 20 feet of an impressive gorge that Euclid Creek cut through 300 feet of Devonian strata. Berea sandstone caps the gorge. Below this massive cliff is the softer Bedford formation, which includes the more resistant Euclid member. The Bedford shales and siltstones grade into darker Cleveland shale. The lower part of the gorge exposes gray Chagrin shale that contains numerous trace fossils, some brachiopods, and a few sponges. Euclid Creek's headwaters are in Beachwood, 450 feet above the lake and about 20 miles south

Devonian-age Chagrin shale forms most of the valley wall in Euclid Creek Reservation in Euclid. The Bedford and Berea formations form the upper part of this gorge.

of it. Commercial development and highway construction upstream has led to flooding problems. Increased runoff after major storms is more than the natural channel can hold. This is not unique to Euclid Creek. Flooding is a concern all along the edge of the Allegheny Plateau.

Cleveland to Elyria

On the west side of downtown Cleveland, the Cuyahoga River meanders across the "Flats," a heavily industrialized floodplain. Between Lakewood and Rocky River, Chagrin shale and siltstone are exposed in the deep valley of the Rocky River near river level. Upstream is Rocky River Reservation, where the overlying Cleveland shale contains a wealth of Devonian fish fossils.

West of Rocky River there are occasionally views of Lake Erie's shoreline from U.S. 6 and Ohio 2. Huntington Reservation, along U.S. 6 in Bay Village, has a groin-protected beach. U.S. 20 follows the Glacial Lake Whittlesey beach ridge from Rocky River to Elyria; along this stretch it is called Center Ridge. Just to the north is the Glacial Lake Warren shoreline, known locally as North Ridge. The Lake Maumee beach forms Butternut Ridge, traversed by Ohio 10 to the south.

A Remarkable Fossil Occurrence

Brittle stars, or ophiuroids, a type of slender-arm starfish, are among the rarest fossils in the world because they rapidly disintegrate after death. Imagine their surprise when paleontologists uncovered thousands of them in 1970 in a thin, silty shale bed just east of the Ohio 3 overpass on I-80, south of North Royalton. The fossil bed gradually thinned east and west from a 1-inch-thick center, stretching for about 18 feet along the slope above the turnpike. How far it extended into the bank is unknown. The unit is Mississippian-age Meadville shale, a member of the Cuyahoga formation. During the time of deposition this area must have been a nearshore shelf where an ocean's tidal currents washed back and forth, carrying food to an established colony of these filter-feeding echinoderms. Only a few parts of crinoids and bryozoans were preserved with the brittle stars, but because they were broken it seems that they must have been washed into the area and were not necessarily living with the brittle stars. Scattered fossil plant fragments amongst the brittle stars indicate that this area was close to the shoreline. Only a few traces of the fossil-rich rock beds poke through the slope wash—eroded rock and soil carried downslope by water—at this location today.

The Black River

The Black River begins as two branches on the Defiance end moraine, the east branch near Lodi and the west branch near Nova on U.S. 224. Both branches meander extensively across till plain to near Laporte, where they flow onto the lake plain. At Ohio 10 both branches veer east and then cut north through the Glacial Lake Maumee shoreline. Presumably, as drainage returned to this area following the retreating Wisconsinan ice sheet, both branches flowed toward the lake basin only to encounter Lake Maumee's beach sand that marked its shore. The branches flowed east along the beach ridge until finding a spot that offered them little resistance, which they cut through near present-day Laporte and about 3 miles west along Ohio 3 before entering Glacial Lake Maumee II. As later lower lake levels developed, the branches continued developing meandering courses northward through lake plain.

Elyria developed where the branches of the river fall some 40 feet over Devonian-age Berea sandstone before joining together. Cascade Park in Elyria is a great site to stand at the juncture of the two branches. You can hike up the west branch past huge slump blocks of sandstone and observe the falls of the west branch, which has a large rock shelter behind it. Elsewhere along the trail there are cross-beds and ripple marks in the sandstone, evidence of its origin, in which currents were moving the sands of a river bottom back and forth. Underneath the resistant sandstone is the crumbly Bedford shale, which continues to be broken down by exposure to the atmosphere. As it weathers away and undercuts the overlying sandstone, blocks of sandstone break off and fall downward. It is in this very manner that the waterfalls have migrated some 4 to 5 miles upstream since their origin. Boardwalks also provide great views of the falls of the east branch, off Lake Avenue, but concrete walls mar the natural face and remind viewers of the area's early use by mills. Just west of the falls there are abandoned sandstone quarries, which in the 1840s supplied markets across the continent with 12- to 700-pound grindstones. The Black River forms a tight loop just north of I-80 and surrounds a wedge-shaped spur of erosion-resistant Devonian-age Chagrin shale. Between Elyria and its

The East Branch Black River drops over Devonian-age Berea sandstone in Cascade Park in Elyria.

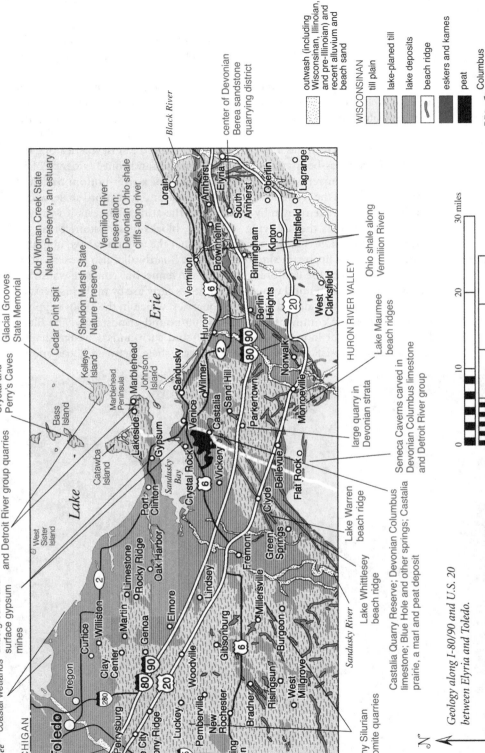

32

Maumee
River

MICHIGAN

Toledo

Oregon

Perrysburg

Lime City

Stony Ridge

Luckey

Pemberville

New Rochester

Woodville

Gibsonburg

Bradner

Risingsun

West Millgrove

Bowling Green

Clay Center

Curtice

Williston

Martin

Genoa

Elmore

Lindsey

Millersville

Burgeon

Green Springs

Fremont

Clyde

Bellevue

Flat Rock

Monroeville

Norwalk

Parkertown

Berlin Heights

West Clarksfield

Lagrange

Pittsfield

Kipton

Oberlin

South Amherst

Elyria

Amherst

Brownhelm

Birmingham

Lorain

Vermilion

Huron

Sandusky

Wilmer

Castalia

Vickery

Sand Hill

Crystal Rock

Venice

Gypsum

Lakeside

Port Clinton

Oak Harbor

Rocky Ridge

Limestone

West Sister Island

Catawba Island

Bass Island

Kelleys Island

Marblehead Peninsula

Johnson Island

Marblehead

Sandusky Bay

Lake Erie

Black River

Maumee River

Sandusky River

underground and surface gypsum mines

coastal wetlands

Devonian Columbus limestone and Detroit River group quarries

Crystal and Perry's Caves

Glacial Grooves State Memorial

Old Woman Creek State Nature Preserve, an estuary

Cedar Point spit

Sheldon Marsh State Nature Preserve

Vermilion River Reservation; Devonian Ohio shale cliffs along river

center of Devonian Berea sandstone quarrying district

Ohio shale along Vermilion River

HURON RIVER VALLEY

Lake Maumee beach ridges

large quarry in Devonian strata

Seneca Caverns carved in Devonian Columbus limestone and Detroit River group

Lake Warren beach ridge

Lake Whittlesey beach ridge

Castalia Quarry Reserve; Devonian Columbus limestone; Blue Hole and other springs; Castalia prairie, a marl and peat deposit

many Silurian dolomite quarries

Geology along I-80/90 and U.S. 20 between Elyria and Toledo.

N

outwash (including Wisconsinan, Illinoian, and pre-Illinoian) and recent alluvium and beach sand

WISCONSINAN

till plain

lake-planed till

lake deposits

beach ridge

eskers and kames

peat

Columbus

0   10   20   30 miles

mouth at Lorain, the Black River flows in a wider valley that is walled by bluffs of Chagrin shale. Near Sheffield the river crosses the Glacial Lake Warren beach ridge. The Black River parallels the coast from Sheffield, passing steel mills before entering Lake Erie.

Lorain County Metro Parks operates several parks—Black River Reservation in Elyria and Lorain, Carlisle Reservation in Lagrange, French Creek Reservation in Sheffield, and Indian Hollow Reservation in Grafton—that offer closer views of the natural history of the Black River. Several abandoned meanders, or oxbows, are flourishing wetlands in these parks, and abandoned sandstone quarries are now ponds in Indian Hollow Reservation. Berea sandstone quarrying began near Grafton in the 1850s; eventually, Cleveland Quarries, the company that owned most of the sandstone quarries in the Amherst-Berea area, controlled it. Discarded grindstones occur in the park.

Elyria to Toledo

From Elyria to Toledo, U.S. 20 and I-80/90 (the Ohio Turnpike) cross relatively flat lake deposits and lake-planed till. Both roads cross the deep valley of the Black River at Elyria. Just north of the turnpike the river twists back and forth across softer Devonian shales after leaving the harder sandstones that underlie lake plain and till south of the turnpike. Although not visible from the routes, Berea sandstone quarries dot the countryside around Amherst and Oberlin. Near Ohio 60, U.S. 20 and the turnpike cross a deep valley the Vermilion River carved into Devonian-age Ohio shale. North of the turnpike, the rest areas on Ohio 2 where it crosses the Vermilion River are good places to observe the river's 50- to 100-foot shale bluffs; there are also numerous large glacial erratics in the woods at these areas. The Vermilion River Reservation, off North Ridge Road south of Vermilion, is yet another good place to view these bluffs. Native Americans once made red paint from the local soil, thus the name "vermilion." The river's headwaters are in Mississippian-age sandstone uplands near Savannah, and it meanders through a postglacial valley it cut across the lake plain to the shore at Vermilion. Vermilion's Showse Park offers a glimpse of the Ohio shale and overlying Wisconsinan till at the wave-cut bluff at Lake Erie.

Oil wells are common around Birmingham, where since 1966 oil companies have pumped petroleum from the Cambrian-age Knox dolomite, 3,895 feet down. Berlin Heights, just west of Birmingham, is situated on an erosional knob of Devonian Berea sandstone. The active Birmingham Quarry lies north of the turnpike near this town, removing Berea sandstone building stone.

I-80/90 crosses the Huron River near the U.S. 250 interchange. In the valley there are low exposures of Devonian Huron shale with characteristic boulder-sized concretions. Along the lakeshore on the west side of Huron, the rounded tops of Ohio shale concretions pop out of the water when lake level is low. The area used to be called Boulder Camp, named after the nearly spherical concretions. The genesis of these concretions is not well understood, but they seem to have formed around small mineral particles, or organic material, embedded in the sediment. The iron carbonate mineral siderite formed small crystals around the original object and added more in concentric shells. This happened while the surrounding muddy sediment was slowly hardening into shale. The layers of shale show upward and downward bends above and below the concretions. The Huron River begins in wetlands near New Haven about 20 miles south of the turnpike and ends in coastal wetlands at Huron. From here to Toledo the lakeshore is a low wave-cut bluff of till and lake sediments with drowned river mouths, coastal marshes, and narrow sandy beaches.

South of Norwalk and U.S. 20 on Ridge Road is the water-filled Bronson Quarry, once a source of Berea sandstone for buildings and bridge piers. The quarry operated sporadically from around 1860 to the early 1900s. Other quarries are scattered on farms northeast and southeast of town. Another important source of building stone was the sandstone quarry at West Clarksfield. Monroeville had an active clay plant making drainage tile from Wisconsinan till into the 1940s. U.S. 20 follows a Glacial Lake Whittlesey beach ridge between Norwalk and Monroeville; the upper sands of the ridge form low sand dunes near the intersection of U.S. 20 and County Road 270. McPherson Cemetery in Clyde resides on the Lake Warren beach ridge. Ohio 101 parallels this ridge to the northeast as far as Castalia.

Between Clyde and Toledo the routes travel across the former lake beds away from shoreline areas. With the exception of recent alluvium in the valleys of Green Creek and the Sandusky River at or east and north of Fremont, the routes traverse wave-planed till or lake sediments.

Grindstones and Building Stone—Quarrying in the Amherst Area

The Devonian-age Berea sandstone that underlies Cuyahoga, Lorain, and eastern Erie Counties is a world-famous building stone. The first commercial use of this stone began at Berea, near Cleveland, in the 1830s, where workers fashioned grindstones. The earliest recorded quarrying in the Amherst area occurred at Brownhelm in 1847. The following year a quarry opened near Amherst. Initially, these quarries produced grindstones and some local building stone. Transportation was a problem because roads were poor and area railroads were not yet built. The Brownhelm operation, being closest to the lake, was able to haul stone blocks to Vermilion and Lorain where it could be shipped to other lake ports.

A quarrying boom began just after the Civil War. Many quarries opened between Amherst and South Amherst where the Berea sandstone was thicker than 200 feet. The sandstone from the Amherst area rapidly became a highly desired building material. As railroads were constructed in the 1850s, the stone was widely distributed across the United States and Canada. The first stone mill built in the area to saw the stone into desirable shapes opened in 1868. In the late 1860s Amherst-area quarries shipped grindstones as far as Cuba, Europe, Australia, and other foreign ports.

**OTHER QUARRY TOWNS OF
NORTHERN OHIO'S SANDSTONE BELT**

Berlin Heights	Grafton	North Olmsted	Pittsfield
Birmingham	Kipton	North Ridgeville	Sheffield
Columbia Station	Lagrange	Oberlin	West View
Elyria			

Cleveland Stone Company No. 6 Quarry at South Amherst around 1907.

By 1870 at least eight companies employing thousands worked the Amherst sandstone, also called the Amherst Stone. This was only the beginning of a thriving industry, for a new surge of quarrying activity began in the late 1800s and early 1900s as larger companies entered the market. The Cleveland Stone Company bought up many small quarries in the late 1880s. The Ohio Quarries Company, formed in 1903, opened several large quarries and mills in the Amherst area. This company opened the Buckeye Quarry at South Amherst, which at 240 feet deep gained fame as the deepest quarry in the world at the time. This quarry supplied over 500 million cubic feet of building stone. The merger of Ohio Quarries and Cleveland Stone Companies to form Cleveland Quarries in 1929 led to this area being unofficially titled the "sandstone center of the world."

Some fifty quarries once operated in the sandstone belt of northern Ohio, from eastern Huron County to Cuyahoga County. Except for a quarry at Birmingham and Kipton and two at Amherst, they all lie disguised today. Be certain to make a stop at the Amherst Historical Society & Sandstone Museum on South Lake Street in Amherst for local stone quarry lore.

A Side Trip to Cedar Point

Cedar Point, north of U.S. 6 and Ohio 2, has been the destination of fun seekers since the late 1800s. From a bathing beach and dance hall it has grown to a great amusement park. The point is an 8-mile-long finger of sand, or spit, that westward-flowing longshore currents built out from the shoreline east of Sandusky. The sand traveled west along the coast and then began accumulating in Sandusky Bay. A corresponding but much shorter spit, Bay Point near

Marblehead, occurs on the Catawba Island side of the bay. Sheldon Marsh State Nature Preserve, off U.S. 6 west of Huron, provides views of the eastern part of the original Cedar Point spit—where it first started forming. Construction for the original road to Cedar Point, which ran along the spit, began at the present preserve in 1913, and portions of it remain as a nature trail. This portion of the road, however, was susceptible to storm damage and was abandoned in 1920 for Chaussee Road, farther west and closer to Sandusky. The spit was breached west of the preserve during a 1972 storm and no longer connects to the western portion, which contains the amusement park.

Sandusky Bay is the largest of many drowned river mouths along the southwestern shore of Lake Erie. Sandusky Bay formed as lake waters submerged the southwestern shoreline after the ice sheets melted and the Canadian landscape slowly rebounded. Canada, being farther north, was released from the glacier's icy burial later. The removal of the heavy ice altogether allowed the land to spring upward, pushing lake waters farther south. The Sandusky River and others were partially submerged and developed drowned river mouths. The mouths of these streams are exceptionally wide and their water levels are the same as the lake's; basically, they are landward extensions of the lake. Old Woman Creek State Nature Preserve, east of Huron, is a good place to observe a typical drowned river mouth blocked by a baymouth bar—a freshwater estuary. The preserve's nature center has information on the local geology.

Erie County Quarries

By the 1870s Sandusky, north of I-80/90, was known for its high-quality building stone. The Devonian-age Columbus limestone occurs in thick and thin beds in this area, and it is a bluish gray stone when freshly quarried. Builders used the "Sandusky bluestone," as it was called by the stone industry, in many churches, schools, and residences across northern Ohio. The Erie County Courthouse in Sandusky, built in 1872, is a good example. By 1910, the city was the leading producer of limestone building stone in the state. The town's main quarries were about 2 to 3 miles south of town, west of present-day U.S. 250. The Sandusky-area quarries eventually fell under the Wagner Quarries Company's ownership and began to produce aggregate more so than building stone.

The easternmost of several limestone quarries near the Ohio Turnpike and U.S. 20 appears at Parkertown. I-80/90 cuts through a distinct ridge of Devonian Columbus limestone at Parkertown; note the low exposures of gray limestone on either side of the interstate. The Parkertown Quarry on the south side of the turnpike exposes 27 feet of Delaware limestone, 51 feet of fossiliferous Columbus limestone, and more than 30 feet of Detroit River group. The quarries in Devonian limestone extend from Sandusky and Kelleys Island south in a 14- to 20-mile-wide band about 125 miles south to Marion, Delaware, and Columbus.

Other quarries in the Columbus and Delaware limestones operated at Castalia, Sand Hill, Venice, and Wilmer. The first major quarrying in the Castalia area began in the 1870s. The Wagner Quarries Company opened a quarry in the Devonian Columbus limestone at Castalia in the early 1900s. It produced building stone in its early years but later turned to crushed stone. The Great Depression of 1929 led to the closure of the Castalia Quarry, but when construction on the Ohio Turnpike began in 1954 it was reopened to supply aggregate. The quarry was deepened into the underlying Detroit River group dolomites at this time. The quarry closed for the second time in 1965 and lay abandoned until the Erie MetroParks took over in 1987, creating the Castalia Quarry Reserve off Ohio 101 southwest of town.

The uppermost strata of the quarry, the Columbus limestone, is quite fossiliferous. Fossils of brachiopods, bryozoans, clams, snails, horn and colonial corals, and occasional nautiloid cephalopods and trilobites can be seen by the careful observer in loose blocks and slabs of limestone along the trails. The rocks provide clues to the animal and plant life that frequented this seafloor around 375 million years ago. The lower strata of the quarry, the Detroit River group, also contains fossils but in smaller numbers and often heavily crystallized with calcite. Elsewhere in the quarry, the bedrock surface shows scratches or striations left as the Wisconsinan ice sheet spread over this region during the last ice advance. Study of the orientation of these markings allows geologists to better understand glacial movements in northwestern Ohio.

Castalia—Springs, Marl, and Tufa

Castalia lies on the edge of the Columbus escarpment—the western edge of eastward-tilted Devonian Columbus limestone—southwest of Sandusky. From U.S. 20, travel north on Ohio 269 from Bellevue to reach Castalia; from I-80/90 head north on U.S. 250, west on Ohio 2, and south on Ohio 101. The Columbus escarpment marks the western edge of the more resistant Columbus limestone, where it contacts the more easily eroded Salina group strata. It stretches some 125 miles as a continuous low ridge from Kelleys Island to just south of Columbus. Although the escarpment's origin is a matter of conjecture, geologists feel that it developed because of extensive erosion in preglacial time, perhaps related to the

Devonian-age Columbus limestone forms the upper walls of a quarry near Castalia, now part of the Castalia Quarry Reserve.

Erigan and Teays drainage systems. The escarpment must have been much more prominent in preglacial time; in places only thin tills lie on top of it. Today the escarpment is often only noticeable as a dip or rise in an east-west highway, but it can be traced by the eye to the north and south.

Flowing, or artesian, springs that remain ice-free year-round made the Castalia area notable to pioneers as early as 1760. An early report of the Ohio Geological Survey mentioned a 3-foot-high fountain pouring forth from the side of a hill in Castalia. Drainage on the escarpment is mainly underground for 40 square miles; note the lack of surface streams atop the escarpment. This karst area is also riddled with large sinkholes, including one southeast of Castalia on Billings Road that measures 1 mile long and 50 feet deep. The road travels through it. Many are irregularly shaped and have rock walls that enclose smaller, more circular depressions. The Brewery and Crystal Rock Caves, which developed in Silurian Put-in-Bay dolomite of the Salina group, are south of Crystal Rock, northwest of Castalia and below the escarpment.

The construction of a mill pond in Castalia in the early 1800s altered the flow of local groundwater, and a spring burst forth from the ground in 1820. The Blue Hole, as it was called, became synonymous with Castalia in the 1920s when the Castalia Trout Club opened it to the public. The hole is about 45 feet deep, and an average of 11 million gallons of water flows from it daily. Tourists flocked to the popular spring in the 1940s and 1950s; often five thousand visitors passed through the gates on a summer Saturday or Sunday. Declining attendance led to its quiet closure in 1990.

The large duck pond in downtown Castalia is fed by five artesian springs and serves as the source of Cold Creek, which flows north from town, falling some 50 feet to Lake Erie. By the early 1820s gristmills operated along Cold Creek, taking advantage of its year-round flow. About 6 miles west of town, off

The Blue Hole at Castalia in the 1920s.

U.S. 6, is Millers Blue Hole in the Millers Blue Hole Wildlife Area, the smallest of the three major springs in this area.

The cool waters of this karst landscape are charged with lime dissolved from the surrounding limestone. As the water warms, calcium carbonate solidifies in the form of flowstone, stalactites, and stalagmites in caves, and marl and tufa in the wet prairie between Castalia and Sandusky Bay. In the early 1800s this wet prairie, also underlain by rich organic sediment or peat, extended from Sandusky to Port Clinton with 7-foot-tall grasses, but now much of it has been developed and/or drained. Tufa forms when calcium carbonate crystallizes on plant stems or leaves. It often encrusts the edges of springs. Some cottages in the area were made of this locally abundant rock. On the northern edge of Castalia, marl, a calcareous mud laid down in water, underlies Resthaven State Wildlife Area to a depth of 7 feet. Large amounts were dredged to manufacture portland cement from about 1900 into the early 1940s along now-abandoned railroads west of town. Cement Street remains near the old pits as a reminder of the former industry. Castalia Portland Cement Company had a plant along the south shore of Sandusky Bay at the community of Bay Bridge. This area still contains abandoned company houses composed of cement blocks and a large concrete storage elevator.

The Castalia karst area is similar in origin to the nearby karst landscapes at Bellevue, Catawba Island, Flat Rock, and Marblehead. Dyes dropped in sinkholes near Flat Rock travel to the Castalia area, indicating that there is a vast underground network of caves. The sinkholes in this region formed as surface limestones and dolomites collapsed into poorly supported caverns that formed as subsurface gypsum dissolved.

Seneca Caverns and the Bellevue-Clyde Area

Seneca Caverns, another of Ohio's commercial caverns, is about 1 mile southwest of Flat Rock, which is southwest of Bellevue and U.S. 20. Boys discovered a crack in the ground between rock layers while hunting in this area in 1872. From 1897 to 1899 the newly christened Good's Cave offered self-guided tours through the uppermost three levels of the caverns. Caverns develop horizontal passageways, or levels, when the water table stays at a certain depth below the ground surface for a long period of time. If the water table fluctuates up or down, this process of lateral exploitation slows down. By this process caverns become larger and deeper. The isolated location of Good's Cave and its narrow muddy passages led its owners to close it. The cave reopened as Seneca Caverns in 1933 after it was enlarged and deepened to a seventh level, 90 feet below the surface.

The caverns formed along joints in the Devonian-age Columbus limestone and extend into the underlying Detroit River group dolomites. The caverns' origins, however, probably began even deeper in Silurian-age Salina group evaporites. The recrystallization of anhydrite (calcium sulfate) into gypsum (hydrous calcium sulfate) in this group pushed the overlying rocks up, cracking them extensively. Water seeping from the surface eventually dissolved the gypsum, creating the caves. If you look closely at the floor and ceiling at Seneca Caverns, you will notice that at places their surface configuration matches like pieces of a jigsaw puzzle, suggesting they were once adjacent layers.

An underground stream occurs at the seventh level, 90 feet down, except during droughts. The caverns extend at least another 100 feet below this level. Seneca Caverns continues to be a cool way to spend a summer afternoon.

Besides Seneca Caverns, other karst features, like sinkholes, swallow holes, and disappearing streams, have developed in the immediate countryside. Speck Creek, northwest of Bellevue, and Schneiders Creek, south of town, are examples of disappearing streams. A disappearing stream looks much like a normal stream; it lies in a channel cut in the surface. The difference is that at some point in its history, the surface stream entered a sinkhole or other connection to the groundwater system. A disappearing stream flows underground through dissolved tunnels and shafts for a good portion of its course. The surface channel of one of these streams is usually dry except during heavy rains, when the underground channel completely fills with water and runoff is forced to follow the original surface channel. Sometimes the surface portion of a disappearing stream will end in a small pool on the surface that fills a swallow hole, a vertical shaft to the water table.

At least six quarries tapped the Devonian bedrock of Bellevue by the 1870s. Initially, the Columbus limestone was the desired building stone that quarry men removed. It was similar in quality to the stone quarried at Sandusky, which was famous throughout the Midwest. If you drive around town, you will see several vintage businesses and homes that were built of this local, gray to buff stone. As the quarries deepened into underlying dolomites, they turned their

Seneca Caverns near Flat Rock developed along joints in Devonian-age Columbus limestone. Note the small stalactites forming as water drips from the ceiling of this narrow passage.

production to crushed stone and lime. Although the larger Bellevue quarries are now filled with water and surrounded by subdivisions, a quarry at Flat Rock still furnishes building stone and aggregate.

A Side Trip to Catawba Island and Marblehead

A short trip north on U.S. 250 and then northwest on Ohio 2 across Sandusky Bay leads to the Marblehead Peninsula and Catawba Island. Catawba Island, actually a peninsula, projects north from the Marblehead Peninsula east of Port Clinton. The Portage River used to flow just north of Port Clinton and between Catawba Island and the Marblehead Peninsula, entering Lake Erie at present-day Gem Beach. The Lake Erie shoreline, at this time, was perhaps ½ mile north of present-day Port Clinton. Waves gradually eroded away the land between the lake and the Portage River, eventually intersecting the river off the coast of Port Clinton. This became the new and present mouth of the Portage River. The eastern part of the river channel filled with sand just east of Port Clinton, making Catawba Island a tombolo. The downstream portion of the former Portage River became a bay that separates Catawba Island from the Marblehead Peninsula. The west side of Catawba Island exhibits a shoreline cliff of Silurian-age Salina group overlain by the Tymochtee dolomite, which is overlain by the Put-in-Bay dolomite. The formations are not easily distinguished, but geologists recognize the contact along Sand Road and near Catawba Island State Park. Because some gypsum in the Tymochtee dolomite was dissolved after the Put-in-Bay dolomite was laid down on top, the bedding of the Put-in-Bay dolomite is deformed at places.

On the Marblehead Peninsula a slight rise on eastbound Ohio 163 at Lakeside marks the edge of harder Columbus limestone bedrock—the Columbus escarpment. Quarried land is visible from Lakeside to the tip of the peninsula. Hartshorn Road, just west of Lakeside off Ohio 163, leads through an abandoned quarry. An excellent place to observe active quarrying is from St. Joseph Cemetery, located on a promontory that juts into the active quarry off Ohio 163 in Marblehead. Quarrying began here around 1835 with a quarry that grew to 110 acres within five months. Initially, quarriers produced Columbus limestone building stone. Considerable quantities reportedly went to Toledo for Miami & Erie Canal construction. At least six pits operated by the late 1800s before lime manufacturing became important. Forty kilns built by the new owner of the pits, Kelleys Island Lime & Transport Company, produced over 100 barrels of lime a day in 1888. Other uses included stone for breakwaters and fertilizer and flux additives.

The Marblehead quarries always had the advantage of ready access to lake shipping. The Lakeside & Marblehead Railroad, a short line connecting Marblehead with Danbury, provided a connection to the mainline of the Lake Shore & Michigan Southern Railroad, which connected Chicago and Buffalo and passed through Sandusky and Toledo. At the historic Marblehead Lighthouse you can observe the fossils embedded in the Columbus limestone, which forms the rocky shore, also called an *alvar*. Glacial grooves, striations, and polished bedrock—all created as the rock- and sediment-ridden Wisconsinan ice sheet scraped across the landscape—are still found on surfaces not disturbed by quarrying, particularly at the east end of the peninsula. The rocks also contain many fossils, including brachiopods, horn and colonial corals, bryozoans, snails, clams, nautiloid cephalopods, and trilobites. Marblehead's name is derived from the limestone outcrops that form the easternmost part of the peninsula. Marble is a metamorphic rock, metamorphosed limestone or dolomite, but people often used the term when speaking about polishable limestone. By the 1980s Marblehead Quarry, which basically comprises all the early quarries of the area, ranked as the largest construction stone quarry of the Great Lakes region. As of 2004 it was still a major producer of aggregate, tapping strata below the Columbus limestone.

Johnson Island, a separate knob of Columbus limestone that juts out into Sandusky Bay, lies off Bay Shore Road on the southeast shore of Marblehead Peninsula. The small island, used as a prison camp for confederate soldiers, also had a limestone quarry. Large blocks of Columbus limestone and Detroit River group dolomite were quarried to build breakwaters. It's now a marina for a housing development.

Kelleys Island—A Groovy Place
Kelleys Island is the largest of several rocky islands in Ohio's portion of western Lake Erie. It is a more resistant part of the Columbus escarpment that crosses—for the most part underneath—the lake. Columbus limestone layers tilt gently toward the east shore of Kelleys Island, on the east edge of the Findlay arch. It's certainly worth a ferry trip, from Marblehead or Sandusky, to explore this historic island and its geology.

Devonian-age Columbus limestone from Marblehead contains horn corals that weather to a white color. This alvar is at Marblehead Lighthouse.

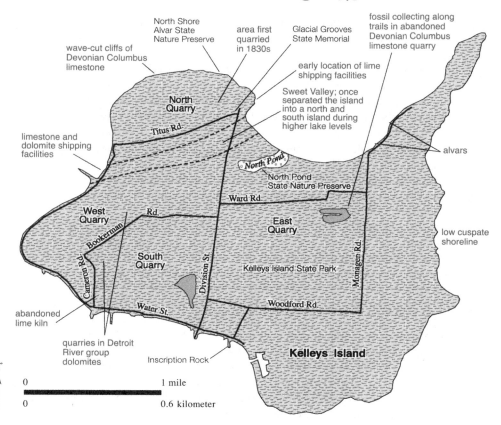

North Shore
Alvar State
Nature Preserve

area first
quarried
in 1830s

Glacial Grooves
State Memorial

fossil collecting along
trails in abandoned
Devonian Columbus
limestone quarry

wave-cut cliffs of
Devonian Columbus
limestone

early location of lime
shipping facilities

North
Quarry

Sweet Valley; once
separated the island
into a north and
south island during
higher lake levels

Titus Rd.

limestone and
dolomite shipping
facilities

North Pond

alvars

North Pond
State Nature Preserve

Ward Rd.

West
Quarry

Rd.

East
Quarry

low cuspate
shoreline

Bookerman

South
Quarry

Division St.

Kelleys Island State Park

Monagen Rd.

Cameron Rd.

abandoned
lime kiln

Water St.

Woodford Rd.

quarries in Detroit
River group
dolomites

Inscription Rock

Kelleys Island

N

0   1 mile

0   0.6 kilometer

Geologic features of Kelleys Island.

Residents of Kelleys Island developed quarries as early as the 1830s, which led to the establishment of the Kelleys Island Lime & Transport Company, the same company that later began operations at Marblehead. The company erected lime kilns at a dock on the north side of the island where stone and lime could be directly transferred to boats. A tremendous volume of this island's stone was used in construction, including the construction of the locks at Sault Ste. Marie, Michigan. Lime manufacturing on the island slowed down by 1909 because of economic reasons, and the Marblehead operations replaced it. Stone for breakwaters was in demand, though, so the Kelleys Island quarries began producing aggregate. A network of trails laces through abandoned quarries south of Ward Road.

The island is a great place to hike and observe Ohio's current wildlife and its fossil inhabitants (check with the state park office for fossil collecting information). If you hike through the quarries, note the steeply tilted rock layers along the quarry walls, the result of quarry blasting. Other abandoned quarries occur near the glacial grooves off Titus Road and on either side of Bookerman Road. At the intersection of Water Street and Cameron Road there is a lime kiln hidden in the trees; other stone ruins of the lime industry become visible when foliage dies back for winter. A quarry deepened into Detroit River group rocks operates north of the abandoned lime kiln. Inscription Rock lies under a shelter in the island's business district. It's a large slab of Columbus limestone that once bore a face of petroglyphs, or carvings. They're virtually unrecognizable today.

North Pond State Nature Preserve, off Ward Road, has an observation tower from which you can get a good view of the island. North Pond, an estuary, has a narrow connection to Lake Erie and fluctuates with lake level. To the west is Sweet Valley, which intersects Division Street between Ward Road and the north shore. During past higher lake levels it divided the island into a north and south island. At low water times, the lake bottom of South Bay, which separates the island from the Marblehead Peninsula, lies partially exposed, adding land mass to the island.

The big attraction on Kelleys Island is Glacial Grooves State Memorial at the north end of Division Street. Deep gouges in Devonian bedrock underlie the soils of the island. Quarrying obviously destroyed many of them, but a small area preserved by the state survived. The exposure at the memorial was greatly enlarged by excavation back into the hillside in the 1970s, where slightly meandering grooves up to 10 feet deep were uncovered. Glacial geologists have several theories concerning the origin of the grooves. Some deduce that the deep grooves were made when ice embedded with rocks and boulders moved through small stream channels that already existed in the bedrock; others suggest they formed as preexisting bedrock ridges were further eroded; others believe they are meltwater channels carved near the margins of an ice sheet; and some believe they are related to the plucking and abrasion of strong flows of sediment-laden glacial meltwater under the ice sheet.

The Devonian-age Columbus limestone at Glacial Grooves State Memorial. A combination of glacial abrasion and meltwater scouring made these grooves and gouges.

Rocky cliffs, 10 to 15 feet high, surround the northern and western margins of the island. The eastern and southern shores are low where the rock layers dip into the lake on the eastern flank of the Findlay Arch. The North Shore Trail in Kelleys Island State Park leads to flat limestone beds along the lake. Here an undercut cliff of Columbus limestone forms the shore; note the tilted blocks of limestone that fell into the lake. This rock beach is an alvar and the home of a unique plant and animal community adapted to the lime-rich soil filling joints in the limestone. North Shore Alvar State Nature Preserve protects a portion of this habitat. East of the nature preserve are more eroded shoreline cliffs of Columbus limestone. The strata at lake level, about 5 to 10 feet below Monagen Road, have thin beds and are more susceptible to wave erosion than the more massive thick beds of limestone that form the top of the shoreline cliff. At places the lake cuts under the overlying rock, causing vertical joints to separate the upper limestone from the mainland; sometimes blocks of it collapse or tilt lakeward.

The Bass Islands

Catawba Island, attached to the mainland, and the Bass Islands in the lake, including South Bass, Middle Bass, North Bass, and several smaller islands are part of an exposed chunk of Silurian-age Put-in-Bay dolomite. These islands mark the edges of an escarpment that is carved in harder Salina group dolomites on the east flank of the Findlay Arch. This escarpment is similar to the Columbus escarpment that exists in Devonian strata farther east at Kelleys Island. The preglacial Erigan River drainage system certainly played a role in the erosion of this escarpment, but its present appearance is related to modifications by the Pleistocene ice sheets and associated meltwater. Analysis of glacial grooves on the lake floor and the islands show that there was a definite east-west trend of ice flow across the island area of western Lake Erie. Each of these islands has cliffs of Put-in-Bay dolomite on the west, and low shorelines on the east. The Bass Islands can be reached by ferry from Port Clinton, on Ohio 2.

Gibraltar Island, which is in Put-in-Bay's harbor along South Bass Island's north shore, has Ohio's most famous sea arch, the Needle's Eye. During the War of 1812, Commodore Perry, a distinguished leader in the U.S. Navy, is said to have posted men at the needle to watch for the British Navy. Needle's Eye can only be observed by boat coming into the harbor of Gibraltar Island. Its opening measures 3 feet wide and 15 feet high; at least half of it is below lake level. Waves eroded this hole through a narrow fin of rock.

South Bass Island is the largest and most noteworthy island of the grouping. Near the ferry dock on the south tip of the island there is an abandoned lime kiln. As with the other islands nearby, lime production was also an early industry here. Karst features dot the island; about fifty caves riddle the island's subsurface. Several of the larger caves once were commercial attractions, and Crystal and Perry's Caves still offer tours. Geologists deduce that the caves formed as dolomite collapsed into cavities that were created when subsurface gypsum dissolved. Once the cavities and fractures opened, groundwater enlarged them, leading to dripstone features like stalactites.

Perry's Cave is the largest cave that has been discovered so far on South Bass Island. Its single chamber measures 208 feet long by 165 feet wide. Soda straws

Silurian-age Bass Islands dolomite exhibits a small upward bend, or anticline, along the south shore of South Bass Island.

dot the ceiling, which is 7 feet at its highest point. A small pool along one of the cave's walls fluctuates with lake level, suggesting that the two are connected.

Crystal Cave, just across the road from Perry's Cave, is certainly atypical. It is lined with large crystals (up to 18 inches long) of bluish white celestite (strontium sulfate) rather than the calcium carbonate dripstone features of a typical cave, like Perry's Cave. This mineral often coats the walls of fractures and cavities in Silurian dolomite of northwestern Ohio, but normally the crystals and cavities are much smaller. The celestite formed when strontium- and sulfate-rich groundwater passed through the Salina group dolomite in preglacial times. Entering Crystal Cave is like walking into a large geode, and it is advertised as the "world's largest geode cave." Reportedly the cave was discovered in 1882, and its owner intended to mine the celestite for use in fireworks manufacturing. In 1897 a new owner deepened the original 3-foot-high cave so people could walk in it and opened it as an attraction at his winery, which it remains today.

Gypsum Mining
The year 1821 marks the first report of "plaster beds" in Ohio, which locals discovered along the north shore of Sandusky Bay. The "plaster" was the mineral gypsum, which came from the Salina group, a deposit of Late Silurian time when a shallow evaporating basin covered the lower Great Lakes area. Gypsum (hydrous calcium sulfate) forms when the mineral salts in seawater become concentrated due to water evaporation during times of warmer and drier climates. Under normal conditions calcite is the dominant mineral to crystallize and settle on the sea bottom; this forms limestone. At other times, as evaporation increases,

dolomite precipitates first, followed by gypsum, and finally halite (sodium chloride). By 1822 a quarry and mill were operating at what was destined to become the town of Gypsum, the only area where gypsum was mined in Ohio.

Initially, people used this soft mineral as fertilizer, and ships of the Portage Plaster Company distributed it throughout the Great Lakes region. In 1860, as the plant grew, it turned its attention to plaster for the construction industry. Other quarries in the Gypsum area opened in the 1880s, and companies began underground mining in 1901. The early mines of Marsh & Company and Granite Wall Plaster Company opened a network of room and pillar mines. From 1912 to 1918 Kelly Plaster Company worked an underground mine south of the U.S. 6 and Ohio 269 junction south of the bay near Crystal Rock. Ohio gypsum production climaxed in 1928 when it produced nearly 563,000 tons. By 1940 two companies controlled the gypsum mines at Gypsum—Celotex Corporation and U.S. Gypsum Company. Celotex terminated its underground production in 1946 and expanded a surface quarry; U.S. Gypsum closed its mine in the late 1970s when it was flooded. Annual gypsum production in the Gypsum area from the 1960s to 2001 averaged around 300,000 tons. In early 2004, the last gypsum producer terminated operations because of flooding problems at the former Celotex quarry. The quarry lies abandoned south of Ohio 2, and water-filled tunnels of four mines underlie the Gypsum area. A few company homes lie along Ohio 2, but the little community of Plasterbed, which once surrounded the U.S. Gypsum plant on the bay, has vanished. A park on Port Clinton's Adams Street preserves a large grindstone that workers used on the north shore of Sandusky Bay to pulverize the gypsum.

Green Springs—Healthy Water

Greenish water fills pools along Flag Creek southeast of Fremont. At least seven springs bubble from the Silurian dolomite near Green Springs. In the late 1800s, 8 million gallons flowed forth daily and attracted the usual health resorts. The year 1868 saw the erection of the first Oak Ridge Hotel, which was followed by other hotels and numerous boarding houses. A modern care facility continues to operate on the old sanatorium grounds. The town advertises these as the largest natural sulfur springs in the world. The springs are most noticeable during winter, when they remain ice free and stink up the north side of town. Not as successful as Green Springs was nearby Sulphur Springs. Here a small community existed in the late 1800s but has since disappeared.

"Beef Hearts" and Celestite—Quarries of the Toledo Area

Due to the relative resistance of its reef-bearing zones to erosion, the wavy upper surface of Silurian-age Greenfield and Lockport dolomites occurs at the surface at numerous areas, giving rise to many small quarries in the early 1800s. Settlers needed stone for foundations, bridge piers, chimneys, and mill dams, and they found it in the Silurian rocks beneath them. They also burned lime for use in mortar and whitewash, and later to make fertilizer. Beginning in the 1850s, stone was in high demand for paving roads through the swampy terrain of northwestern Ohio in order to make them passable in all kinds of weather.

The development of quarries sometimes gave rise to towns like Lime City, and sometimes quarries opened to supply towns that already existed, like Genoa and Woodville. Many towns were named after local geology, notably Gypsum, Lime City, Limestone, Rocky Ridge, and Stony Ridge. In 1870 Genoa topped the list of lime shippers along the Lake Shore & Michigan Southern Railroad, producing over 12,000 tons in that year alone. At least five quarries in town tapped the Silurian bedrock. Gibsonburg's lime industry began in the early 1870s with a couple of small quarries and simple kilns but grew to a number of large lime plants by 1900. Quarries consumed farmland northeast, northwest, and southeast of town. Rocky Ridge had a lime kiln as early as 1870 and several quarries by 1876. Clay Center, Fremont, Luckey, Millersville, New Rochester, Pemberville, Risingsun, and West Millgrove also had quarries.

The quarries in this region are well-known to rockhounds for spectacular mineral specimens and large fossil clam molds. Collectors came from around the world in the mid-1900s to poke around the crystal-lined cavities of Clay Center Quarry for beautiful specimens of blue celestite, barite, calcite, galena, iridescent pyrite, sphalerite, and fluorescent brown fluorite in the Silurian Lockport dolomite. The town began as a small farming community in 1871, at that time founded in the center of Clay Township along the Lake Shore & Michigan Southern Railroad. A small quarry opened to provide stone for road building. The first lime kiln fired up in 1874. By 1891 the Clark and Toledo White Lime Companies operated adjacent quarries and lime plants. These companies became part of the Kelleys Island Lime & Transport Company, which also had plants at Marblehead and on Kelleys Island, in 1905. In 1955 the company sold its Ohio properties and sought stone suitable for flux in northern Michigan. Crushed stone remains the major product of this small community, but the wealth of collectible minerals has diminished.

Woodville's quarries have radiating crystals of white celestite. Embedded in the dolomite there are large fossils that quarry workers called "beef hearts." They are internal molds of large reef-dwelling clams (*Megalomoidea canadensis*) that reached lengths of nearly 1 foot. A number of other fossils occur in Woodville's quarries, including *Favosites*, or honeycomb coral; *Trimerella* and *Pentamerus*, brachiopods; and various gastropods. The fossils are generally internal molds and often heavily dolomitized or covered with crystals. The original reef limestone, deposited in shallow water, was very fossiliferous, but later when dolomite replaced the limestone many fossils were destroyed and distorted. Genoa and Gibsonburg quarries also yielded brown fluorite. Lime City's quarries had clear blue celestite crystals. Each quarry had its own distinct assemblage of minerals that often varied in color, transparency, or crystal shape. Veteran collectors could tell what quarry a particular specimen came from just by examining it. Most of the quarries are now abandoned, filled with water, and inaccessible to collectors. In many the mineral-rich horizons were surpassed years ago and only exist high on quarry walls today.

As of 2004 only a few quarries continued to operate, including those at Clay Center, Genoa, Lime City, Millersville, and Woodville. Most of them lie abandoned, though a few continue to serve communities in other ways. Sometime around 1930 the quarry along U.S. 20 in Stony Ridge became a swimming quarry; another quarry just northeast of town followed suit. In 1944 Genoa purchased an abandoned quarry from the Kelleys Island Lime & Transport

Internal molds of large reef-dwelling clams (Megalomoidea) *are common in many quarries tapping Silurian-age dolomites.*

Celestite, calcite, and fluorite are commonly found in cavities of Silurian-age dolomite, such as this specimen from Clay Center.

Company and converted it into a park. In 1975 the Sandusky County Park District purchased Gibsonburg quarry land, which eventually became White Star Park. The old Doherty Quarry on the south side of Luckey also became a swimming hole.

Brick and tile manufacturing was another major use of local mineral resources, namely glacial, lake, and floodplain clays. By the 1880s brick plants operated in Bellevue, Clyde, Elmore, Fremont, Green Springs, and Lindsey; tile plants operated at Bates Station (Rossford) and Perrysburg.

Prospectors drilled for oil and gas in the Silurian Clinton horizon (Brassfield formation) and the Ordovician Trenton strata some 1,200 to 1,400 feet below the surface. People drilled thousands of prospect wells between 1885 and 1900 in this area. Initially, production ranged from 25 to 100 barrels daily per well. The more-successful wells were centered around Bradner, Curtice, Gibsonburg, Martin, Pemberville, Williston, and Woodville. Pemberville wells produced 8,000 barrels per day in 1892, and people continued drilling into the 1920s. The Kirkbride well west of Gibsonburg, drilled in 1894, was a classic gusher. It sprayed oil across a ½-mile radius, and tremendous amounts of it were wasted. Estimates of initial production ranged from 10,000 to 40,000 barrels per day. A blast of nitroglycerin released oil from the Klondike well in present-day Oregon in 1897. This was another big one, but it is difficult to estimate its initial production because of spotty records. The Toledo area became a refinery center by the 1890s, and refineries shipped petroleum products widely by ship and rail. A number of towns had short-lived supplies of natural gas from local wells, including Burgoon and Oak Harbor. Some of the wells still pump, but their yields are low.

THE OHIO VALLEY—
A PRODUCT OF THE ICE AGE

The Ohio River, stretching 451 miles from East Liverpool to the Indiana line west of Cincinnati, forms the southern boundary of Ohio. Where the river incises the Allegheny Plateau between Rome and East Liverpool, it is noted for its many higher valleys and terrace remnants related to preglacial drainage, including the Teays system. Commonly these high-level valleys are 200 feet or more above the level of the present-day Ohio River. From Ripley west to the Indiana line, the valley contains till and outwash of pre-Illinoian and Illinoian age where ice lobes advanced as far south as northern Kentucky. All of the types of bedrock you encounter in this chapter are discussed in more detail in preceding chapters, including their origins and composition.

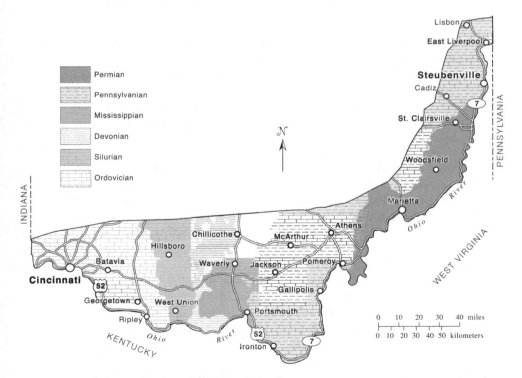

Highways and geology of the Ohio River valley. —Modified from Ohio Department of Natural Resources, Division of Geological Survey publication

History of the Ohio River

Southwest of Ohio, the present-day Ohio River flows through the valley the preglacial Ohio River originally developed, which had a source in southwestern Ohio or southeastern Indiana; geologists aren't certain. Most of the Ohio section of the river valley began developing in a landscape that the north-flowing preglacial Teays River system carved. Interpretations of this buried drainage network vary. Some geologists feel that the Teays system drained most of central, south-central, and southwestern Ohio. Others suggest that southwestern Ohio was drained by the southwestward-flowing preglacial Hamilton River that flowed from Hamilton southwest into southeastern Indiana. Geologists debate whether it was the Hamilton or the Teays that drained the area between Hamilton and Dayton. In eastern Ohio, between East Liverpool and New Martinsville, West Virginia, the preglacial drainage was northward through the preglacial Pittsburgh River, which connected with the Erigan River where Lake Erie now lies.

When the Ice Age began, perhaps as early as 2 million years ago, preglacial drainage was permanently changed. Geologists cannot be certain of when the first ice moved across Ohio or how far south it reached because each advance stripped away evidence of its predecessor. Pre-Illinoian till on the west side of Cincinnati and in the neighboring hills of Kentucky tells us that ice sheets reached southwestern Ohio and Cincinnati at least as early as 240,000 years ago.

When the pre-Illinoian ice sheet advanced to the Cincinnati area, the preglacial Manchester River was flowing from near present-day Manchester into Indiana. It followed a broad loop north of Cincinnati to Hamilton, where it joined the preglacial Hamilton River, which then joined the preglacial Ohio River in Indiana. From Manchester to California, just southeast of Cincinnati, the Manchester River set the course for the future Ohio River. At that time the valley was much shallower, and all of its tributaries were much shorter than they are now. Tremendous volumes of meltwater flowed down this ancestral portion of the Ohio River as the pre-Illinoian ice melted. The north-flowing preglacial Teays River system was blocked, so impounded water flowed south, seeking new and lower outlets through the Allegheny Plateau. From 240,000 to 230,000 years ago the Ohio Valley region underwent the so-called Deep Stage of erosion, when drainage carved deeply into the valley-bottom sediments and then the underlying bedrock. Wide, deep valleys were the result, and the topographical relief of these was greatest during this time of ice-free conditions.

Beginning about 230,000 years ago Illinoian ice moved into southwestern Ohio. This ice sheet found the landscape to be quite different from the pre-Illinoian glaciers, mainly due to the deep downcutting and weathering that had taken place during the Deep Stage. The ice sheet spread into the Cincinnati region from the northeast and sent a thin finger of ice far down the present-day Great Miami River to the Ohio River and beyond. By the end of Illinoian time, about 125,000 years ago, the northward loop of the old Manchester River channel from Cincinnati to Hamilton had been abandoned because it was clogged with sediment. The Manchester River carved a new channel through the neck of the meander at the Ohio River's present location in Cincinnati. During Illinoian time the deeply carved valleys of the Deep Stage were filled with outwash, greatly reducing topographical relief. Ice would not return to the Ohio Valley, and the present course of the Ohio River was then established. Between 125,000 and 117,000 years ago Ohio was ice free, and the Ohio River began to take on its present appearance.

Wisconsinan ice advanced as far as the northern suburbs of Cincinnati, and at one point a tongue of ice extended down the Mill Creek valley to within a couple miles of the Ohio River. The fluctuations of the Wisconsinan ice sheet led to the formation of several terraces cut into outwash sediments within the Ohio Valley. The terrace at the highest elevation (about 50 feet above present river level at Cincinnati) correlates with this farthest advance of the ice sheet to just north of Cincinnati around 23,000 years ago. Periodic meltings of the Wisconsinan ice sent tremendous volumes of meltwater and outwash down the Great Miami River valley; note the large sand pits at Elizabethtown. Buildup of outwash in the Ohio River led to damming of smaller tributaries and the formation of lakes in their headwaters.

Wisconsinan ice retreated from southwestern Ohio and returned only to the Union City moraine far to the north of the Ohio Valley at Greenville. More outwash was carried to the Ohio Valley, mainly down the Great Miami and Scioto Rivers, and with later erosion formed a lower terrace about 10 feet above present river level. Although there were many fluctuations of the ice sheet during the last part of Wisconsinan time, they had little effect on the Ohio River because of the great distance between them. Beginning about 13,000 years ago, the Ohio River began cutting away the outwash deposits and developing its modern floodplain. The river formed and is forming its floodplain by lateral deposition of sediments and by the periodic settling of sediments from floodwaters that submerge the floodplain.

As the river flows along, its erosional and depositional behavior adjusts in response to the bottom topography or shape, amount and particle size of its sediment load, strength of currents, narrowness of the channel, and friction with different rock strata in the channel and valley walls. The river sometimes follows cracks or joints in the rock, forming relatively straight courses, and at other places, where the rock is uniformly hard, the river forms broad, sweeping meanders.

Typically, the fastest current is in the center of a relatively straight channel, with sluggish water near the banks. However, when the river encounters some resistance to flow and is forced to meander, the strong currents swing to the outside of the meander and erode the outside bank. The bank here often drops off into the river and the water might be quite deep. The opposite side of the meander, the inside, accumulates sediment in sandbars and gravel bars because the water is barely moving. Over hundreds of years, this constant erosion causes the meander to move in the direction of erosion, becoming a tighter U shape.

The bar of sand and gravel that fills the middle of the meander is called a *point bar*. The point bar may have arcuate depressions either filled with water or holding wetlands that mark former edges of a point bar. Point bars along the Ohio River in southwestern Ohio are often quite extensive and are usually referred to as *bottomland*; they are less prominent in southeastern Ohio because of the more erosion-resistant bedrock of the Allegheny plateau. (Good examples are across the river from North Bend and Sciotoville and between Aberdeen and Manchester.) Waterfalls and rapids formed where the river crossed harder, erosion-resistant layers of rock, like sandstone, as it carved through the Allegheny Plateau. The waterfalls and rapids are gone now due to the construction of a series of dams, which have raised the water level, but town names like Letart Falls remind one of the early conditions.

Through a combination of meandering and flooding the river developed its present wide, flat-bottomed valley. The area just above normal water level forms the floodplain. Although the flatness of this terrain makes it ideal for town sites and farming, the very nature of its formation makes it susceptible to complete submergence during floods.

The Ohio River has always been an important artery for travel and commerce. Native Americans had villages on bluffs and high terraces along the river, above the reach of floodwaters. Meriwether Lewis traveled along the upper portion of the Ohio River in 1803 on his way to meet William Clark in Louisville, where they launched their famous journey west. On the way, he stopped at Big Bone Lick, in the Kentucky hills southwest of Cincinnati, to collect fossil bones for President Thomas Jefferson. Boatloads of scientists and educators made their way to Cincinnati, Louisville, Marietta, New Harmony, and other centers of learning west of the Alleghenies in the early 1800s.

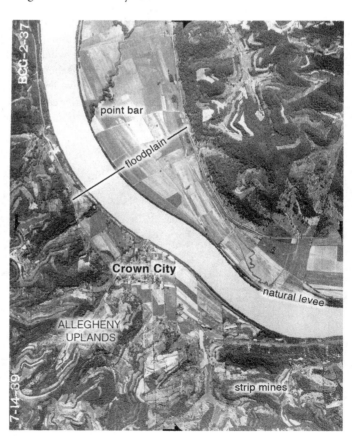

Features of the Ohio River valley at Crown City.

Travel on the Ohio River was slow because of its shallowness and the accumulation of debris, so in the 1820s government crews began clearing snags. The first recorded dam on the river was built in 1838 to divert water around one side of Browns Island near Steubenville. The U.S. Army Corps of Engineers began recording water levels along the Ohio River in 1869. Dredging of sandbars began in the late 1870s. Between 1885 and 1929 the U.S. Army Corps of Engineers built a series of dams and locks along the Ohio, Allegheny, and Monongahela Rivers. These were replaced with new structures between 1938 and 1976. Today nine locks and dams control navigation along Ohio's portion of the river. Flood control on the Ohio River is accomplished by controlling discharge from its major tributaries with dams and reservoirs, like those of the Muskingum Watershed Conservancy District and the Miami Conservancy District.

Notable Floods

Rivers and streams develop the size and shape of their valleys over thousands of years. All the water that moves through the system, from the headwaters of tiny tributaries to the main river, impacts the valley. Since water flows rapidly in hilly areas, such as those in the uplands of southwestern Ohio and throughout the Allegheny Plateau of southeastern Ohio, the stream channels are straight, narrow, deep, and generally V-shaped. There are many small streams in these areas, and water from precipitation is efficiently and quickly moved to lower elevations. These streams have developed relatively quickly in geologic terms. Resistant beds of rock crossing their channels cause waterfalls, which slowly move upstream, sometimes disappearing when the hard rock is surpassed. Over time the streams mainly cut deeper gashes into the hillsides.

At lower elevations water courses have less slope, so water moves more slowly. The discharge from several tributaries can add up to make a much larger volume of water that these streams have to handle; thus they become wider at

Submerged New Richmond during the 1907 flood; waters reached 65.2 feet during this flood and submerged Ohio Valley communities.

NOTABLE FLOODS OF THE UPPER OHIO RIVER

Year	Events
1832	Winter storms in January dumped snow on the upper Ohio Valley. Four days of rain in February caused it to rapidly melt. East Liverpool, Wellsville, and other eastern Ohio communities were inundated. As the flood moved down the Ohio Valley, the Cincinnati area experienced a 15-foot rise in water level. Few riverside frame structures survived.
1862	In January, floodwaters rose 4 feet above Spring Grove Avenue in Cincinnati.
1865	A spring flood devastated parts of East Liverpool, Wellsville, and other upper valley towns; riverside industry, including all the potteries, came to a standstill.
1882, 1883, 1884	After river level dropped to an all-time low of 1.7 feet at Cincinnati in September 1881, serious floods returned in 1882, 1883, and 1884. In February 1884 the river crested at 71.1 feet at Cincinnati, the second highest level ever (a significant flood in Cincinnati is one that rises above a 52-foot depth on a flood gauge). It was also a record flood in the upper valley; high watermarks survived for many years on the older buildings from Steubenville to East Liverpool.
1913	Two notable floods—one in January and the other in March and April—wreaked havoc across Ohio and neighboring states. Four days of torrential rain (some 9 to 11 inches) between March 23 and 27 fell on ground already saturated with spring melt that had been weakened by the earlier January flood. The first flood crested at 61.2 feet at Cincinnati and the March flood reached 61.8 feet. Regional flood losses totaled over $10 billion and 527 people died.
1918	Floodwaters carried large chunks of ice that damaged many steamboats docked at Cincinnati. The steamboat industry never recovered.
1937	Water levels of the Ohio River reached 79.9 feet at Cincinnati, which still reigns as the highest level ever recorded. One-fifth of the city was submerged. Upstream at Portsmouth the flood set a record flood level for that town at 74.23 feet. The flooding continued for nineteen days, causing $500 million in property damage and 385 deaths from Pittsburgh, Pennsylvania, to Cairo, Illinois. This flood led the U.S. Army Corps of Engineers to build a dam across the mouth of Mill Creek in downtown Cincinnati to prevent it from backing up floodwaters into the Ohio Valley.
1945	The dam on Mill Creek failed in a flood, so crews constructed floodwalls and levees but did not rebuild the dam.
1948, 1955, 1964, 1996, 1997, 2000	Flood levels in Cincinnati approached or exceeded 60 feet.

the expense of deepening, and the channels become more U-shaped. Because of periodic rises when water level fills and spills out of the channels, these streams form floodplains to hold the excess water.

Water may cover the Ohio River floodplain in spring as tributaries discharge meltwater from the past winter's snowfall. Ohioans expect annual spring floods, which rejuvenate the valley bottoms. Floodplain soils are among the most fertile soils because of this rejuvenation. Water levels also rise when weather systems stall over the drainage basin, mainly during spring and fall. Now and then flood levels exceed the average river level during annual flooding and last for days or weeks at a time. It's these major floods that locals remember. Because early Ohioans insisted on settling on floodplains, major floods can be disastrous today, destroying life and property. Floods are gauged by how far water rises above a typical level, which is an average water level of the stream or river over several years of record. This level differs according to local lay of the land, stream geometry, and the number of streams draining a given area.

Early records of floods are scanty; many came from Native Americans who pointed out watermarks high on floodplain trees to settlers. One of the earliest written records of a flood recounted one that occurred at the mouth of the Licking River in 1773, across from the future site of Cincinnati. Just two months after its settlement in November 1789, the small village of Columbia, situated at the junction of the Little Miami and Ohio Rivers, was under water. This early flooding led settlers to site Cincinnati across the Ohio River, opposite the mouth of the Licking River, where terraces provided higher ground than that along the Little Miami River. Floods move downstream as a wave, slowly rising and cresting before lowering to normal water depth, and downstream locations are often the most severely affected by floodwaters.

U.S. 52
Sybene—Indiana Line
179 miles

U.S. 52 extends across 117 miles of unglaciated Pennsylvanian- to Ordovician-age rock between Sybene and Ripley and traverses hilly Illinoian terrain and Ordovician bedrock between there and the Indiana line. U.S. 52 now bypasses many towns, but you can follow the old routes that parallel the river more closely. The highway crosses the necks of many point bars where the river meanders toward Kentucky, and it parallels rocky cliffs where the channel cuts into Ohio.

Chesapeake to Portsmouth

Coal mining has always been common around Coal Grove; slope mines in Pennsylvanian-age Pottsville and Allegheny group seams dot the edge of the surrounding hills. Abandoned clay pits lie above the highway just east of town, and an old brick and tile plant once sat next to the river in Coal Grove. Downstream is Ironton, the gateway to the Hanging Rock Iron Region. On the east edge of town look to the north for the remains of the Ironton Portland Cement plant. U.S. 52

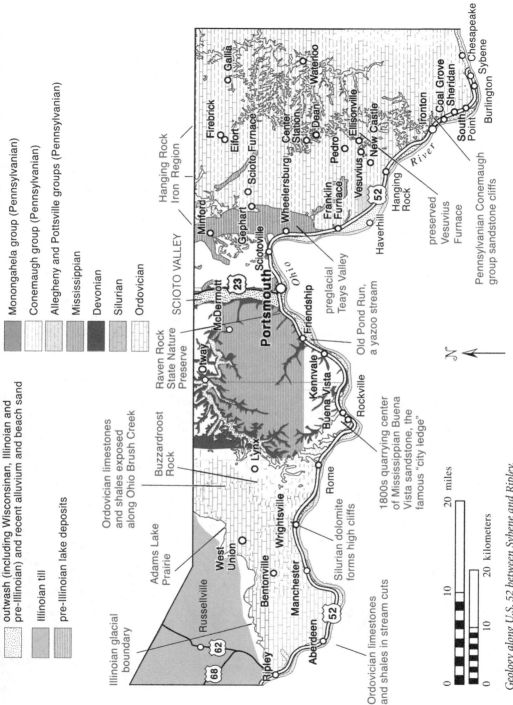

Legend:

- Monongahela group (Pennsylvanian)
- Conemaugh group (Pennsylvanian)
- Allegheny and Pottsville groups (Pennsylvanian)
- Mississippian
- Devonian
- Silurian
- Ordovician

- outwash (including Wisconsinan, Illinoian and pre-Illinoian) and recent alluvium and beach sand
- Illinoian till
- pre-Illinoian lake deposits

Illinoian glacial boundary

Ordovician limestones and shales exposed along Ohio Brush Creek

Adams Lake Prairie

Raven Rock State Nature Preserve

Buzzardroost Rock

SCIOTO VALLEY

Hanging Rock Iron Region

preserved Vesuvius Furnace

Pennsylvanian Conemaugh group sandstone cliffs

preglacial Teays Valley

Old Pond Run, a yazoo stream

1800s quarrying center of Mississippian Buena Vista sandstone, the famous "city ledge"

Silurian dolomite forms high cliffs

Ordovician limestones and shales in stream cuts

Towns and locations: Ripley, Aberdeen, West Union, Russellville, Bentonville, Manchester, Wrightsville, Rome, Rockville, Buena Vista, Kennvale, Lynx, Otway, McDermott, Portsmouth, Friendship, Sciotoville, Minford, Gephart, Wheelersburg, Haverhill, Franklin Furnace, Pedro, Vesuvius, New Castle, Ellisonville, Center Station, Dean, Waterloo, Scioto Furnace, Eifort, Firebrick, Gallia, Hanging Rock, Ironton, Coal Grove, Sheridan, South Point, Burlington, Chesapeake, Sybene

Ohio River

Routes: 68, 62, 52, 23

Geology along U.S. 52 between Sybene and Ripley.

Scale: 0 — 10 — 20 miles; 0 — 10 — 20 kilometers

hugs cliffs of Pennsylvanian sandstone as it skirts the city. Above Hanging Rock, slump blocks of sandstone still make the town's name appropriate.

Pine Creek follows a twisting path as it enters the Ohio River near Wheelersburg. Several meander scars and oxbow lakes are evidence of changes in its channel. Wheelersburg sits in the valley of the preglacial Teays River, which drained Ohio before the Ice Age. Note the large gap in the plateau heading to the north as you pass through town, evidence that this was a major drainage system in the past. The valley is filled with Minford clay, which settled out of a lake that formed when the Teays River was blocked by pre-Illinoian ice to the north. The bedrock changes to Mississippian-age Logan formation sandstones and shales near Haverhill and Cuyahoga formation near Sciotoville, although both formations are buried by alluvium along U.S. 52. Bricks and building stone made from these rocks paved the way for early industrial growth in Portsmouth, at the mouth of the Scioto River.

Charcoal-Fired Furnaces in the Hills

Stretching 75 miles north of Ironton and south into Kentucky is the Hanging Rock Iron Region, the first major region of iron manufacturing west of the Allegheny Mountains. The iron ore lay on top of the Pennsylvanian-age Vanport limestone and thus became known as the Limestone ore, but it was also called the Ferriferous, Baird, and Franklin ore. Both the limestone and ore vary in thickness, as is typical of Pennsylvanian deposition because of the fluctuating courses of streams flowing away from the Appalachians. Streams of this time moved across the alluvial plain, carving new channels as their previous

Typical Pennsylvanian-age sandstone, coal, and shale at Coal Grove. Due to recurrent rock falls and slides, the cliff has been cut back from the road. Pottsville group strata are near the road level and Conemaugh group rocks lie above.

channels were filled in. This movement caused the sandstone, shale, coal, limestone, and clay beds across the alluvial plain to thin and thicken in places. It is often difficult to distinguish different Pennsylvanian sandstones because a layer that normally separates them may have been removed or thinned.

The limestone forms prominent cliffs on the sides of hills in this region. The iron ore averages 10 to 12 inches thick, while the Vanport limestone averages 6 feet. The ore, usually only 25 to 40 percent iron, is a mixture of siderite, an iron carbonate, and limonite, an iron oxide. The siderite was difficult to melt, so the limonite became the desired ore. Workers stripped back the overlying rock and sediment layers to get at the ore, often cutting benches back into the hillsides. The limestone was used as flux in the smelting process. Smelters also used a few thinner layers of kidney ore, a nodular iron deposit that occurred above the Limestone ore. In areas where the ore thickened, miners developed drift mines.

Geologists speculate that the iron accumulated in wetlands of Pennsylvanian time—the same wetlands where coal was accumulating. Rivers and streams carried the iron, derived from the weathering of the Appalachians, to present-day eastern Ohio. It accumulated as iron carbonate, or siderite, in swamps and other wetlands. In places it formed layers that averaged 1 foot thick. Where the carbonate ore was later exposed by erosion it oxidized, turning into hydrous iron oxide, or limonite. Carbonate ores were hard to smelt in the early furnaces, so mining occurred where limonite was present.

The necessary requirements of successful iron manufacturing in the early 1800s occurred in this southern Ohio region—sufficient ore, limestone for flux, timber for making charcoal, sandstone to build furnaces, and water to provide power. The first Ohio charcoal-fired furnace in the region, Union Furnace, was erected near Kellys Mills (later Union Furnace) in Hocking County in 1826. It produced around 3 tons per day, somewhat below the average of the time of 8 to 12 tons. Workers cast the molten iron into rectangular masses called *pigs*, which were sold to manufacturers. Occasionally the furnaces cast pots, skillets, and other implements. The first furnace in Scioto County was Franklin Furnace, located in the town now known by this name. It operated from 1827 to 1860. During its heyday it produced an average of 10 tons of pig iron daily. Miners pulled ore directly from the hills just above the Ohio River floodplain. In 1828 the nearby Franklin Junior Furnace fired up, operating until 1876. The Pine Grove, Little Etna, Hecla, Mt. Vernon, Buckhorn, Lawrence, and Center Furnaces were in operation by 1836. In 1837 Vesuvius Furnace became the first hot blast furnace in the nation.

The earliest furnaces were cold blast furnaces, in which outside air was blown up through the mixture of iron ore, limestone flux, and fuel with some form of bellows. The hot blast technique involved heating air before injecting it into the furnace. Hot blast iron was more malleable; cold blast iron was more durable. In the Hanging Rock Iron Region, hot blast furnaces typically produced about 3,000 tons of iron annually and used less fuel than cold blast furnaces, which produced about 2,000 tons of iron per year.

By the 1870s forty-six charcoal-fired furnaces dotted the hills. Company towns of a few hundred people rose up around the furnaces. England favored Hanging Rock pig iron. It played an important part in the Crimean War of the

Vesuvius Furnace was typical of the many charcoal-fired iron furnaces of the Hanging Rock Iron Region.

mid-1850s. Furnaces in the Hanging Rock region also cast cannon and armor plating for the *Monitor* and other Civil War gunboats.

Just as demand for iron was increasing, other signs pointed to the imminent death of charcoal-fired furnaces. The once-verdant forests had been rapidly depleted, driving up the cost of charcoal. Higher-grade Lake Superior iron became widely available as more and more railroads were built across the country. Coke- and coal-fired furnaces, which were being built in the area and elsewhere, far surpassed the efficiency of the old furnaces. By 1900 most of the region's furnaces were out of blast. The last charcoal-fired iron in the Hanging Rock region flowed from the Jefferson Furnace in Jackson County in 1916.

Because of the sturdy construction of stone furnace stacks and their isolated locations, nineteen stacks in the region survive today. Nature, however, rapidly reclaimed wood-frame casting houses, engine houses, forges, and other associated structures. A visit to the Ohio Historical Society's reconstructed Buckeye Furnace near Wellston is a must if you would like to witness the operation of a typical charcoal furnace. The furnace industry also lives on in names of communities like Franklin Furnace and in road names such as Haverhill Ohio Furnace Road, between Ironton and Wheelersburg.

Other Mineral Industries of the Hanging Rock Iron Region
When the furnaces were in blast, miners paid little attention to the relatively thin coals of the Hanging Rock region—iron was the most precious commodity. Furnace companies usually obtained all the coal they needed somewhere on their properties, so large coal mines generally didn't exist. Sheridan, off

U.S. 52, was a short-lived mining center in the 1860s and 1870s. Miners removed the easily accessible part of a deeply buried 3- to 4-foot seam of Middle Kittanning coal. Scattered small mines in the Waterloo area mined Upper Freeport coal during the late 1800s.

CHARCOAL-FIRED IRON FURNACES OF THE HANGING ROCK IRON REGION

County/Location	Furnace Name	Date Built	Current Status
Gallia/Gallia	Gallia	1847	Little evidence
Hocking/Haydenville	Hocking	1856	Gone
/Logan	Logan	1853	Gone
/Union Furnace	Union	1854	Gone
Jackson/Milton Township (southeast of Wellston)	Buckeye	1851	Restored stack; replica of furnace complex
/remote location	Cambria	1854	Partial stack
/Jackson	Cornelia	1853	Gone
/Jackson	Diamond (formerly Salt Lick)	1856	Gone
/south of Mabee Corner	Jackson	1836 or 1838	Gone
/west of Oak Hill	Jefferson	1854	Partial stack
/Keystone Furnace	Keystone	1848	Partial stack
/Berlin Crossroads	Latrobe	1854	Gone
/north of Clay	Limestone	1855	Partial stack
/Rempel	Madison	1854	Partial stack
/near Blackfork Junction	Monroe	1856	Partial stack
/near Petrea	Young America	?	Gone
/Buckhorn	Buckhorn	1833	Partial stack
/Center	Center (Centre)	1836	Gone
/Etna	Etna	1832	Partial stack
/Ironton	Grant	1869	Gone
/Hecla	Hecla	1833	Gone
/LaGrange	La Grange	1836	Partial stack
/south of Bartles	Lawrence	1834	Gone
/Coal Grove	Monitor	1868	Gone
/east of Buckhorn	Mount Vernon (Vernon)	1833	Little evidence
/west of Wilgus	Oak Ridge	1856	Stack in good shape
/Olive Furnace	Olive	1846	Partial stack
/Pine Grove	Pine Grove	1828	Little evidence
/north of Olive Furnace	Pioneer	1857	Partial stack
/west of Pine Grove	Union	1826	Gone
/Lake Vesuvius Recreation Area	Vesuvius	1833	Preserved stack
/Blackfork	Washington	1853	Partial stack
Scioto /Bloom	Bloom	1832	Gone
/south of Bondclay	Clinton	1832	Gone
/east of Powellsville	Empire	1846	Gone
/Franklin Furnace	Franklin	1827	Gone
/Harrison Furnace	Harrison	1853	Little evidence
/east of Lyra	Howard	1853	Gone
/Junior Furnace	Junior	1832	Gone
/Ohio Furnace	Ohio	?	Little evidence
/Scioto Furnace	Scioto	1828	Gone
Vinton/Oreton	Eagle	1852	Gone
/southwest of Hamden	Hamden	1854	Gone
/Lake Hope State Park	Hope	1854	Preserved stack
/Richland Furnace State Forest	Richland (formerly Cincinnati)	1854	Partial stack
/Vinton	Vinton	1854	Partial stack; Belgian coke ovens
/Zaleski	Zaleski	1858	Gone

Ohio 650 north of Hanging Rock winds through a Lower Kittanning coal-field of the Hanging Rock region, what geologists called the New Castle Field. Beginning in the 1860s companies pulled the coal from underground mines centered around New Castle and Vesuvius. These mines supplied all the heating and manufacturing needs of Hanging Rock and Ironton. The coal averaged about 3½ feet thick, and miners mined over 400 tons daily. Today, east and west of the highway, strip-mining operations have scraped off the tops of the hills. New Castle and Vesuvius are ghosts of once-bustling towns, and the old Iron Railroad (later the Detroit, Toledo & Ironton Railroad), which served the mines, lies abandoned.

Although originally Hanging Rock was supposed to be the terminal of the Iron Railroad, the floodplain was too narrow and the community was subject to flooding. In 1849, a better site was platted 3 miles east, and with the opening

SOME CLAY AND COAL INDUSTRIES OF THE IRONTON-PORTSMOUTH AREA

Town	Industry
Center Station, Culbertson, Dean, Eifort, Ellisonville, Firebrick, Lawrence Furnace (south of Bartles), Pedro, and Pioneer Furnace (north of Olive Furnace)	There were strip and drift mines in Upper Mercer, Clarion, and Lower Kittanning coals of the Pottsville and Allegheny groups in these towns in the early 1920s.
Ironton	At least five brick plants used Pennsylvanian Anthony, Portsmouth, and Sciotoville clays of the Pottsville group. Thousands of pavers came from the kilns daily and were shipped to locations in the eastern United States. Plants closed and merged until by 1940 only two remained open.
Oak Hill, Portsmouth, Scioto Furnace, and Sciotoville	These towns had firebrick and/or refractory plants that used Sciotoville clay in the late nineteenth and early twentieth centuries.
Pine Grove	Mining of Pennsylvanian Pottsville group Sciotoville clay began here in 1864. This plant became the Scioto Fire Brick Company in 1871, at the time one of the largest firebrick plants in the United States. Workers dug the clay from pits north of town and at Gephart.
Sciotoville	Pennsylvanian Allegheny group Lawrence clay was used to make terra-cotta around 1859. Manufacturers also used it to make yellowware and stoneware, as a bonding agent for molding sands, and in refractories.

Blast furnaces like this one at Hanging Rock replaced the earlier charcoal-fired furnaces.

of the Iron Railroad, which was chartered to haul iron from the furnaces in the hills of the plateau to the Ohio River, Ironton became an iron-shipping port of the 1860s.

Ironton Portland Cement Company began operations in 1902 near Coal Grove. It made cement from the Pennsylvanian-age Vanport limestone, which the company mined from the surrounding hillsides. It used the Lower Kittanning coal above the Vanport limestone to fire its kilns for the first few years and clay above and below the limestone as an additive for the cement mixture. About 1905 the company switched from coal to natural gas. In 1914, the company opened a shaft mine to the Mississippian Maxville limestone about 510 feet below Coal Grove. This limestone, which was between 70 and 100 feet thick at Coal Grove, replaced the more or less depleted Vanport limestone as the primary ingredient in its cement product. To make portland cement, the limestone is mixed with clay or shale and ground to a fine powder. The mixture then is heated in rotating kilns to a temperature of 2,400 degrees Fahrenheit and then ground again. The company supplied cement for federal dams and locks on the Big Sandy, Ohio, and Kentucky Rivers; Norfolk & Western Railway bridges and tunnels; and various buildings throughout the region.

Portsmouth to Manchester

In the late 1700s the mouth of the Scioto River was about 1 mile farther west than it is today; it has meandered to its present position. In 1796 several settlers began developing a town on the west side of the river, which became known as Alexandria. In a short time it became the county seat. Portsmouth began developing around 1803 on high ground on the east side of the river, across from Alexandria. Initially, the new plat attracted few people, but after the flood of 1805 inundated Alexandria, Portsmouth began to grow. Continued flooding problems caused settlers to abandon Alexandria. The construction of several charcoal-fired furnaces in the area and the opening of the Ohio & Erie Canal in 1832 thoroughly established Portsmouth as a major commercial center. Floodwaters continued to plague the area, inundating Portsmouth as the city spread down from the higher ground. The 1884, 1913, 1937, and 1945 floods hit the town hard, washing away hundreds of frame homes and businesses and flooding the basements of thousands of structures. Between 1946 and 1949 workers constructed a 7½-mile-long floodwall and levee adjacent to downtown Portsmouth to protect the city from water that could rise as high as 77 feet; the normal flood level is about 50 feet.

The Scioto River once had three meander loops just north and west of the U.S. 23 bridge over the Ohio River, a typical pattern of most of the Ohio River's tributaries. Builders had to cut a short channel across the neck of the most southerly meander when constructing the Ohio & Erie Canal; the "neck" is the part of a meander a river eventually cuts through when it creates an oxbow lake. The loops just west of town were also cut off later, drastically altering the mouth of the Scioto. Only meander scars mark the early channels. The river now flows much farther east than it did during the early days of Portsmouth. Slab Run follows the lower course of the most westerly channel. Look for it on the south side of U.S. 52 after crossing the present channel of the Scioto River. The southernmost extant lock on the canal is at West Portsmouth along Ohio 73. Builders used Mississippian-age Buena Vista sandstone to make the lock.

Just west of Careys Run Pond Creek Road, between Portsmouth and Friendship, is Raven Rock State Nature Preserve. In this preserve there is a ridge of Mississippian-age Cuyahoga formation that forms a 500-foot-high, prominent lookout. The preserve, accessible by special permit from the Ohio Department of Natural Resources Division of Natural Areas and Preserves, offers a great view of the mouth of the Scioto River and the Ohio River valley. On a clear day you can see 14 miles of the meandering Ohio River. There are many legends that involve the landform, from a lifesaving leap by Daniel Boone to escape Shawnee captors to burial caves of the Ancient Ones of Native American lore hidden beneath rockslide debris. One delicate natural arch of sandstone graces the ridgetop. It is 15 feet long and only 14 inches thick at its narrowest point. Two other arches occur on the cliff face below. These three small arches formed from the differential weathering of the Cuyahoga formation sandstones, probably where sand grains were weakly cemented together.

About 9 miles southwest of Portsmouth, near Kennvale, a classic yazoo stream, Old Pond Run, flows between the highway and the Ohio River. Sediments deposited by the Ohio prevent the tributary from entering it until the main channel of the Ohio swings to intersect Old Pond Run. Downriver, Buena Vista and Rockville gave birth to the Buena Vista freestone industry in the early 1800s. Most quarries were up Lower Twin Creek and Rock Run. Soft Devonian-age shales underlie the floodplain and U.S. 52. Between Rome and Wrightsville,

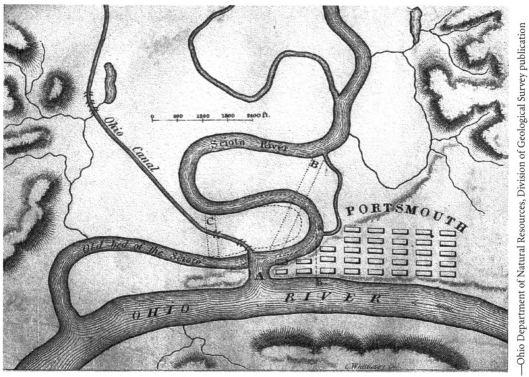

—Ohio Department of Natural Resources, Division of Geological Survey publication

The Scioto River's mouth has undergone a number of changes in Portsmouth. The 1838 map (top) depicts the new mouth of the river that was dug when the Ohio & Erie Canal was constructed. The bottom map shows the area some 150 years later.

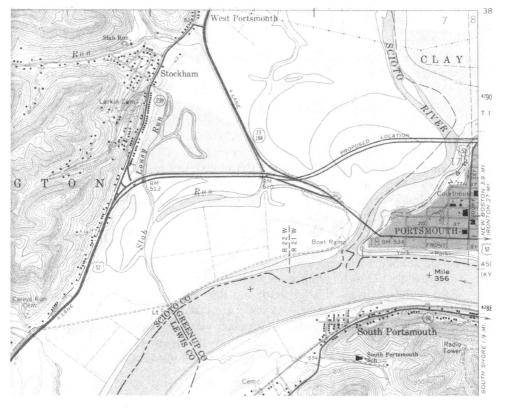

—United States Geological Survey 7 ½-minute Friendship quadrangle

Ordovician-age Grant Lake limestone forms thick beds in this quarry at Manchester.

Ohio Brush Creek, which exposes fossiliferous Ordovician limestones and shales, enters the Ohio River. Note small Brush Creek Island, which sits in the Ohio near the mouth of the creek. It developed as this large tributary carried great amounts of sediment out of the uplands and dumped them in the Ohio; since the island occurs on the inside of a meander, the Ohio hasn't washed it away. Middle Silurian dolomites cap the hills in Ohio Brush Creek valley.

About 1 mile northeast of Wrightsville off Ohio 247, 10- to 30-foot cliffs of dolomite border a narrow ridge along Cummins Creek in the Whipple (Robert A.) State Nature Preserve. Small sinkholes dot the ridgetop. From Wrightsville to the Indiana line, U.S. 52 lies on upper Ordovician strata, much of it covered by Illinoian and pre-Illinoian till. Every exposure reveals marine fossils.

A quarry on Broadway Street in Manchester exposes shale and limestone of Ordovician-age Grant Lake limestone. Brachiopod and bryozoan fossils are abundant. Silurian Bisher and Peebles dolomite from Lynx and West Union found wider use as building stone than the Ordovician strata. Wisconsinan glaciers did not reach this far south and Illinoian glaciers only affected the northwestern corner of Adams County, so upland soils are thin and nutrient poor. The underlying bedrock is in many places the easily eroded Silurian Noland formation, mainly shales. Short-grass prairies developed on mounds of rounded shale of this formation, particularly around West Union, certainly before Wisconsinan glaciation worked over northern Ohio and perhaps much earlier. Prairie vestiges remain at the Adams Lake Prairie, Chaparral Prairie, and Johnson Ridge State Nature Preserves, around West Union.

Buena Vista's City Ledge

Near the Scioto and Adams County line is Buena Vista, which was an important sandstone quarrying center during the 1800s. The first evidence of good building stone in the area came in 1814 when Joseph Moore, a stonecutter, erected a house using local stone in the valley of Rock Run between Buena Vista and Rockville. He then marketed this stone, which he had taken from slump blocks in the hills. Rafts carried large shipments to Cincinnati. In 1831 actual quarries opened on the hilltops to supply stone for the Miami & Erie Canal, which workers were building through Cincinnati.

A problem these hilltop quarries faced was getting stone to the river for shipment. At first quarries used oxen to drag the stone; later, they slid it down chutes, hauled it in wagons, and moved it on incline railways. Until the erection of the first stonecutting mill in 1847, they shipped stone in large, rough blocks. By the late 1860s at least four companies operated around Buena Vista; others had quarries between town and Portsmouth.

The sandstone that made Buena Vista famous was from the Mississippian-age Cuyahoga formation (probably the Black Hand and Buena Vista members) that outcropped on the hilltops. Operators of the Loughry quarries in Rockville eventually discovered a 3- to 4-foot-thick layer of sandstone near the base of the Cuyahoga formation that proved to be of the highest quality because it had held up well after ten years of use along the canal. By 1843 operations focused their attention on this unit that they called the "city ledge," so named because so much of it went to Cincinnati. Although the stone industry has left Buena Vista, the wide, flat terrace left by the quarrying of this unit is still noticeable about 260 feet up the hillside north of U.S. 52.

Buena Vista quarries lasted into the early 1900s, gradually losing business to other quarries located along or near railroads. The depletion of good stone led these quarries to market inferior stone, which slowly tarnished the name of the Buena Vista product. Concrete was also becoming a favored building material in the early twentieth century. The last company to operate at Buena Vista was the Buena Vista Freestone Company, which closed in 1909. The 1913 Flood nearly wiped out the remaining community. Other important early quarries were along Careys Run and Pond Run, north of U.S. 52 and west of Portsmouth. One of the first steam-powered stone sawmills west of the Alleghenies operated near Portsmouth from 1847 to 1884. Most of these quarry-related operations closed in the early 1900s.

A Side Trip to Buzzardroost Rock

To get to Buzzardroost Rock, a cross section of Silurian-age rocks, follow Ohio 136 north out of Manchester past Ordovician-age outcrops. Reddish shales of the Ordovician Drakes formation underlie the Silurian-age Brassfield formation in roadcuts at Bentonville. From here continue on Ohio 41 north, traversing a rolling karst landscape that developed as groundwater dissolved parts of the Brassfield limestone surface. At its intersection with Ohio 125, Ohio 41 climbs onto the Silurian Bisher dolomite. On the south side of West Union there are abandoned quarries in the Bisher formation. Go east on Ohio 125 in West Union and follow the road 7 miles to the bridge over Ohio Brush Creek. The trailhead to Buzzardroost Rock is ½ mile north on Weaver Road.

This scenic spot offers a grand view of the Ohio Brush Creek valley and a cross section of Silurian rocks. At the trailhead, Ohio Brush Creek runs over fossiliferous Brassfield limestone. As the trail rises, the Noland formation (mainly shale) forms the gentle slopes that fall away from it. Near the top of the hill, slump blocks of fossiliferous Peebles dolomite appear. Buzzardroost Rock is some 500 feet above Ohio Brush Creek and typifies the knobby Silurian landscape. The Silurian dolomites underlying this area are crisscrossed by joints that extend deep into the rock layers; they are ready paths for surface water to percolate through to join a main stream that has incised into the bedrock. As the descending water dissolves and wears away the rock on the walls of the joints, the joints widen and eventually isolate vertical columns of rounded dolomite—the knobs. Knobs like this generally occur at the edge of the Silurian, or Niagaran, escarpment, or along streams that have cut deeply into the Silurian bedrock.

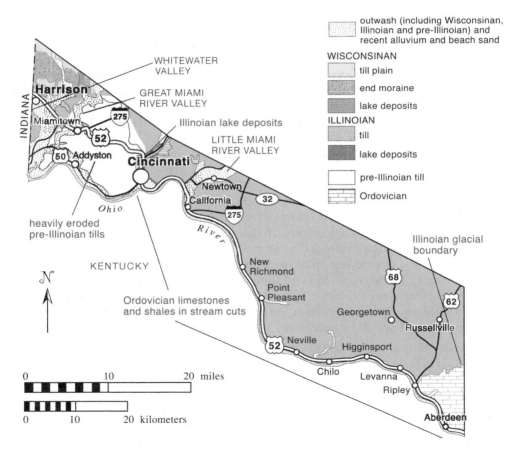

Geology along U.S. 52 between Ripley and the Indiana line.

Aberdeen to the Indiana Line

West of Aberdeen, Big Threemile Creek, another yazoo stream, parallels U.S. 52, blocked by sediments from entering the Ohio River. From Ripley to the Mill Creek valley in downtown Cincinnati, U.S. 52 lies on Illinoian till. Since the tills of this region weren't covered by Wisconsinan ice and have been exposed to streams for thousands of years longer than younger tills, the landscape is rugged, with many deeply incised valleys. Many of these valleys have exposures of Ordovician limestones and shales. The limestones often form steplike waterfalls up the valleys; for example, like those in Kope Hollow on the east side of Levanna, the namesake of the Kope formation. All exposures of the bedrock are very fossiliferous. Brachiopods and bryozoans are the dominant fossils, but clams, snails, horn corals, burrows, borings, and occasional trilobites can also be found.

Between Chilo and Neville the Captain Anthony Meldahl Dam and Locks cross the river. Farther downstream Point Pleasant was once an important quarrying center, supplying Cincinnati with much building stone. This is one of the few places in southwestern Ohio where the Point Pleasant formation is not buried under younger strata. The thick bedding made it a favored building stone; most Ordovician strata had too much shale and only thin beds of limestone to make it very useful for building. This bluish gray limestone came from

The fossiliferous Ordovician-age Kope formation, composed of limestones and shales, outcrops along the Ohio Valley east and west of Cincinnati. This typical exposure occurs along U.S. 50 in Addyston.

beds at the Ohio River's edge, where it was easily loaded onto flatboats for shipment. The thinner slabs were used locally as flagging and for making fences. Across the state the Point Pleasant formation is best known as a subsurface unit usually referred to as the Trenton limestone, which was a major source of oil and natural gas.

The Ohio River valley is relatively narrow, about ¾ mile wide, around New Richmond. Bluffs underlain by Ordovician-age strata form the northern edge of this town, which spreads along the narrow floodplain at this slight bend in the river channel. The Kentucky side of the valley, the outside of the meander, has only a cutbank. The construction of the dam and lock just downstream of town has alleviated some of the flooding problems that this community has experienced. California sits south of where the highway crosses the wide Little Miami River valley. The river, which only occupies a small part of the valley bottom, is flowing in the valley the preglacial Manchester River carved where it swerved north of present-day Cincinnati toward Hamilton. The Ohio River channel between the mouth of the Little Miami River and the Kentucky, Ohio, and Indiana border was not carved until sometime around Illinoian time. U.S. 52 crosses a small patch of Illinoian lake sediment and outwash that forms a higher terrace on the east side of the Little Miami River valley. A lower flat area west of the road is a remnant of the Wisconsinan terrace. Meltwater from the Wisconsinan ice sheet resting at its terminal moraine about 15 miles to the north probably eroded away most of the Illinoian outwash here, leaving only scattered patches. Just north of the U.S. 52 bridge across the Little Miami River is an oxbow lake.

In downtown Cincinnati, on the east side of the Mill Creek valley, U.S. 52 joins I-75 and heads north to its junction with I-74. On the west side of the Mill Creek valley U.S. 52/I-74 is underlain by pre-Illinoian till; except for occasional building excavations and stream cuts the material is buried under commercial and residential development. The pre-Illinoian till ends on the east side of the Great Miami River valley, which is filled with Wisconsinan outwash. This valley was a major meltwater channel throughout most of the Ice Age. At Miamitown, Wisconsinan outwash sand and gravel was dug for fill during the construction of I-74. Ordovician limestones and shales are visible in the hillsides. About 2 miles west of where I-275 leaves I-74 and heads south, U.S. 52/I-74 crosses a wide valley filled with Wisconsinan outwash, through which the Whitewater River flows south of the highway. The Whitewater River leaves the valley and flows west into Indiana near the southern limits of Harrison. North of U.S. 52/I-74 the wide valley is only occupied by the Dry Fork Whitewater River. This marks an abandoned portion of the preglacial route of the south-flowing Hamilton River. The valley once extended northeast to the present Great Miami River valley but was blocked by a lobe of the Wisconsinan ice sheet, which deposited part of the Hartwell moraine in the former valley. The headwaters of the Dry Fork Whitewater River parallel this end moraine to the northwest.

OHIO 7
East Liverpool—Chesapeake
239 miles

Ohio 7 traverses the unglaciated Allegheny Plateau and Ohio Valley. Between East Liverpool and Bellaire, the new Ohio 7 bypasses most towns, but farther south it resumes its old route. The typical cyclothems—repeating layers of sandstone, shale, limestone, clay, and coal—of Pennsylvanian age are visible from East Liverpool to Tiltonsville. The youngest Paleozoic rock in the state, which is Permian in age, forms the hills from Yorkville to Chester; however, to most people it's indistinguishable from the Pennsylvanian rocks. From Minersville to Chesapeake the route crosses more Pennsylvanian strata.

East Liverpool to Steubenville

Immediately south of East Liverpool there is a long roadcut in Pennsylvanian strata that shows what underlies most of the surrounding hills. The rocks are inaccessible, however, and stopping is ill-advised. The lowest part of this cut, the closest to Wellsville, exposes interfingering beds of Pottsville group sandstones and shales, probably the Homewood and Tionesta members. A lens-shaped deposit of sandstone is visible, having been deposited in a scoured-out channel, but it is extremely difficult to trace strata very far across this exposure. Overlying the Pottsville strata, closer to East Liverpool and higher up the cliff face, are Allegheny group marine limestones and shales between the Brookville coal horizon and the Lower Kittanning coal. It is difficult to distinguish individual members. This cut is typical of the interfingering, thickening and thinning nature of Pennsylvanian cyclothemic strata. Without a few key fossils in the various layers, it is very difficult to decipher the history of events.

Upper Pottsville group and lower Allegheny group rocks form a long roadcut just south of East Liverpool.

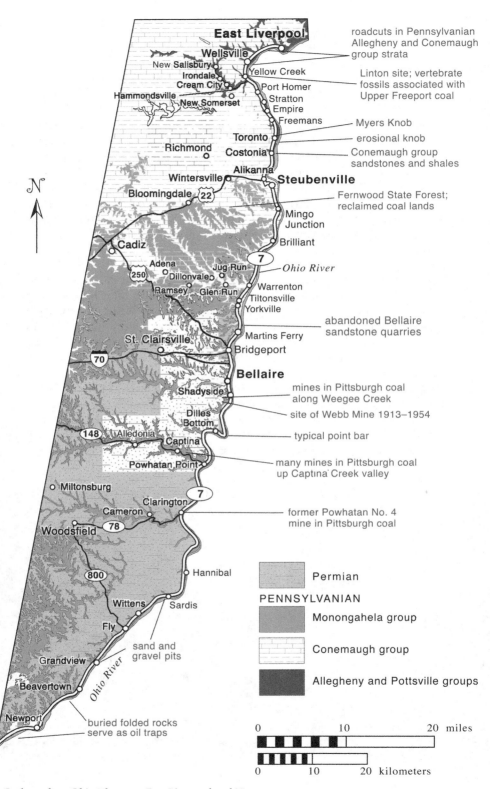

roadcuts in Pennsylvanian
Allegheny and Conemaugh
group strata

Linton site; vertebrate
fossils associated with
Upper Freeport coal

Myers Knob
erosional knob

Conemaugh group
sandstones and shales

Fernwood State Forest;
reclaimed coal lands

Ohio River

abandoned Bellaire
sandstone quarries

mines in Pittsburgh coal
along Weegee Creek

site of Webb Mine 1913–1954

typical point bar

many mines in Pittsburgh coal
up Captina Creek valley

former Powhatan No. 4
mine in Pittsburgh coal

sand and
gravel pits

buried folded rocks
serve as oil traps

Permian

PENNSYLVANIAN

Monongahela group

Conemaugh group

Allegheny and Pottsville groups

East Liverpool
Wellsville
New Salisbury
Irondale
Cream City
Hammondsville
New Somerset
Yellow Creek
Port Homer
Stratton
Empire
Freemans
Toronto
Richmond
Costonia
Alikanna
Wintersville
Steubenville
Bloomingdale
22
Mingo
Junction
Brilliant
Cadiz
7
Adena
Jug Run
Dillonvale
Ramsey
Glen Run
Warrenton
Tiltonsville
Yorkville
250
St. Clairsville
Martins Ferry
Bridgeport
70
Bellaire
Shadyside
Dilles
Bottom
148
Alledonia
Captina
Miltonsburg
Powhatan Point
Clarington
7
Cameron
78
Woodsfield
800
Hannibal
Wittens
Sardis
Fly
Grandview
Beavertown
Newport
Ohio River

| 0 | | 10 | | 20 miles |

| 0 | | 10 | | 20 kilometers |

Geology along Ohio 7 between East Liverpool and Newport.

Prospectors sought oil near Little Yellow Creek, north of Wellsville, in the 1860s, but successful yields were not encountered and they abandoned the wells. In 1899 oil was struck in the same area, but it proved to be a small pool and dried up by 1910.

A solitary abandoned pottery kiln lies just below the highway in Wellsville. Four Wellsville brick plants, and plants at New Salisbury and Yellow Creek, used Lower Kittanning clay in the 1920s. This clay occurs below the base of the roadcuts south of town. Note the large blocks of sandstone at the Columbiana and Jefferson County line, just south of Wellsville, that have fallen from high on the hillside. Rockfalls are prevalent along this highway wherever thick sandstone layers are underlain by easily eroded shale or clay. For 1½ miles south of Wellsville, the highway cuts through strata of the Allegheny and Conemaugh groups. Six coal seams lie in the 180-foot-high roadcuts west of the road, although slumping and weathering have taken their toll, making it difficult to discern them from other strata. Upper Freeport coal was mined for many years in the vicinity of Wellsville. It lies high on the hillside in the first large roadcut south of town and marks the boundary between Allegheny and Conemaugh group rocks. Fossiliferous coal balls, calcareous nodules containing fossil plant fragments, weather out of strata associated with the Freeport coals just south of the railroad bridge on Ohio 7 in Yellow Creek, just south of Wellsville; massive Mahoning sandstone lies above.

At Port Homer the Peerless Clay Manufacturing Company had a slope mine in the Lower Kittanning clay it used to make sewer pipes. The plant operated east of Ohio 7 until 1971 but is long gone. The towers of the W. H. Sammis Power Plant are a major landmark between Wellsville and Steubenville. The plant was previously the site of the Northern Sewer Pipe Company. The New Cumberland Dam and Locks are in Stratton.

Stratton's first clay plant opened around 1895. At first it made bricks but later produced sewer pipe. The town of Empire had clay plants on the north and south ends of town. Plants using Lower Kittanning clay also operated at Calumet and Freemans. Allegheny sandstones were quarried for pulp stone at Empire and west of New Somerset. The Upper Freeport sandstone was the source rock at Empire, and as the limits of its quarry were reached the company drove drift mines into the hillside and continued mining the sandstone underground. Today the underground section serves as a home for bats.

At Toronto five companies mined Lower Kittanning clay to manufacture sewer pipe, building blocks, and firebricks. Myers Knob, above the Union Cemetery just west of Ohio 7, provides a scenic view of the Ohio Valley. On the south edge of town the highway swings around a large hill that is separated from the Pennsylvanian escarpment by an abandoned channel of the Ohio River. During Wisconsinan time this was a bedrock island in the Ohio River. At some point sandbars cut it off and the river assumed its present course. The city of Toronto is located on a terrace of Wisconsinan outwash. Ohio 7 passes along the west side of the city at the edge of the former river bluff of Pennsylvanian strata. Across the river at Weirton, West Virginia, there is a large filled-in meander loop that is now an industrial site with railroad yards. A roadcut at the Costonia exit is composed of Conemaugh group shales and sandstones. A Costonia company once made firebricks using Bolivar clay of the Allegheny group. The town name Pottery Addition hints at an early industry of the Steubenville area.

At Alikanna, just north of Steubenville, a roadcut exposes 7 feet of concretion-bearing Pennsylvanian Brush Creek shale. Concretions—nodular rocks of harder material—weather from the shale and accumulate at the base of the slope. Splitting the concretions often reveals veins of sphalerite, barite, calcite, quartz, and other minerals and sometimes a fossil at the center.

A Side Trip to Irondale and Other Mining Towns

Up Yellow Creek and Ohio 213 lie a number of communities that were once important to the clay, coal, and iron industries. Hammondsville had mines in the Lower Kittanning coal and a pulp stone quarry in the Lower Freeport sandstone. Coke manufacturing was a prevalent industry around 1900. Leave Ohio 213 at Cream City, which once had mines in the Middle Kittanning and Lower Freeport coals and a plant of the McLain Fire Brick Company, and head northwest to Irondale. In the 1870s, Irondale's furnace, rolling mill, coal washer, and extensive mines played an important role in the local economy. Obviously its name came from early successes in iron production. A more important claim to fame, however, was the modification of a sheet-steel mill that led to the first American tin product in 1890, a product that met the standards of imported tin from Wales. The mill's gone now, but the community survives. In the 1920s Banfield Clay, McClain Fire Brick, and East Ohio Sewer Pipe companies used Pennsylvanian Clarion and Lower Kittanning clays to produce bricks and sewer pipe. Upper Freeport coal was used to fuel the kilns. Further northwest there are more kilns at New Salisbury, the turnaround point.

Rare Vertebrate Fossils

One of the more famous coal mines in this area was the Diamond Mine at Linton, which is now a ghost town just north of Yellow Creek. In 1853 miners began drift mining Upper Freeport coal, which reached a thickness of 9 feet locally. Although the mine was known for its high-quality coal, it was fossils that brought it worldwide acclaim. In 1856 John Strong Newberry, Ohio's state geologist from 1869 to 1882, is credited with having discovered a fossil-bearing, 4- to 6-inch-thick bed of cannel coal, or spore-bearing coal, which formed the floor of the mine. Newberry was mainly interested in the numerous fish remains, but others, including Edward D. Cope of Philadelphia, a well-known paleontologist perhaps best known for dinosaur discoveries out West, described significant amphibian and reptile remains from the mine up until its closure in 1892.

The mine reopened from 1917 to 1921 and then closed permanently. Through the years at least ten taxa of invertebrates, including small worms, millipedes, and crustaceans, and forty taxa of vertebrates, mainly fish, have been uncovered; in a number of cases the fossils of some species are the only examples that have ever been found anywhere. The Linton location ranks as the most prolific Pennsylvanian vertebrate fossil locality in the world.

The most common fish fossil at the Linton mine is the coelacanth (*Rhabdoderma elegans*), numbering in the thousands of specimens, followed by other bony fish. The coelacanth was a lobe-finned, jawed fish that first appeared in

Devonian time. Lungfish and sharks are less common. Scales and teeth are common, but complete impressions of fish are uncommon. The amphibians consisted of eel-like, salamander-like, and froglike creatures. Reptiles are rare but included one sail-back reptile, or pelycosaur (*Archaeothyris* species), and two taxa of captorhinomorphs, very early primitive reptiles.

The peaty Linton sediments accumulated in an oxbow lake on a delta plain some 310 million years ago. The Linton fossils were preserved by pyrite. The bones, teeth, and spines of these early vertebrates were buried in the bottom sediments shortly after the death of the animals. In many cases the bones were separated after the flesh decayed and were redistributed by water currents or scavenging animals, thus many fossils are far from complete. Eventually, chemicals moving through the sediments precipitated in the pore spaces of the organic remains and gradually replaced molecule for molecule the original hard parts. At Linton, the organic sediments accumulating in the still waters of an abandoned river meander were rich in sulfur and iron. These ions joined together to make the brassy colored mineral pyrite, which concentrated in the former bones and preserved them. Unfortunately, some of the pyrite changes to a less stable form of iron sulfide when exposed at the surface and the fossils crumble.

Scientists continue to unravel the biologic and geologic history of the Linton area making use of nearby roadcuts. Although most of the Linton specimens wound up at museums outside Ohio, the Cleveland Museum of Natural History and Orton Geological Museum at Ohio State University have representative collections.

Steubenville Mining and Coal Balls

Pennsylvanian Conemaugh group rocks form the bedrock of Jefferson County. Along the Ohio River from Alikanna to Steubenville, miners widely mined Lower Freeport coal, the so-called Steubenville shaft coal, into the early 1900s. While boring a water well in 1829, workers first discovered the seam 225 feet below the city. Geologists originally estimated that the seam was 11 feet thick. Since there was plenty of Pittsburgh coal in the surrounding hills, however, nobody thought that sinking deep shafts to get at the seam was reasonable.

Twenty years later, though, the city was demanding more coal than surrounding mines could steadily supply, especially when muddy road conditions slowed wagon travel. In 1857 workers sank a shaft on upper Market Street. Disappointment ran through the city when workers found that the seam was

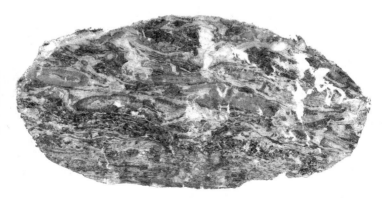

When cut into thin slices, a typical coal ball often shows the internal parts of Pennsylvanian-age plants that grew in eastern Ohio wetlands.

The Plum Run Mine near Steubenville tapped the important Pittsburgh seam from around 1905 to 1929.

only 4 feet thick, but work continued and they soon discovered that the vein thickened in places to 8 feet. Thirteen shafts eventually tapped this high-grade coal that ranged from 60 to 200 feet down. By the 1920s only three remained open. Northwest of town miners found the coal along Island Creek and soon developed strip mines. Shaft and strip mines in the Pittsburgh coal operated near Bloomingdale and between Steubenville and Mingo Junction. Quarriers pried the Buffalo sandstone from the ground on the north edge of town near the Ohio River until 1925. The sandstone was used as pulp and building stone.

A short trip west of town on U.S. 22 quickly leads to major strip mines. Strip-mined land lies south of Wintersville, and there is reclaimed land in Fernwood State Forest south of town. Fossiliferous Ames limestone bounds the hill at Reeds Mill. Miners have extensively stripped the hills south of Bloomingdale.

After swamp deposits are covered by overlying sediments they eventually harden to coals and organic-rich shales. Occasionally, these buried swamp deposits are selectively mineralized by waters that are rich in calcium, magnesium, and iron carbonates, forming erosion-resistant masses called *coal balls*. Coal balls are a rich source of plant fossils, which are often preserved in living position. Because of the precipitation of calcite that forms the harder, compact globular masses, the fossils have not been flattened and have not lost their structure, as usually happens with Pennsylvanian-age plant fossils. By cutting thin sections through coal balls, paleobotanists can study the internal organs and cells of many plants that lived in these ancient swamps.

Nine miles west of Steubenville along U.S. 22 there are Conemaugh group exposures, which in 1975 were discovered to yield coal balls. Some twenty-five species of Cordaites (early conifer-like plants), ferns, horsetails, lycopods (scale trees), and seed ferns have been identified in these areas. The coal balls weather out of two thin coal seams above the road and collect in a roadside ditch.

Steubenville to Bridgeport

The hills west of Brilliant are riddled with mines in the 5-foot-thick Pittsburgh seam, the thickest persistent coal seam in southeast Ohio. One of the oldest mines in the Pittsburgh coal opened at Warrenton in 1884. Companies began strip-mining the area around 1913. West of Warrenton, along Short Creek, there were many other mining communities, including Adena, Dillonvale, Glen Run, Jug Run, and Ramsey. Most of the mines in the Pittsburgh coal in this area were abandoned by the 1920s. South of Brilliant along Salt Run was an early 1800s saltworks.

South of the Jefferson and Belmont County line near Tiltonsville, Ohio 7 lies on the youngest rocks of Ohio—the Permian-age Dunkard group, a series of sandstones, shales, and limestones much like the underlying Pennsylvanian Monongahela group strata. Permian rocks form the hilltops from Yorkville to Chester, 138 miles to the southwest. Belmont County continues to be the state's leading producer of bituminous coal, which mainly comes from the Pittsburgh and Sewickley seams of the Monongahela group. Underground mining is still the dominant form of extraction. Most of the communities in northeastern Belmont County and along the Ohio River once had mines in the Pittsburgh coal; much of this seam was depleted by the early 1960s. The Siebrecht Quarry near Martins Ferry was the source of Bellaire sandstone that builders used to construct the suspension bridge at Wheeling, West Virginia. Quarriers pulled Permian Washington formation limestones from a quarry northwest of Martins Ferry.

Bellaire-Area Coal and Sandstone

The first recorded coal mining in Belmont County occurred north of Powhatan Point around 1804. This was the Pittsburgh coal, a thick and high-quality coal destined to become Ohio's most important seam. Until 1830 it served mainly local domestic uses. After 1830, coal mined at Bellaire was shipped down the Ohio and Mississippi Rivers on flatboats to fuel the furnaces of sugar companies in Louisiana. Steamboats gradually replaced the flatboats and became major consumers of coal as well. Bellaire became an important refueling point for them. Mines along Shadyside's Wegee Creek opened around 1912 and also provided steamboats with coal. The largest mine in the area was the Webb Mine, opened in 1913 by the Toledo Mining Company. In its first year, miners dug 17,800 tons of coal, which came from the Pittsburgh seam 90 feet down. The mine prospered, employing around seven hundred men by the 1920s and shipping more than seven thousand carloads of coal per year. The mine closed in 1954, but its remains are visible along Township Road 811. North American Coal Company, which operated a mine at Businessburg west of Dilles Bottom, mined the remaining reserves of the Webb Mine at some point.

The stone-arch railroad viaduct stretching across downtown Bellaire is made of Pennsylvanian-age Bellaire sandstone. This unit is 10½ feet thick in the Robinson Quarry along McMahon Creek on the south edge of town. Parker & Sons Cement Works opened in 1858, north of Barnesville about 25 miles west of Bellaire, and provided 11,000 barrels of natural cement for the bridge.

The Fishpot limestone member of the Monongahela group from a local drift coal mine was the source of this natural rock cement. Several quarries pried Pennsylvanian sandstones from the hills in this area, and Permian clays were the raw materials for brick factories. Bellaire also supported glass factories, and the Bellaire Glass Museum has a lot of interesting displays and information about this former industry.

Bellaire to Marietta

South of Shadyside, Ohio 7 loops across a large point bar. A power plant rests on the edge of this growing sand deposit in a town called Dilles Bottom. Note the corresponding steep bluffs on the West Virginia shore, where the river has carved deeply into the landscape. Two miles farther south there is another meander, but now the point bar is on the West Virginia side, and the steep bluffs on the Ohio side adjacent to Ohio 7. South of this meander, at Powhatan Point, companies began developing coal mines in the late 1800s.

Coal seams that were so prevalent in Belmont County to the north were deeply buried below Permian strata in Monroe County, which starts south of Powhatan Point. The seams that were exposed were too thin and sporadic to be of much value except for local use in the past. The first records of coal use in the area date to 1840, and area production climaxed between the 1870s and 1900. A lapse of sixty years followed. In 1966, with improvement in mining technology, deep shaft mines opened and their production far exceeded the earlier attempts at commercial mining. The Powhatan No. 4 Mine, operating near Clarington from 1971 to 1999, was one of the largest coal mines in Ohio and at one time mined over 3 million tons of Pittsburgh coal per year and employed one thousand miners.

The Allegheny Plateau is heavily eroded by streams along this stretch of Ohio 7. Streams flow through short, deep cuts into the Ohio River. South of Hannibal is the Hannibal Locks and Dam. At Wittens the floodplain widens and borrow pits in it provide gravel and sand. Monongahela group rocks flank the highway between Wittens and Beavertown and a lock and dam lies in the river at Beavertown. Conemaugh group rocks form the bedrock at Newport, and Permian-age rocks cap the hills to the north. The hill just northwest of Newport was once an island. An abandoned meander of the Ohio River loops north of Newport, and the deepening of this meander at one time cut into the underlying Conemaugh strata. The route skirts the edge of Wayne National Forest from Newport to Marietta. Although only noticeable from geologic mapping, several anticlines and synclines of the West Virginia Valley and Ridge Province, part of the folded Appalachian Mountains, dive under the Ohio River and southernmost Ohio along this stretch. The folded strata trapped oil at certain spots, including the area just west of Newport. Wells tapped the Newell Run oil field around 1890. Three hundred wells operated in the area by 1900. Production initially came from the Conemaugh group rock called the Cow Run sandstone; later, wells pumped oil from the deeper Devonian-age Berea sandstone. Some still operate today.

A Side Trip up Captina Creek

At Powhatan Point you can leave Ohio 7 on Ohio 148 for a trip up Captina Creek to a number of mining and quarrying communities. Early settlers hacked small farms out of this rugged wilderness in the early 1800s. Settlers knew that Pittsburgh coal lay under Bellaire and northern Belmont County, but nobody knew if it occurred south and west of that region. In 1882, however, miners discovered a 6-foot-thick seam 68 feet below the ground. The coal appeared in a shaft they sank at Captina Station on the new Bellaire & Southwestern Railway, a narrow-gauge railroad. Geologists mapped the area and showed that the Pittsburgh seam extended throughout southern Belmont and northern Monroe Counties. The Captina Mine, which developed around the shaft, operated until 1927, eventually succumbing to labor and water inflow problems, not a lack of coal.

In the late 1940s the Pennsylvania Railroad became interested in Captina Creek valley for its remaining Pittsburgh coal. Mines that had been operating for a long time elsewhere were running out of coal. In 1952 the railroad began constructing a rail line up the valley from Powhatan Point, approximating the route of the old narrow-gauge railroad, which had been abandoned in 1931. Huge earthmoving equipment cut into the hills, redistributing over 1 million cubic yards of soil and rock in what became a $3 million project. The hills didn't give in easily, and landslides were common. At Captina the new rail line used the old mine dump as fill. The line was built in segments, extending to new mines like the Norton No. 3 and Powhatan No. 6 near Alledonia. The Powhatan No. 6 Mine supplied a Cleveland power plant with coal. The Century and Powhatan No. 6 Mines remain the only underground mines still in operation as of 2004; most coal is now removed by stripping.

A Side Trip to Woodsfield

Ohio 78 west from Clarington is yet another scenic trip into the Allegheny Plateau uplands. The route initially follows the twisting course of Sunfish Creek to near Cameron. Monongahela group rocks—sandstones and shales—form cliffs, rock shelters, and slump blocks along the deep valley as well as in similar tributary valleys of Sunfish Creek. Most of the terrain along this route has been dissected by water, and the relief in this area may reach 400 to 500 feet. Exposures of the Permian-age Greene formation near Cameron yielded numerous fossil lungfish teeth and bones; bones of two early amphibians, 3-foot-long *Eryops* and *Diploceraspis* (the latter with a unique crescent-shaped skull); and skeletal parts of the 10-foot-long *Dimetrodon* and 5-foot-long *Edaphosaurus*, sail-back reptiles. These creatures were inhabitants of the broad alluvial plain that covered eastern Ohio in Permian time, with its meandering streams, swamps, and marshes.

About a mile east of Woodsfield, along the abandoned grade of the Ohio River & Western Railway, lies a heavily overgrown quarry in the Pennsylvanian Gilboy sandstone of the Monongahela group. This quarry was the main source of building stone in the area until it closed in the early 1900s. The Permian-age Mannington sandstone was used to a lesser extent, coming from quarries in the Miltonsburg area, to the north off Ohio 145.

Marietta Grindstones, Building Stones, and Bricks

As early as 1819, Marietta settlers realized that some local sandstones made great grindstones, especially the Lower and Upper Marietta and Hundred

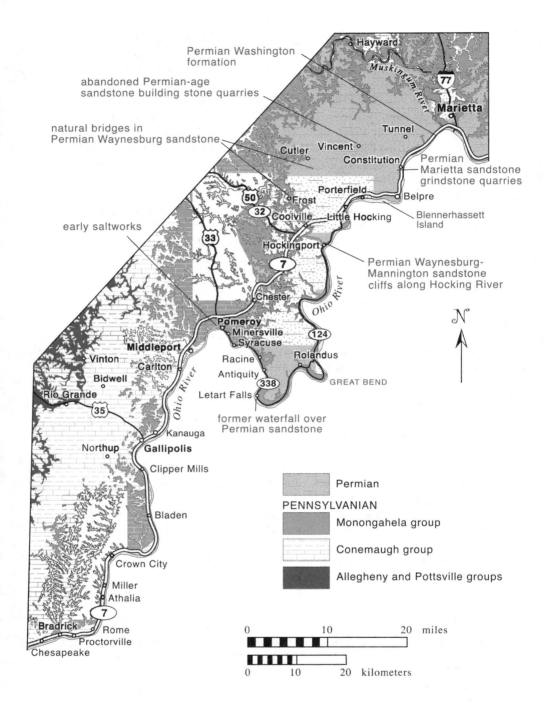

Permian Washington
formation

abandoned Permian-age
sandstone building stone quarries

natural bridges in
Permian Waynesburg sandstone

early saltworks

Hayward

Muskingum River

77

Marietta

Tunnel

Cutler   Vincent
         Constitution

Permian
Marietta sandstone
grindstone quarries

50   Porterfield
Frost   Belpre
32 Coolville Little Hocking
Hockingport

Blennerhassett
Island

33

7

Chester

Ohio River

Permian Waynesburg-
Mannington sandstone
cliffs along Hocking River

N

Pomeroy
Minersville
Syracuse

124

Racine   Rolandus
Antiquity
Middleport   338

GREAT BEND

Vinton
Carlton
Bidwell
Rio Grande   Letart Falls

Ohio River

former waterfall over
Permian sandstone

35

Kanauga

Northup Gallipolis

Clipper Mills

Bladen

Crown City

Miller
Athalia

7

Bradrick   Rome
Proctorville
Chesapeake

	Permian
PENNSYLVANIAN	
	Monongahela group
	Conemaugh group
	Allegheny and Pottsville groups

0   10   20   miles

0   10   20   kilometers

Geology along Ohio 7 between Marietta and Chesapeake.

sandstones of the Permian-age Washington formation. The best grindstones are made from sandstones that have angular quartz grains of equal size; are somewhat porous; and have thick, uniform beds. A quality grindstone maintains an abrasive surface that doesn't dull. Quarriers removed the stone in 3-foot-thick circular blocks from the quarry face using a channeler. Workers then finished the suitable stone disks at a mill. The abandoned quarries in the hills above Marietta still contain circular scars in the walls and discarded grindstones. The Hall Grindstone Company was one of the last producers of natural grindstones in Ohio. Their abrasives went to saw, axe, file, and knife manufacturers.

Sandstone quarries operated throughout the hills west of Marietta, particularly around Constitution and Tunnel. Grindstones also came from Hayward, up the Muskingum River. The last plant in this region closed in the early 1970s. Other sandstones, like the Permian-age Waynesburg sandstone, were cut and used for bridge piers. The abandoned Marietta, Columbus & Cleveland Railroad made wide use of this stone for trestles; some of them are still standing. Locally, people used other sandstones for trim, fireplaces, and tombstones.

Marietta is the oldest settlement in Ohio and the first settlement of the Northwest Territory, having developed around Campus Martius, a fort that was built between 1788 and 1791. The original log buildings had brick chimneys that settlers fired from local Permian clays. Brick making was well established by 1797, and Marietta is the only place in the state where Permian shales were used to make them. Acme Brick and Marietta Shale Brick Companies mined the Creston Red shale that occurred between the Washington coals in the early 1900s.

Marietta to Pomeroy

Just south of Marietta near the intersection of Ohio 550, the Permian-age Washington formation outcrops. A partial amphibian skull, found in the redbeds here on the west side of Marietta in 1969, turned out to be a new species of labyrinthodont, *Trematops stonei*. Labyrinthodonts were not unlike salamanders, but generally they stood somewhat taller and had a more rounded body. Many were small but some reached lengths of several feet.

In the hills above Constitution there are abandoned grindstone quarries in Permian Marietta sandstone. Belpre sprouts from a large point bar. Blennerhassett Island formed as excess alluvium the Ohio carried accumulated downstream of the meander and the mouth of the Little Kanawha River of West Virginia. North of Porterfield on Ohio 339 is Vincent, which was a quarrying center for local Permian Upper Marietta sandstone for building purposes. More sand and gravel operations are evident where the Ohio River floodplain widens.

Between Belpre and Coolville, Ohio 7 and U.S. 50 are coterminous and leave the Ohio River valley. The meandering course of the sediment-choked Little Hocking River enters the Ohio River at Little Hocking. To stay along the Ohio River leave Ohio 7 and follow Ohio 124 and 338 between Little Hocking and the eastern edge of Pomeroy.

High cliffs of Permian Waynesburg-Mannington sandstone rise above the Hocking River at Hockingport on Ohio 124. The Washington coal caps the sandstone, and a little below river level lies the Pennsylvanian-age Waynesburg coal. Downstream of Hockingport the Ohio River makes a broad swoop

A building stone quarry in Permian-age sandstone near Vincent around 1907.

called the Great Bend opposite Ravenswood, West Virginia. Pennsylvanian-age bedrock underlies this narrow promontory; sandstone outcrops along the northeast side across from Ravenswood. On the southwest side of Great Bend, Ohio 7 follows a flat terrace of Wisconsinan-age outwash. Lake sediments below this outwash mark the path of a tributary of the preglacial Marietta River that was just to the north. Two more lock and dam complexes govern flow at this point.

At Great Bend continue on Ohio 338. Letart Falls developed around a waterfall where the Ohio River crossed over resistant Permian sandstone. Before dams, people winched boats through a rocky channel around the falls. The damming of the Ohio submerged the falls, and now people wonder about the origin of the name. The Racine Locks and Dam, built in the 1950s, contain rock quarried from Cedar Hill, behind the cemetery in Letart Falls. Note the slump blocks against the rocky cliffs north of Ohio 338 in Antiquity. Supposedly the town received its name from a large, smooth rock 4 miles downstream that was etched with ancient engravings, the so-called rock of antiquity. The namesake rock was submerged when the dams were built. The 1848 discovery of salt in deep wells drilled along the river led to the opening of saltworks at Antiquity. Saltworks became associated with coal mines in the area because the coal was used to fuel the boilers that distilled the brine. From here to past Middleport, coal mines dotted the hills above the river. A saltworks and a 190-foot-deep mine in the Pomeroy coal operated in Antiquity in the mid- to late 1800s. Syracuse and Minersville were the sites of coal mines, saltworks, and brick plants in the nineteenth century. The operations shipped coal and salt down the Ohio into the early 1900s. West of Pomeroy, Ohio 7 heads into the part of the Allegheny Plateau where Pennsylvanian-age rocks form the bedrock.

Belpre Vertebrate Fossils

A large overgrown roadcut of Washington formation strata near the intersection of Ohio 7 and U.S. 50 near Belpre was the source of Permian-age amphibian and reptile remains that were uncovered in the 1960s and 1970s. The fossils, mainly broken and waterworn isolated teeth and bones, came from a thin lens-shaped deposit of conglomerate, the fill of a former stream channel. *Eryops* and *Diploceraspis* were among the more common amphibian fossils paleontologists found. The sail-back reptiles, or pelycosaurs, *Dimetrodon* and *Edaphosaurus* were also common. Paleontologists know that these animals lived around ponds or small lakes more than 250 million years ago. The Cleveland Museum of Natural History participated in a number of excavations at this Belpre site and has fossils on display.

Scenic Rock Bridges

A short jaunt north on Ohio 144 from Coolville on Ohio 7 or Hockingport on Ohio 124 leads to one of the largest natural arches in the state—Ladd Natural Bridge. The arch, carved in Permian-age Waynesburg sandstone, is part of a state nature preserve on County Road 26 north of Frost. You need a permit, which is free from the Division of Natural Areas and Preserves, to enter the preserve. The bridge is 40 feet long, 12 to 16 feet wide, and 50 feet high. It formed as an ephemeral stream cascading from the lip above a large rock shelter widened

—indstone quarry in Permian-age Hundred sandstone in the hills above Marietta around 1920. —Ohio Department
Natural Resources, Division of Geological Survey photo

Ladd Natural Bridge near Frost is carved in Permian-age Waynesburg sandstone.
—Ohio Department of Natural Resources, Division of Geological Survey photo

a vertical fracture that was behind the face of the cliff. Eventually the crevice captured the water flow and the rock mass—now the bridge—separated from the rest of the cliff. A stream now runs under the bridge rather than over it. Qualey Natural Bridge, which also developed in the Waynesburg sandstone, is off Township Road 238, ½ mile north of Ohio 555 and southeast of Cutler.

Pomeroy

Salt seeps along Leading Creek at Middleport became the site of a salt well in 1822. Salt production shifted to Pomeroy around 1850 and thirteen saltworks eventually opened. The Pomeroy brines came from ancient seawater trapped in pores of the underlying Pennsylvanian, Mississippian, and Devonian sandstones. In later years production was focused on the Black Hand and Berea sandstones 300 to 1,600 feet below the surface. The brines were highly concentrated compared to other Ohio locations. The wooded slopes above town and numerous coal seams provided the saltworks with plenty of fuel, and the river provided ready transportation. By the 1870s one of the Pomeroy plants ranked as the largest in the state. Increased rock salt production in Michigan, New York, and later northeastern Ohio and declining salt prices led many Pomeroy operations to close. In 1905 only six plants were still in production. Bromine and calcium chloride, two economically important by-products of salt distillation, kept these Pomeroy plants in business. They shipped these chemicals throughout the lower Ohio Valley and as far west as the Mississippi Valley. The last salt plant in Pomeroy closed in 1972.

The first records of coal mining in the area date to around 1819; Pomeroy became a coal mining center as early as the 1830s. Shaft mines in the Pomeroy coal dotted the uplands into the mid-1900s. Strip-mining later removed coal from uplands west of town, including the deeper Pittsburgh coal.

The Pomeroy sandstone became a desired building stone, and quarriers took it from the hills above town. Look for it in stone buildings in town and the wall along the downtown parking lot. A notable rockfall of this sandstone occurred in winter 1971 behind Pomeroy's post office, which was built below a rocky escarpment. This cliff has a base of soft shale that is prone to undercutting. Joints behind and parallel to the cliff face also allow water to seep into the sandstone, setting in motion freeze-thaw erosional processes. Several hundred cubic yards of large blocks crashed through the back wall of the post office and onto the parking lot, luckily during early morning hours.

Pomeroy to Chesapeake

Ohio 7 bypasses downtown Pomeroy and Middleport. On the west side of Pomeroy the floodplain narrows and slump blocks of Pomeroy sandstone rest at the roadside. Strip mines lie to either side of the Pomeroy bypass. Between Middleport and Chesapeake, Ohio 7 stretches across narrow river terraces edged by rocky cliffs and wide point bars; the edges of the uplands are far from the roadside. The huge James M. Gavin Power Plant at Cheshire once received its coal supply from the former Meigs No. 2 and No. 31 drift mines in the Clarion coal of the Allegheny group east and north of Wilkesville. A 10-mile-long overland conveyor belt transported the coal to the plant at 1,000 feet per minute. The mines closed around 2002–03. The company, forced to use coal with a lower sulfur content, now receives shipments on the Ohio River.

From Middleport to Carlton many drift and shaft mines in the Pomeroy coal lie abandoned north of the highway. Middleport now covers the site of Coalport, where entrepreneurs attempted to ship coal downriver as early as 1804. Middleport had two brick plants in the 1890s, and both used silt from the Ohio River floodplain. You can see these dark bricks, sometimes containing rounded pebbles, throughout the community. Gallipolis once had a drain tile and brick plant that utilized floodplain clays and also used Pomeroy sandstone of the Monongahela group in prominent structures, including many of the original buildings at the state institute, a hospital for people with epilepsy.

The wide valley of Raccoon Creek meets the Ohio north of the Gallipolis Lock and Dam. The upstream community of Northup sits about 50 feet above the valley bottom on a Wisconsinan-age meltwater terrace. Farther up Raccoon

Creek near Rio Grande is Daniel Boone's Cave, which he supposedly occupied from 1791 to 1792. Scattered coal mines dot this region but are nowhere near as common as they are farther northeast.

The Pennsylvanian strata west of Rio Grande on U.S. 35 contained sufficient iron ore to support the construction of a charcoal furnace in 1847. The Gallia Furnace was built on Dirty Face Creek near Gallia and grew into a community of 136 people by 1880. Each year 3,000 tons of ore was made into wagon wheels and machinery. A narrow-gauge railroad connected the extensive coalfields with the furnace. Local Vanport limestone served as a flux. The furnace shut down in the early 1900s and little trace of it remains.

Watch for occasional roadside exposures of Pennsylvanian-age sandstone at Clipper Mills, Bladen, and Crown City as Ohio 7 heads to the most southerly part of the state. Between Crown City and Athalia, Connellsville sandstone forms 50-foot-high cliffs. Between Crown City and Miller, signs warn of rockfalls that occur in sandstone cliffs at the edge of the road. From this constricted section of the Ohio Valley the floodplain widens to 1½ miles wide just east of Proctorville, where the river begins to loop westward and eventually northward, completing a huge meander that began back in Crown City. Rock slides and rockfalls in Pennsylvanian-age sandstones occur along the road from Bradrick to Chesapeake. Meandering Symmes Creek meets the Ohio across from Huntington, West Virginia, east of where Ohio 7 has been rerouted at the edge of the cliff of Pennsylvanian sandstones. Mines in Upper Freeport and Wilgus coals once dotted this valley. Between Burlington and South Point, what is now U.S. 52 cuts through the Pennsylvanian bedrock of the plateau. This bedrock is also visible in the ramps of the bridge leading to I-64 and Huntington, West Virginia. On U.S. 52 across from Ashland, Kentucky, an immense cut in the sandstone escarpment along the Ohio River has pushed the rock back from the roadside, reducing rock-slide hazard. This stretch of the roadway has been plagued by rock slides.

Coal Grove (*see* discussion in the U.S. 52 road guide) was an active coal mining area at one time. From here to Portsmouth and northward into the Allegheny Plateau lies the Hanging Rock Iron Region, where Pennsylvanian iron ores sparked the early part of the iron boom in the Ohio Valley.

Conemaugh group shales, sandstones, and clays are exposed along roads heading out of the Ohio Valley near Gallipolis. Iron within the sediments leads to their reddish color—the so-called redbeds. They are notoriously weak and often lead to landslides.

A major tributary of the Ohio River joins it at Point Pleasant, West Virginia, across the river from Kanauga. Pennsylvanian-age Buffalo sandstone forms the piers of the railroad bridge.

GLOSSARY

aggregate. Sand, gravel, and crushed stone added as filler material to mortar or concrete or used alone as a construction medium.

anticline. An upward bend or fold in rock strata caused by compressive forces; often associated with mountain building.

aquifer. Sediment or rock that readily yields groundwater.

arch. An upward bend, or fold, in rock layers; also called an *anticline*.

artesian water. Groundwater that is overlain by low-permeability sediment or rock that forces the water to rise above the normal water table.

arthrodire. An extinct group of heavily armored, jawed fish, or placoderms, that were characteristic fish of Devonian time.

arthropod. An animal with jointed legs, an external skeleton, and a segmented body. Insects, spiders, crabs, and lobsters are common members of this diverse group.

basement. Ancient Precambrian rocks underlying Ohio's Paleozoic strata.

basin. A region of the earth's crust that is downwarped in a circular manner; sedimentary strata incline and increase in thickness toward the center of a basin.

beach ridge. A curving ridge or rise of sand marking the shoreline of a major proglacial lake. They are common in northern Ohio, where roads sometimes follow them.

bed. A layer of sedimentary rock, or stratum.

bedrock. Solid rock that formed in place and was not transported to its location. The rock underlying loose surface sediment or soil.

bituminous coal. A sedimentary rock that formed as a result of the compression and gentle heating of peat. It is the most common kind of coal.

bluestone. A stone-industry term for building stones that appear bluish gray, at least until they have weathered.

bog. A poorly drained wetland underlain by acid-rich plant debris and spongy peat.

bone bed. A thin, concentrated layer of whole and fragmental fossils, including fish bones and teeth. Bone beds are distinctive in some parts of Ohio's Devonian-age strata.

brachiopod. A sea-dwelling animal that superficially looks like a clam but is not one. Each brachiopod valve is split into symmetrical halves. Their fossils are common in Ohio's Paleozoic marine rocks. Inarticulate brachiopods lack hinge teeth.

breakwater. An engineered structure built offshore but parallel to a shoreline and meant to reduce the force of waves.

brine. Groundwater enriched in salt, or sodium chloride.

bryozoan. A tiny sea-dwelling colonial animal that builds a variety of calcareous structures. Their fossils are common in Ohio's Paleozoic marine rocks.

building stone. Any rock quarried in various blocks or slabs and used to construct bridges, buildings, and monuments.

buried valley. A former river valley filled with sediment and not necessarily tied to present drainage.

calcite. The mineral form of calcium carbonate; it is common in limestone and is the main cementing material in many types of sedimentary rock. It also occurs as pore and fracture fillings, often exhibiting six-sided crystals in Ohio's Paleozoic rocks.

cannel coal. A waxy coal mostly composed of fossil plant spores and pollen.

carbonate bank. A warm, shallow part of the sea where the bottom sediments are composed of minerals rich in calcium. Also called a *carbonate platform*.

cephalopods. A group of sea-dwelling animals that includes nautiloids, squids, and extinct ammonoids. Tentacles surround a cephalopod's head. Nautiloids and ammonoids have cone-shaped or coiled shells; squids lack them.

channeler. A quarry machine used to score a quarry face so building stone can be removed in large blocks.

chert. A microscopically fine form of quartz. It typically occurs as nodules or thin layers in sedimentary rocks. It was the main raw material for Native American arrowheads.

clay. A sedimentary material composed of different types of weathered silicate minerals that are less than .0039 millimeter in diameter.

coke. Baked bituminous coal.

colonial coral. A sea-dwelling invertebrate of the cnidarian group composed of thousands of individuals amassed in a large rocky skeleton.

concretion. A mass of mineral matter formed around a nucleus that differs in composition from the rocks that enclose it.

conglomerate. A sedimentary rock composed of rounded pebbles or cobbles cemented together in a mass.

conodont. Brownish fossils that are barely visible and look like minuscule jaws.

crinoid. A stalked echinoderm with many arms and a body enclosed in solid plates of calcite. They are common fossils in Ohio's Paleozoic rocks.

cross-bedding. Layers of sediment or rock within larger layers. They are typically inclined in relation to other horizontal layers and are indicative of flowing or turbulent water.

crystal. A naturally occurring geometric form that minerals develop. Crystals may be cubes, parallelograms, pyramids, and other shapes and range in size from microscopic to a few feet.

cyclothem. A repetitive series of sedimentary layers, including sandstone, siltstone, shale, and limestone, that are associated with coal seams of late Paleozoic time. They were caused by changes in the depositional environment.

delta. An accumulation of sediment deposited where a river enters a lake or ocean.

dimension stone. Building stone cut to specific dimensions.

dolomite. The mineral form of calcium magnesium carbonate. The name also refers to sedimentary rocks made of dolomite.

dolomitization. The chemical transformation of limestone into dolomite by the addition of magnesium-rich fluids.

dragline. A huge power shovel used in strip mines.

drainage tile. Drain pipe, in a variety of sizes and designs, made from locally available clay. It was placed into excavated channels in farm fields to drain water.

drift mine. A type of underground mine with an entrance that is in the side of a hill.

dripstone. A general term for mineral features in a cave that formed by dripping water.

drowned river mouth. Where rising lake level has flooded the mouth of a river and created a bay.

echinoderm. Any of a group of sea animals that typically have five-part symmetry and a spiny outer skeleton. Starfish, brittle stars, sand dollars, and sea urchins are common types today; cystoids, edrioasteroids, and crinoids were common types during Paleozoic time.

edrioasteroid. An encrusting echinoderm common in some Ohio Ordovician-age rocks. It is a cross between a brittle star and small sand dollar and commonly occurs on brachiopod shells.

end moraine. A ridge of till that builds up along the edge of an ice sheet when it is at a standstill, when melting equals ice accumulation.

epicenter. The point on the surface closest to an earthquake's point of origin.

erratic. A glacially transported rock that generally differs in type from local bedrock. Erratics are often rounded and found in till.

escarpment. A long ridge of varying height that marks the upslope edge of a resistant, tilted rock layer.

esker. A winding ridge of sand and gravel (outwash) that was deposited in a meltwater tunnel or channel carved into an ice sheet.

fault. A fracture in rocks along which bedrock on one side slides past the bedrock on the other side.

fen. A poorly drained wetland underlain by alkaline plant debris and marl.

firebrick. A type of brick that resists deformation at high temperatures. Made from fireclay, a type of clay that won't deform at high temperatures, firebrick is often used to line furnaces.

flagstone. A sedimentary rock that readily splits into thin slabs.

flash flood. A rapidly developing, short-term flood that occurs in hilly regions.

flint. A darker colored variety of extremely fine-grained quartz. This mineral is Ohio's state gemstone and was used widely by the first Ohioans.

floodplain. The part of a valley floor that is underwater during floods.

flowstone. A smooth deposit of calcium carbonate precipitated from groundwater as it trickles down the walls of caves; it also forms mounds around springs.

flux. A material used to lower the melting temperature of ore.

formation. A body of rock distinctive enough to be recognizable from one place to another. Formations are simply rock units described by their composition, so they may vary in age from location to location.

freestone. A type of sandstone that breaks easily.

glacial erratic. *See* erratic.

glass sand. A high-purity silica sand used in glass manufacturing.

gneiss. A metamorphic rock that often has obvious light and dark bands of minerals. It occurs in Ohio's basement and as glacial erratics in Ohio.

granite. An igneous rock composed of quartz, feldspar, mica, and accessory minerals that forms deep within the earth. In Ohio, granite and granodiorite occur in the basement or as glacial erratics.

grindstone. Sandstone wheels or blocks used for grinding grain or sharpening tools and implements.

groin. A narrow structure built in the water perpendicular to the shoreline in order to trap sand.

ground moraine. *See* till plain.

ground sloth. An extinct mammal (*Megalonyx jeffersoni*) whose bones are found in Ohio's Pleistocene deposits. These toothless mammals were related to modern tree sloths; however, they were much larger.

groundwater. Water that moves slowly through subsurface geologic material. We retrieve groundwater from wells.

group. Two or more formations that form a mappable unit.

gypsum. The mineral form of hydrous calcium sulfate. The name also refers to sedimentary rocks made of gypsum. It is used to make plaster and wallboard.

halite. The mineral form of sodium chloride. It commonly occurs as clear to colored crystalline masses. In large masses it is known as *rock salt.*

honeycomb weathering. Raised circular ridges of resistant mineral matter that occur on some sedimentary rock surfaces. The surfaces resemble honeycombs.

horn coral. Paleozoic corals with a cone-shaped calcareous skeleton. A cross section of a skeleton reveals a radial growth pattern. Also called *rugose corals.*

hydraulic cement. Cement that sets under water.

ice lobe. A tongue of ice that sticks out from the advancing or receding edge of an ice sheet; typically, it is controlled by underlying topography.

ice sheet. An immense glacier that covers a significant landmass and has ice moving out in various directions; it is not greatly influenced by underlying topography.

igneous rocks. Rocks formed from the cooling of molten mineral matter. In Ohio they are found in the basement or as glacial erratics at the surface.

industrial sand. Sand or crushed sandstone of high purity.

interlobate. Occurring between adjacent ice lobes and consisting of a mixture of tills and outwash.

jetty. An engineered structure that extends into open water on either side of a channel to prevent sand deposition from closing the channel off.

joint. A crack in rock along which little or no movement has taken place.

kame. A hill of sand and gravel (outwash) that was originally deposited on or at the edge of a glacier or in depressions or crevasses at the ice's surface. As the ice melted, the outwash was lowered onto the landscape.

kame terrace. A flat level of outwash that accumulated between an ice lobe and a valley margin.

karst. A hilly landscape of caves and sinkholes that develops on and in some dissolving limestone formations.

kettle. A depression where a block of ice, left behind in glacial sediments, finally melted; some kettles fill with water, forming kettle lakes.

ledge. Refers to eroded outcrops of jointed Pennsylvanian-age sandstone and conglomerate that caps uplands in northeastern Ohio.

lime. Calcium oxide, often made from limestone.

limestone. A sedimentary rock mainly composed of calcite. It typically forms in shallow, warm seas or lakes.

limonite. A general name for hydrous iron oxide that often occurs as a yellowish brown powder and is associated with minerals that contain iron.

lobe. *See* ice lobe.

loess. A deposit of windblown silt.

longshore currents. Water motion that parallels a coastline and is generated by waves that approach the shore at an angle. Longshore currents carry sediment.

longwall method. A mining method in which a coal seam is entirely removed along a face, leaving no pillars. This method allows for a higher recovery of coal than the room and pillar method.

lycopods. A group of plants that is represented by club mosses today. Lycopods were common during Pennsylvanian time, when some reached heights of 100 feet.

magnitude. In reference to earthquake activity, it is the amount of energy released in a given quake.

marl. A calcareous mud precipitated in lakes and streams. It is dredged to make portland cement.

mastodon. An extinct elephant (*Mammut americanum*) that favored forests and lived in North America until about 10,000 years ago.

meander. A curving or looping stretch of stream channel.

meander scar. An abandoned, dried-up stream meander.

meltwater. Water derived from a melting ice sheet. Meltwater erodes valleys and distributes outwash.

member. A subdivision of a formation based on distinctive rock characteristics.

metamorphic rocks. Rocks that were changed in appearance and sometimes composition due to exposure to high temperature, high pressure, and/or chemically active fluids when they were buried or came into contact with magma. In Ohio they are found in the basement or as glacial erratics at the surface.

mineral. A naturally occurring, solid inorganic chemical element or compound. There are around three thousand minerals.

misfit stream. A small stream that appears out of place and incapable of carving the wide valley it occupies. Misfit streams flow in preglacial channels and/or channels that glaciers and meltwater carved.

molding sand. Sand or crushed sandstone used in the casting industry.

monocline. A local change in the dip, or angle of incline, of sedimentary rock layers.

moraine. *See* end moraine, till plain, and terminal moraine.

mudstone. A sedimentary rock that is composed of clay-sized particles and has no evident layering.

natural arch. A rock bridge that spans a valley or occurs near cliffs. It can have a number of origins.

natural gas. Subsurface gas formed from the breakdown of organic matter and held in traps associated with rocks with lower permeability.

nodule. *See* concretion.

ostracod. A minute arthropod that has a shell composed of two valves and mainly lives in water. Their fossils are common in some Paleozoic rocks.

ostracoderm. An early Paleozoic jawless fish.

outcrop. A surface exposure of bedrock.

outlier. An erosional upland remnant of younger bedrock surrounded by older, lower-elevation bedrock.

outwash. Sand and gravel washed from glacial deposits by meltwater.

oxbow lake. An abandoned stream meander filled with water.

paver. A hardened, thick brick used for street paving.

peat. A black to dark brown sediment composed of broken-down plant matter; it is often found in marshes and swamps.

pelycosaur. An extinct late Paleozoic reptile; some had a notable fin down their back. It is commonly called a *sail-back reptile*.

pig iron. Crude iron pulled directly from a blast furnace.

point bar. A flat to rolling mass of sand deposited on the inside of a meander.

pothole. A circular depression or hole eroded in rock by swirling sediments in a stream.

prairie. Flat to gently rolling land that generally lacks trees and is dominated by grasses. It may be wet or dry.

pressed brick. Bricks that can be made just about any color or shape with special presses. They are mainly used in building walls and for decorative use.

proglacial. A term applied to any temporary landscape feature, such as a lake, associated with the edge of an ice sheet.

pulp stone. Sandstone or siltstone used as rollers in the making of paper.

quartzite. A metamorphic equivalent of sandstone.

reclaimed land. Refers to strip-mined land that has been restored.

redbeds. Sedimentary beds stained red to maroon by iron compounds.

refractory. Heat-resistant ceramic material.

Richter scale. A mathematically calculated expression of earthquake magnitude.

roadcut. Where a road cuts through the landscape, exposing subsurface sediment and bedrock.

rockfall. A relative free fall of rock from a cliff face, cave ceiling, or arch.

Rockingham ware. A type of yellowware pottery.

rock salt. A sedimentary rock composed of halite. It is particularly common in Ohio's Late Silurian strata.

rock shelter. A recess in a cliff face that formed as softer rock layers were undercut. A rock shelter is overhung by more-resistant rock.

room and pillar method. A mining method in which coal or ore is removed in large masses while some material is left as pillars to support the mine roof.

sand. Weathered mineral grains, most commonly quartz, .0625 to 2 millimeters in diameter.

sandstone. A sedimentary rock composed of sand-sized particles, mainly quartz, that are held together by some natural cement.

schist. A mica-rich metamorphic rock showing foliation. In Ohio it is found in the basement and as glacial erratics.

scolecodont. Minute fossils of the jaw apparatus of a type of Paleozoic marine worm. They are usually black to dark brown.

seam. A layer, or stratum, of coal.

sedimentary rocks. Sediments that have hardened into solid rocks. They typically occur in layers. Sedimentary rocks form the bedrock of Ohio.

shaft mine. A type of underground mine with a vertical or near-vertical passageway for an entrance.

shale. A common sedimentary rock mainly composed of clay that shows obvious layering.

siderite. The mineral form of iron carbonate. Sometimes it is found in concretions.

silt. Weathered mineral grains larger than clay but smaller than sand, ranging between .0039 and .0625 millimeter in diameter.

siltstone. A sedimentary rock composed of silt.

sinkhole. A surface depression or hole. Sinkholes form as surface materials sink into an underlying area where limestone has dissolved or where the ceiling of a near-surface cave collapses.

slope mine. A type of underground mine with an inclined passageway for an entrance.

soda straw. An early form of a stalactite that is small and hollow like a soda straw.

spit. A bar of sand that projects into a lake; they form as sand travels along the shore.

spring. Seeping or flowing water where the water table intersects the surface.

stalactite. An icicle-like mass of calcium carbonate precipitated from dripping water and hanging from the ceiling of a cave.

stalagmite. A stubby mound of calcium carbonate on the floor of a cave below a stalactite.

stoneware. A type of pottery that is dense and appears vitreous after being fired at high temperatures.

strata. Layers, or beds, of sedimentary rock.

strip mine. A mine worked from an open pit, so named because overlying material is stripped to expose material of value.

swallow hole. A sinkhole where a stream goes underground.

syncline. A U-shaped downward bend of rock layers caused by compression or sinking.

terminal moraine. A ridge of till marking the farthest advance of an ice sheet. A type of end moraine.

terrace. A flat shelf or plain above the floodplain of a stream. Terraces are remnants of old floodplains that formed when the channel was at a higher elevation.

terra-cotta. A reddish brown ceramic material used mainly for architectural ornamentation and vases.

tidal flat. The shoreline area between normal tide levels.

till. A chaotic deposit of sediments of all sizes deposited by glacial ice.

till plain. Till deposited on the landscape by glaciers. Across glaciated Ohio it varies from flat to gently rolling terrain. Also called *ground moraine.*

trace fossils. Evidence—like tracks, trails, and burrows—of living organisms that they left in sedimentary rocks.

travertine. A general term for calcium carbonate found around springs and caves.

trilobite. A common arthropod of the Paleozoic seas that is now extinct. Trilobites had hard-shelled skeletons that were divided into three lobes. One variety, *Isotelus maximus,* is Ohio's state fossil.

tufa. A form of calcium carbonate that often encrusts plant stems and leaves around a spring.

type area. The geographic location where rock strata or fossil taxa are first described and named. It is usually an area that exhibits all the characteristics of the stratum or species.

unconformity. An old erosional surface that was buried by younger sedimentary rocks. It is a gap in the geologic record.

underclay. A layer of clay beneath a Pennsylvanian-age coal seam. Almost every seam has an underclay.

valley train. Outwash that was deposited by torrents of meltwater, filling a valley beyond the reach of a glacial ice sheet.

weathering. The chemical and mechanical breakdown of geologic material at the surface of the earth due to exposure to the elements.

whetstone. Sandstone or siltstone used for sharpening tools.

white ware. A whitish ceramic material, including white granite and porcelain ware.

yazoo stream. A tributary that flows parallel to a main channel before joining it.

yellowware. A type of pottery that involves a second firing of more-porous material to form its yellowish glaze.

ADDITIONAL READING

I urge you to consult the many publications of the Ohio Department of Natural Resources (ODNR) Division of Geological Survey and its predecessors. More-recent publications are available by mail or over-the-counter at the publication sales office in Fountain Square in Columbus. Many information sheets, class-room materials, and maps are available at the survey's Web site (http://www.ohiodnr.com/geosurvey/index.html). Out-of-print publications are available at many public, college, and university libraries across the state.

The Ohio Academy of Science's *Ohio Journal of Science*, available at many libraries, is a good source of articles on Ohio geology as are publications of the Ohio Biological Survey; the Ohio Historical Society and its predecessor, the Ohio State Archaeological and Historical Society; the Cincinnati Museum of Natural History and Science; and the Cleveland Museum of Natural History. County historical books often contain information on scenic geologic fea-tures and mineral resources as well. Individual county maps or a compilation of county maps, like Delorme's *Ohio Atlas & Gazetteer*, will guide you as you explore the Buckeye State. Below are a few selected publications of interest.

Adams, I. and S. Ostrander. 2002. *Ohio: A Bicentennial Portrait 1803–2003*. San Francisco: BrownTrout Publishers, Inc.

Adams, I. and J. Fleischman. 1995. *The Ohio Lands*. San Francisco: BrownTrout Publishers, Inc.

Bowell, D. F., compiler. 1980. *Inventory of Ohio's Lakes*. Columbus: ODNR Division of Water.

Brockman, C. S. 1998. *Physiographic Regions of Ohio*. Columbus: ODNR Division of Geological Survey.

Carlson, E. H. 1991. *Minerals of Ohio*. ODNR Division of Geological Survey, Bul-letin 69.

Crowell, D. L. 1995. *History of the coal-mining industry in Ohio*. ODNR Division of Geological Survey, Bulletin 72.

Davis, R. A., editor. 1992. *Cincinnati Fossils, an elementary guide to the Ordovician rocks and fossils of the Cincinnati, Ohio, region*. Cincinnati: Cincinnati Museum of Natural History.

Directory of Ohio's State Nature Preserves. 1996. Columbus: ODNR Division of Nat-ural Areas and Preserves.

Drahovzal J. A., D. C. Harris, L. H. Wickstrom, D. Walker, M. T. Baranoski, B. Keith, and L. C. Furer. 1992. *The East Continent Rift Basin: a new discovery*. ODNR Division of Geological Survey, Information Circular 57.

Feldmann, R. M., A. H. Coogan, and R. A. Heimlich. 1977. *Field Guide Southern Great Lakes*. Dubuque, Iowa: Kendall/Hunt Publishing Company.

Feldmann, R. M. and M. Hackathorn, editors. 1996. *Fossils of Ohio*. ODNR Division of Geological Survey, Bulletin 70.

Folzenlogen, R. 1990. *Hiking Ohio: Scenic Trails of the Buckeye State*. Glendale, Ohio: Willow Press.

Frank, G. W., editor. 1969. *Ohio Intercollegiate Field Trip Guides 1950–51 to 1969–70*. Kent, Ohio: Kent State University Printing Service.

Goslin, C. R. 1976. *Crossroads and Fence Corners Historical Lore of Fairfield County*. Lancaster, Ohio: The Fairfield Heritage Association.

Groene, J. and G. Groene. 1994. *Natural Wonders of Ohio: A Guide to Parks, Preserves and Wild Places*. Castine, Maine: Country Roads Press.

Hough, J. L. 1958. *Geology of the Great Lakes*. Urbana, Illinois: University of Illinois Press.

Howe, H. L. 1900. *Historical Collections of Ohio*. Vols. 1 and 2. Cincinnati: C. J. Krehbiel and Company.

Hunt, T. S. 1881. *Coal and Iron in Southern Ohio: The Mineral Resources of the Hocking Valley*. Boston: S. E. Cassino, Publisher.

Krolczyk, J. C. (compiler) and V. Childress (editor). 2001. *Gazetteer of Ohio Streams*. 2nd ed. ONDR Division of Water, Water Inventory Report 29.

Lafferty, M. B., editor. 1979. *Ohio's Natural Heritage*. Columbus: The Ohio Academy of Science.

La Rocque, A. and M. F. Marple. 1955. *Ohio fossils*. ODNR Division of Geological Survey, Bulletin 54.

Larsen, G. E. 1994. *Regional bedrock geology of the Ohio portion of the Lima, Ohio-Indiana 30 x 60 minute quadrangle*. ODNR Division of Geological Survey, Map 7.

Miller, L. L. 1996. *Ohio Place Names*. Bloomington and Indianapolis: Indiana University Press.

Murdock, E. C. 1988. *The Buckeye Empire: An Illustrated History of Ohio Enterprise*. Northridge, CA: Windsor Publications, Inc.

Noble, A. G. and A. J. Korsok. 1975. *Ohio—An American Heartland*. ODNR Division of Geological Survey, Bulletin 65.

Ohio Atlas & Gazetteer. 1989. Freeport, Maine: DeLorme Mapping Company.

Ohio Geology. A biannual publication of the ODNR Division of Geological Survey.

Ostrander, S., editor. 2001. *The Ohio Nature Almanac: An Encyclopedia of Indispensible Information About the Natural Buckeye Universe*. Wilmington, Ohio: Orange Frazer Press.

Pavey, R. R., R. P. Goldthwait, C. S. Brockman, D. N. Hull, E. M. Swinford, and R. G. Van Horn, compilers. 1999. *Quaternary geology of Ohio*. ODNR Division of Geological Survey, Map 2.

Peacefull, L., editor. 1990. *The Changing Heartland: A Geography of Ohio*. Needham Heights, Massachusetts: Ginn Press.

Peacefull, L. 1996, editor. *A Geography of Ohio*. Rev. ed. Kent, Ohio: Kent State University Press.

Ramey, R. 1990. *Fifty Hikes in Ohio*. Woodstock, VT: Backcountry Publications.

Ruchhoft, R. H. 1984. *Backpack Loops and Long Day Trail Hikes in Southern Ohio*. Cincinnati: Pucelle Press.

Sanders, R. E. 2000. *A Guide to Ohio Streams*. Columbus: Ohio Chapter of the American Fisheries Society.

Schumacher, G. A. 1993. *Regional bedrock geology of the Ohio portion of the Piqua, Ohio-Indiana 30 x 60 minute quadrangle*. ODNR Division of Geological Survey, Map 6.

Shrake, D. L. 1997. *Regional bedrock geology of the Marion, Ohio 30 x 60 minute quadrangle*. ODNR Division of Geological Survey, Map 9.

Smyth, P. 1979. *Bibliography of Ohio Geology 1755–1974*. Ohio Division of Geological Survey, Information Circular 48.

Swinford, E. M. and E. R. Slucher. 1995. *Regional bedrock geology of the Bellefontaine, Ohio 30 x 60 minute quadrangle*. ODNR Division of Geological Survey, Map 8.

Trautman, M. B. 1977. *The Ohio Country from 1750 to 1977—a naturalist's view*. Ohio Biological Survey, Biological Notes 10.

Tribe, I. M. 1986. *Little cities of black diamonds: Urban development in the Hocking coal region, 1870–1900*. Athens, Ohio: Athens Ancestree.

Van Tassel, C. S. 1901. *The Book of Ohio*. Bowling Green and Toledo, Ohio: C. S. Van Tassel.

INDEX

Page numbers in italics refer to illustrations.

Mark J. Camp, a native of Toledo, began his teaching career at Earlham College in Richmond, Indiana, in 1974. He joined the faculty at the University of Toledo in 1976 and teaches courses in introductory geology, paleontology, the geology of national parks, and Ohio geology. He received his MS in geology from the University of Toledo and PhD in geology from the Ohio State University. His recent research, focused on the history of Midwest geological studies, the use of building stones in historical buildings, quarry and mine development, and the architecture of railroad depots, reflects his long-standing interest in late nineteenth- and early twentieth-century history. He is the author of *Roadside Geology of Indiana*, *Railroad Depots of Northwest Ohio*, and *Railroad Depots of West Central Ohio*.